Lecture Notes in Computer Science 1355

Edited by G. Goos, J. Hartmanis and J. van Leeuwen

Springer
*Berlin
Heidelberg
New York
Barcelona
Budapest
Hong Kong
London
Milan
Paris
Santa Clara
Singapore
Tokyo*

Michael Darnell (Ed.)

Cryptography and Coding

6th IMA International Conference
Cirencester, UK, December 17-19, 1997
Proceedings

 Springer

Series Editors

Gerhard Goos, Karlsruhe University, Germany

Juris Hartmanis, Cornell University, NY, USA

Jan van Leeuwen, Utrecht University, The Netherlands

Volume Editor

Michael Darnell
University of Leeds, Department of Electronic and Electrical Engineering
Leeds LS2 9JT, UK
E-mail: eenmd@elec-eng.leeds.ac.uk

Cataloging-in-Publication data applied for

Die Deutsche Bibliothek - CIP-Einheitsaufnahme

Cryptography and coding : ... IMA international conference ... ;
proceedings. - Berlin ; Heidelberg ; New York ; Barcelona ; Budapest
; Hong Kong ; London ; Milan ; Paris ; Santa Clare ; Singapore ;
Tokyo : Springer

6. Cirencester, UK, December 1997. - 1997
(Lecture notes in computer science ; 1355)
ISBN 3-540-63927-6

CR Subject Classification (1991): E.3-4, G.2.1, C.2, J.1

ISSN 0302-9743
ISBN 3-540-63927-6 Springer-Verlag Berlin Heidelberg New York

© Springer-Verlag Berlin Heidelberg 1997
Printed in Germany

Typesetting: Camera-ready by author
SPIN 10661337   06/3142 – 5 4 3 2 1 0   Printed on acid-free paper

Preface

The fact that this is the 6th book in a series on "Cryptography and Coding" produced over the past decade is a clear indication that interest in these two aspects of information transmission and processing remains at a high level and, indeed, is progressively increasing.

Since the inception of this series of books containing edited papers presented at a corresponding series of international conferences on "Cryptography and Coding", sponsored and organised by the Institute of Mathematics and its Applications (IMA), successive organising committees have sought to emphasise the links and commonality between the underlying mathematical bases of cryptography and error control coding. These linkages have tended to expand over the years to encompass other elements of information transmission and processing, such as multiple-access coding, image processing, synchronisation and sequence design for a range of applications.

This algorithmic broadening has been matched implementationally by the increasing availability of a wide variety of high-performance digital signal processing (DSP) devices for incorporation into practical systems. The era of both algorithmic and implementational multi-functionality is now with us, whereby flexible digital architectures can host multiple and interconnected adaptive DSP algorithms – an environment which provides both a stimulating challenge for mathematicians and engineers and also a vehicle for investigating the synergy and commonality between what have previously been regarded as separate disciplines. It is an objective of the 6th IMA International Conference on "Cryptography and Coding", to be held at the Royal Agricultural College, Cirencester, from 17th to 19th December 1997 and at which the papers in this book will be presented, to foster such interaction.

It is also my pleasant task to place on record my appreciation of the help and support of the members of the conference organising committee, namely Paddy Farrell, Mick Ganley, John Gordon, Sami Harari, Chris Mitchell, Fred Piper, and Mike Walker. I also wish to express my sincere thanks to Pamela Bye, Adrian Lepper, and Hilary Hill of the IMA for all their help with both the organisation of the conference and the publication of this volume. Finally, my particular thanks go to my colleague, Ahmed Al-Dabbagh, for his invaluable assistance in editing and preparing camera-ready copy to a very tight schedule.

October 1997

Michael Darnell

Contents

The Theory and Application of Reciprocal Pairs of Periodic Sequences

Ahmed Al-Dabbagh and Michael Darnell

Institute of Integrated Information Systems
School of Electronic & Electrical Engineering
University of Leeds
Leeds, LS2 9JT
UK
Email: A.Al-Dabbagh@leeds.ac.uk
M.Darnell@elec-eng.leeds.ac.uk

Abstract. Recently, there has been a substantial interest shown in the development of the concept of inverse filtering in which the inverse $\tilde{s}(t)$, of a given sequence $s(t)$ is such that the convolution between $s(t)$ and $\tilde{s}(t)$ is a perfect impulse. In the context of this paper, the sequences $s(t)$ and $\tilde{s}(t)$ will be termed a *reciprocal pair of sequences*. The authors develop the theory of reciprocal pairs of periodic sequences via the problem of linear time-invariant (LTI) system identification. Initially, the classical correlation-based method of system identification is considered which makes use of test signals with impulsive autocorrelation functions. Analysis of the effects of a non-impulsive autocorrelation function is addressed and a method for its compensation is presented. A new more general formulation of the identification problem is then proposed; it is demonstrated that the classical correlation-based approach is a special case of this more general formulation. Employing the generalised approach and inverse filtering, we show that it is possible to obtain accurate estimates of LTI impulse responses using test signals with non-impulsive autocorrelation functions. The performance is demonstrated under noise-free conditions.

1 Introduction

Pseudo-randomness, in the context of bipolar binary sequences, is associated with sequences that exhibit the following properties, [1]:

1. the numbers of +1s and −1s in any given period are equal,
2. runs of consecutive +1s or −1s frequently occur, with short runs being more probable than long runs;
3. the periodic autocorrelation function (ACF) of the sequence is *ideally* impulsive, ie a peak at zero shift and zero everywhere else.

Considerable research has been carried out over the past few decades to find and develop useful binary sequences which possess these characteristics. Nevertheless, most sequences can only approximate these idealised properties, especially in respect of the ACF being only quasi-impulsive. The search for useful pseudorandom (PR) sequences has also been extended to include non-binary and complex sequences.

PR sequences find application in many areas of system design, for example :

1. *Digital Communication Synchronisation* : A PR sequence may be used in the design of pre-, mid- and post- ambles within a data packet in order to facilitate synchronisation of the receiver to the data packet, [2]. Of particular interest is the impulsive nature of the ACF profile of the PR sequence. The receiver correlates its input with the known PR sequence and decides on a 'synch point' when the correlator output is a maximum. Any deterioration in the quality of the ACF on a noisy channel can cause the receiver to synchronise to the wrong synchronisation epoch.

2. *System Identification* : The use of PR sequences as test signals in the impulse response estimation of an unknown linear-time-invariant (LTI) dynamic system, is well established [3]. Again, the ACF of the test sequence should ideally approximate an ideal impulse. If the ACF of the test signal is not impulsive, then errors in the estimation of the impulse response are introduced. This application is studied in depth in this paper.

3. *Code-Division Multiple Access* : The design of a code-division multiple access (CDMA) communication system requires a set of PR sequences, each of which should ideally possess an impulsive ACF to aid in bandwidth expansion, data detection, synchronisation and other aspects of system design. In addition, all the sequences in the set should ideally be uncorrelated for zero cross-talk. Such conditions can not be realised in practice and, as a consequence, one can only find a limited set of sequences which possesses approximately the above characteristics. System performance will suffer if the required correlation properties are not met.

Recently, there has been a substantial interest shown in the development of the concept of inverse filtering, which can be summarised as follows [4]: the inverse, $\tilde{s}(t)$, of a given sequence, $s(t)$, is such that the convolution between $s(t)$ and $\tilde{s}(t)$ is a perfect impulse. In the context of this work, the sequences $s(t)$ and $\tilde{s}(t)$ will be called *a reciprocal pair of sequences*. The value of this concept is that it can potentially introduce more flexibility into the design of sequences and widen their application. This flexibility could be used to overcome the limitations which arise from ideal correlation characteristics being only partially realised. Inverse filtering has already been proposed for data detection [4]; it has also been propsed for code and code/time division-multiple-access types of multi-access communications [5].

In this paper, the authors analyse the problem of system identification in some depth. Initially, the relevant theory of LTI system characterisation is reviewed in section (2). We then consider the correlation-based LTI system identification procedure in section (3). Particular emphasis is given to identification using test signals with non-impulsive ACFs; and a closed-form expression is derived which exactly characterises the error in the impulse response estimation when the test signal ACF is non-impulsive. A technique for reducing this estimation error is discussed.

We then develop a concept of generalised system identification in section (5), where we also show that the classical correlation-based method of section (3) is simply a special case of our new formulation. Futhermore, the relationship between this new approach and the concept of inverse filtering is specified. Identification using the generalised approach is then demonstrated using Gold, Kasami and trajectory-derived (TD) sequences.

Finally, section 6 summarises the main contributions of the paper and draws overall conclusions.

2 System Characterisation

A *linear* dynamic system is one in which the principles of scaling and superposition apply for arbitrary inputs and outputs; that is, if x_1 and x_2 are two inputs to a system which produce the corresponding outputs y_1 and y_2, then the application of $c_1x_1 + c_2x_2$ at the system input will give an output of $c_1y_1 + c_2y_2$, where c_1 and c_2 are two scalars [6]. Furthermore, a system is called *time-invariant* if an input x applied at time t_1 gives the same output when applied at time t_2, for all $t_1 \neq t_2$; alternatively, a time shift in the input signal causes a corresponding a time shift in the output signal [7].

Consider the general linear dynamic system shown in figure 1.

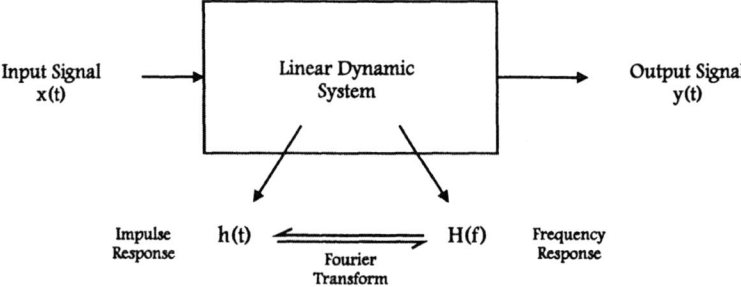

Fig. 1. General Linear Dynamic System.

For this system, the relationship which links its unit impulse response, $h(t)$, its output, $y(t)$, and its input, $x(t)$, may be expressed in the form of the integral

$$y(t) = \int h(u)x(t-u)du = h(t) \otimes x(t) \tag{1}$$

where u is a dummy time variable. The integral of equation 1 is termed a *convolution* integral and the process of convolution is denoted by the symbol $\otimes$. The function $h(t)$, which characterises the physical properties of the system, is also called the *weighting function* of the system. The impulse response describes the system output as a function of time when an impulse is applied at its input. The Fourier transform of the impulse response is the transfer function, or frequency response, $H(f)$, of the system and characterises it in the frequency-domain.

In the context of this paper, *system identification* will be taken to describe the process obtaining a measurement of the *impulse response* (or equivalently the frequency response) of a given unknown linear system.

3 Classical Approach to System Identification

The classical approach to system identification makes use of PR sequences whose ACF approximates to a Dirac delta function. The input of the unknown system is perturbed by a PR test signal. Then, the corresponding system output is crosscorrelated with an exact replica of the test signal to give an estimate which is directly proportional to the impulse response of the unknown system. The arrangement for the classical identification approach is shown in figure 2.

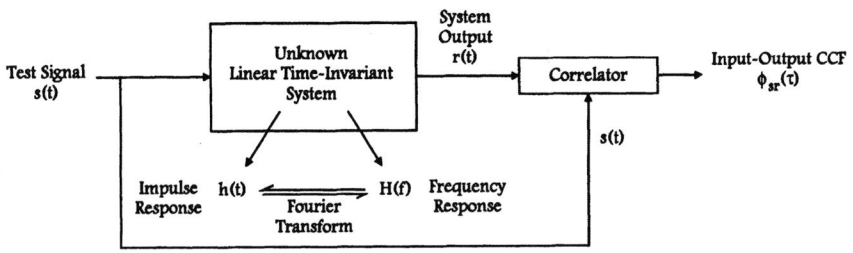

Fig. 2. Arrangement for Classical System Identification.

When the ACF of the test signal is a perfect Dirac impulse function, accurate identification is possible. Any deviation from the ideal ACF case results in the introduction of estimation errors.

Theoretically, an infinitely long segment of Gaussian white noise possesses such a perfect impulsive ACF [8]. Clearly, this requirement cannot be met in practice, and hence the use of pseudo-noise (PN) test signals is more appropriate. PN mimics the statistics of random noise, with the added advantage of being completely deterministic, ie it can be regenerated exactly whenever desired. PN is generated by means of PR sequences and a considerable amount of research has been carried out over many years in order to find PR sequences with useful correlation properties [1]. The application of various classes of PR sequences to system identification has also been investigated thoroughly and shown to give accurate estimates of desired impulse responses [9].

3.1 Ideal and Non-Ideal Sequences

A sequence, $s(t)$, is termed *ideal* if its ACF is an ideal Dirac delta function, ie zero for all time shifts except the zero shift. Such an ACF can be written in the form

$$\phi_{ss}(\tau) = \delta(\tau) = \begin{cases} 1; & \text{when } \tau = 0 \\ 0; & \text{otherwise} \end{cases} \tag{2}$$

where τ is a time delay variable. If $\phi_{ss}(\tau)$ does not strictly conform to the above definition, it is then called *quasi-ideal*. In this case, the ACF off-peak values may be non-zero and $\phi_{ss}(\tau)$ can, in general, be expressed in the form

$$\phi_{ss}(\tau) = \delta(\tau) + g(\tau) \tag{3}$$

where the $\delta(\tau)$ function represents the main ACF peak and $g(\tau)$ is a function describing the sidelobe structure ($\tau \neq 0$) of the ACF as a function of delay τ. The term *quasi-impulsive* is also sometimes used to denote an ACF with non-zero, but small, sidelobes.

For any given sequence, $s(t)$, the corresponding power spectral density (PSD), $\Phi_{ss}(f)$, is the Fourier transform of the sequence ACF [10]; taking the Fourier transform of 3 thus yields

$$\Phi_{ss}(f) = 1 + G(f) \tag{4}$$

where $G(f)$ is the Fourier transform of $g(\tau)$. Note that, when the sequence is ideal, $g(\tau) = 0$ and 4 becomes the Fourier transform of the ideal case described by equation 2.

3.2   Identification Using Ideal Sequences

Referring to figure 2, the correlator output, or the input-output crosscorrelation function (CCF), $\phi_{sr}(\tau)$, can be shown to be a function of both the unknown system impulse response, $h(t)$, and the ACF of the test signal [11, 12], ie a convolution of the form

$$\phi_{sr}(\tau) = \int h(u)\phi_{ss}(\tau - u)du \tag{5}$$

where u is again a dummy time variable.

If the input sequence is ideal, then its ACF is a Dirac delta function and the correlator output of equation 5 is directly proportional to the impulse response of the unknown system by the sifting property of the delta function, ie

$$\phi_{sr}(\tau) = h(\tau) \tag{6}$$

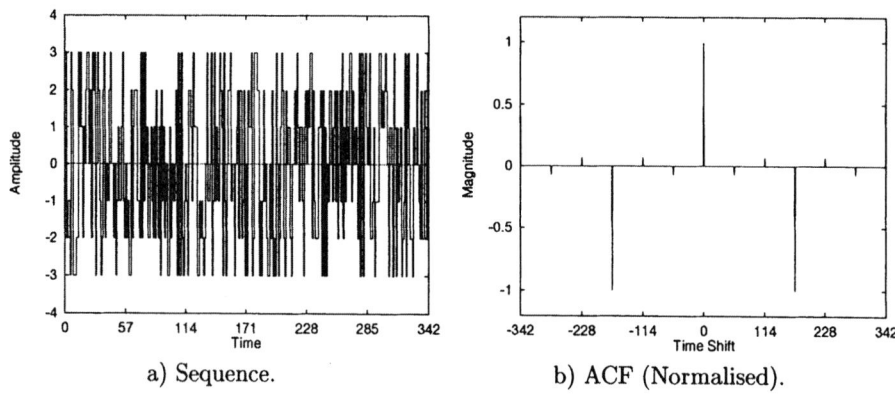

a) Sequence.   b) ACF (Normalised).

Fig. 3. 342-Digit 7-Level Test Signal and its ACF.

Figure 3 shows a periodic 342-digit, 7-level, test signal obtained from the corresponding m-sequence by bipolar level transformation [1], together with its periodic

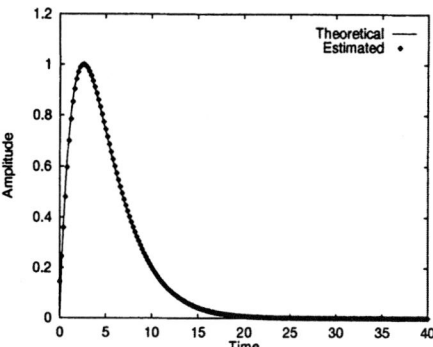

Fig. 4. Classical Identification with an Ideal Test Signal.

ACF. In order to mimic an ideal sequence ACF, the test system response has been confined to the delay interval ±57 digits during which the ACF is ideal. The test system impulse response was chosen to be $h(t) = t \exp(-0.4t)$. The identification of the system was carried out using the classical method of figure 2. The results given in figure 4 show that accurate identification is achieved with an ideal ACF.

3.3 Identification with Non-Ideal Sequences

The majority of PR sequences of practical interest do not possess the ideal ACF described by equation 2. If the ACF of the test signal is non-ideal and only approximates to a Dirac delta function, then the presence of non-zero ACF sidelobes will introduce a measurement error with the classical method. In such a situation, the ACF may be written in the form of equation 3, which when substituted into 5, using the result of 6, gives

$$\phi_{sr}(\tau) = h(\tau) + \underbrace{\int h(u)g(\tau - u)du}_{\text{Error Term}} \qquad (7)$$

The error term in equation 7 quantifies the estimation error exactly when the ACF of the test signal is of a quasi-impulsive type and can be seen to be dependent on the sidelobe function $g(\tau)$ of that ACF.

The 342-digit 7-level transformed m-sequence of figure 3 was next used to estimate the impulse response of a known system with an impulse response of $h(t) = 0.1t \cdot e^{(-0.04t)}$, where the system response now spans a delay interval which encompases the ACF sidelobes at ±57 and ±114 digit delays. The result of the identification in figure 5 clearly shows the estimate error due to the sidelobes of the ACF of the test signal.

3.4 Compensation of Estimation Error

Recently, it has been shown to be possible to compensate for the error in the estimated impulse response when non-ideal sequences are used as the basis of test

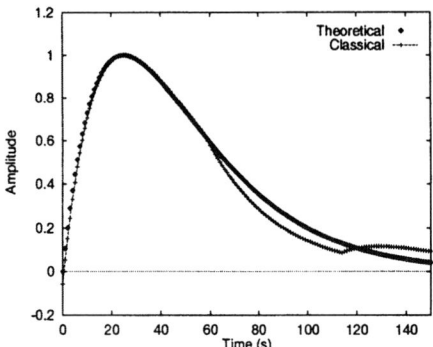

Fig. 5. Identification Using the Non-Ideal 7-Level Test Signal.

signals [13]. Consider the Fourier transform of equation 7:

$$\Phi_{sr}(f) = H(f) + H(f) \cdot G(f) \tag{8}$$

where the convolution in the time domain has been transformed to multiplication in the frequency domain and $H(f)$ is the Fourier transform of $h(t)$. Factorising equation 8 gives

$$\Phi_{sr}(f) = H(f)(1 + G(f)) \tag{9}$$

which can be rearranged to give

$$H(f) = \frac{1}{1 + G(f)} \Phi_{sr}(f) \tag{10}$$

If $G(f)$ does not equal -1 for any f, then equation 10 suggests that the unknown system frequency response, $H(f)$, can be completely determined. Transforming equation 10 back into the time domain yields

$$h(\tau) = q(\tau) \otimes \phi_{sr}(\tau) \tag{11}$$

where

$$q(\tau) = \mathcal{F}^{-1} \left[\frac{1}{1 + G(f)} \right] \tag{12}$$

where the symbol $\mathcal{F}^{-1}[]$ denotes inverse Fourier transform. Equation 12 implies that the complete process of impulse response estimation and error compensation can be performed in the time domain, as shown in figure 6, where the convolver effectively implements a *compensation filter* whose sole task is to compensate for the sidelobes of the ACF; the impulse response of this filter is specified by equation 12.

For the 342-digit level transformed 7-level m-sequence of figure 3, the compensation sequence was computed using equation 12 and is shown in figure 7(a). The identification of the same system as in section (3.3) was carried out using the arrangement of figure 6. The *compensated* impulse response is shown in figure 7(b).

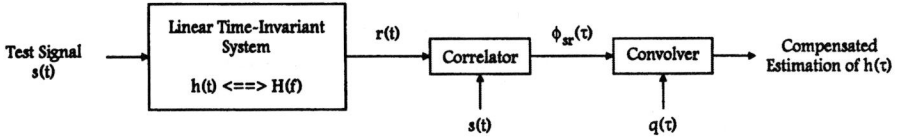

Fig. 6. Compensated System Identification.

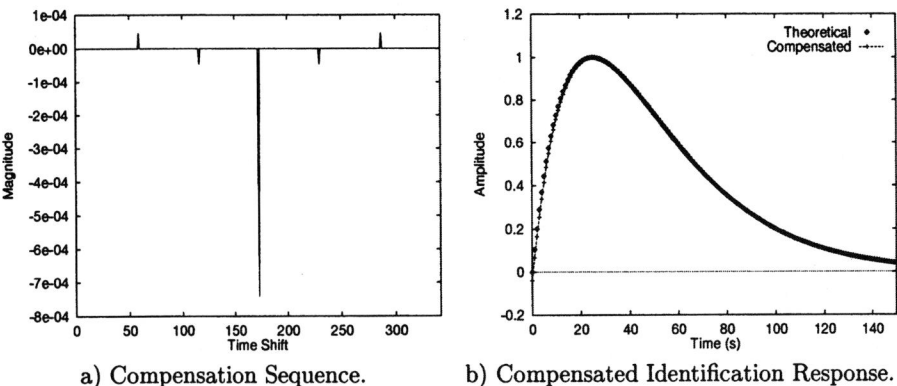

a) Compensation Sequence.   b) Compensated Identification Response.

Fig. 7. Compensated Identification Using the 342-Digit 7-Level m-Sequence.

4 Generalised System Identification

In a practical system, the operations of correlation followed by convolution, shown in figure 6, can be coalesced into one convolution (or correlation), resulting in a less complex procedure. If the correlator-and-convolver combination is replaced with a single correlator, then a *mis-matched filter* is effectively being used; alternatively, if the combination is replaced by a single convolver, then an *inverse filter* is being used.

Suppose that an inverse filter is used for the purpose of identification. Then, the arrangement of figure 8 can considered to be a *generalised* form of system identification which will also encompass that of the classical method. Let $h(t)$ and $H(f)$ be respectively the impulse and frequency responses of the unknown system, as shown in figure 8.

For the purpose of system identification, the system input is perturbed with a test signal, $s(t)$, to obtain a corresponding output response, $r(t)$, Since waveform convolution in the time-domain corresponds to spectral multiplication in the frequency domain, the spectra of the input and output of the unknown system are related by

$$R(f) = H(f) \cdot S(f) \tag{13}$$

where $S(f)$ and $R(f)$ are the Fourier transforms of the input and output signals respectively. By applying an *identification filter*, $\bar{S}(f)$, to the system response, $R(f)$, it will be shown that the this filter output, $z(t)$, will give an estimate of the unknown system impulse response; such filtering is shown in figure 8 as a

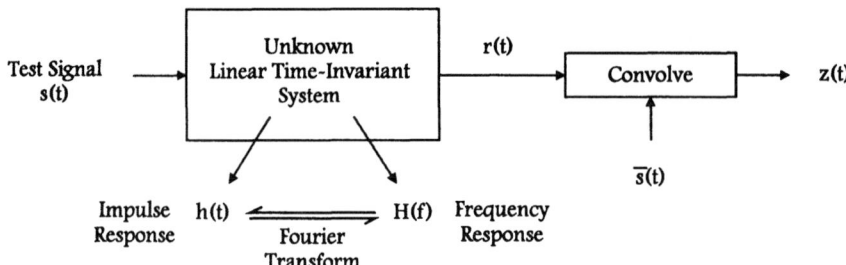

Fig. 8. Generalised System Identification.

convolution operation. The output spectrum of the identification filter is therefore

$$Z(f) = H(f) \cdot S(f) \cdot \bar{S}(f) \tag{14}$$

It is evident from equation 14, that if the relationship

$$S(f) \cdot \bar{S}(f) = 1 \tag{15}$$

holds, then the output of the identification filter will provide an estimate of the system frequency response. The function $\bar{s}(t)$, the inverse Fourier transform of $\bar{S}(f)$, will be called the *reciprocal signal* of $s(t)$ [5, 4]. The inverse Fourier transform of equation 15 is thus

$$s(t) \otimes \bar{s}(t) = \delta(t) \tag{16}$$

since spectral multiplication inverse transforms to waveform convolution.

Equations 15 and 16 are equivalent and define the explicit relationships which characterise a signal and its reciprocal in the frequency and time domains, respectively, to form a *pair of reciprocal sequences*. Furthermore, equations 15 and 16 show that, in general, *system identification can be realised if the unknown system response is convolved with the reciprocal of the applied test signal.*

It is important to note that the classical method of identification is a special case of the above general result, and is obtained by constraining the convolver input, $\bar{s}(t)$, in figure 8 to be the time-reverse of the input test signal. That is, if

$$\bar{s}(t) = s(-t) \tag{17}$$

then, the convolution of equation 16 becomes

$$s(t) \otimes s(-t) = \delta(t) \tag{18}$$

This is the equivalent to the condition derived previously, ie

$$\phi_{ss}(\tau) = \delta(\tau) \tag{19}$$

where $\phi_{ss}(\tau)$ is the ACF of the test signal $s(t)$; this is the pre-requisite for the classical method in which the ACF of the test signal is required to be impulsive.

5   System Identification with Arbitrary Signals

Section (3) dealt with the classical approach to system identification in which
equation 15 was satisfied in one specific way by making the reciprocal sequence
$\bar{s}(t)$ equal to the time-reverse of the test sequence $s(t)$. As a consequence of the
classical approach, a limitation has been implicitly imposed on the choice of test
signals, namely to those with an impulsive ACF.

In section (4), it was established that the restriction on the ACF of the test
signal can be relaxed if the problem of system identification is formulated in a
more general manner and that equation 15 need only be satisfied in order to allow
accurate system identification. In this case, the response of the unknown system
to the applied test signal is convolved with the reciprocal of the test signal. The
only necessary theoretical condition to be met is that the reciprocal of the test
signal itself must be finite, ie

$$-\infty < \frac{1}{S(f)} < +\infty \tag{20}$$

Consequently, the output of the convolution in figure 8 will directly be proportional to the impulse response of the unknown system. It is clear that condition 20
translates to the requirement that the applied test signal must not contain any
nulls in its spectrum. It can be seen that the importance of this result lies in the
fact that an arbitrary test signal can be used, providing the constraint of equation 20 holds to a good approximation. Thus, the 'library' of potential test signals
is greatly enhanced in comparison with those appropriate only for the classical
method.

5.1   Computation of the Reciprocal Signal

Equation 15 can be used to find the reciprocal signal $\bar{s}(t)$ directly for an arbitrary
test signal $s(t)$. Setting $S(f) = \alpha_S(f) + j\beta_S(f)$ and $\bar{S}(f) = \alpha_{\bar{S}}(f) + j\beta_{\bar{S}}(f)$, then
it can simply be shown that

$$\alpha_{\bar{s}}(f) = +\frac{\alpha_s(f)}{\alpha_s(f)^2 + \beta_s(f)^2} \tag{21}$$

and

$$\beta_{\bar{s}}(f) = -\frac{\beta_s(f)}{\alpha_s(f)^2 + \beta_s(f)^2} \tag{22}$$

The restriction of equation 20 applies again to the denominators of equations 21
and 22.

5.2   Example-1 : Gold Sequences

Gold sequences are an important class of periodic binary sequences of length
$2^m - 1$ (m integer) which have found wide application in many areas, especially

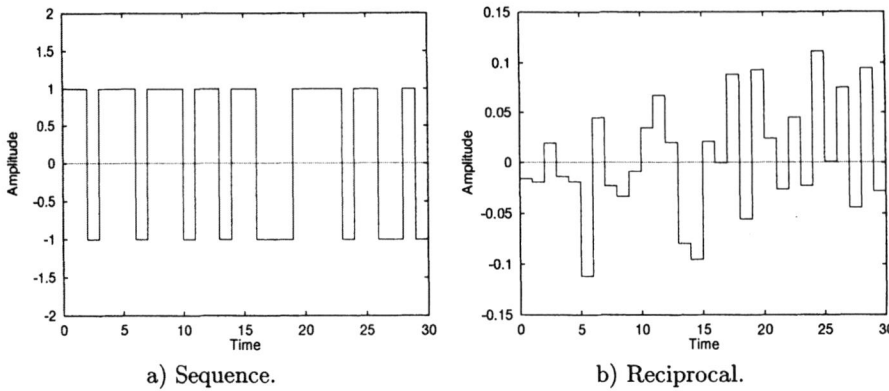

a) Sequence. b) Reciprocal.

Fig. 9. 31-Bit Gold Sequence and Its Reciprocal Sequence.

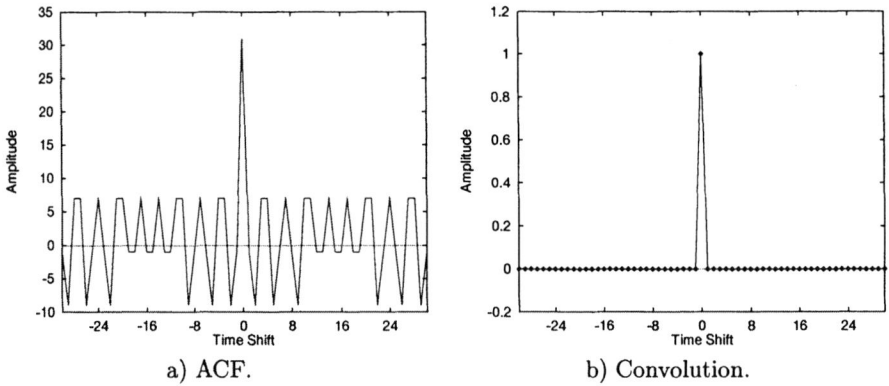

a) ACF. b) Convolution.

Fig. 10. ACF of the Gold Sequence and the Convolution with its Reciprocal Sequence.

CDMA. They were first described in the 1960s by Gold [14, 15]. Although Gold sequences provide favourable crosscorrelation properties, their ACFs are not ideally impulsive.

Figures 9(a) and 9(b) show a typical 31-bit Gold sequence [1] and its reciprocal sequence obtained using the method of section (5.1). Its ACF and the convolution between the sequence and its reciprocal are given in figure 10. It is clear from these two plots that the original ACF is of a quasi-impulsive type. However, the convoution of the test signal with its corresponding reciprocal signal is an ideal impulse.

The Gold sequence was first used in the identification of a known dynamic system with an impulse response given by $h(t) = \exp[-5t]\sin(35t)$ using the classical method; the results are shown in figure 11(a). The identification of the same system was then repeated using the generalised technique; the results of this identification are given in figure 11(b). It can be seen from these identification results that the quasi-impulsive profile of the test signal ACF results in the introduction of large estimation error using the classical method. Despite the imperfection in the ACF of the test signal, however, accurate system identification is shown to be

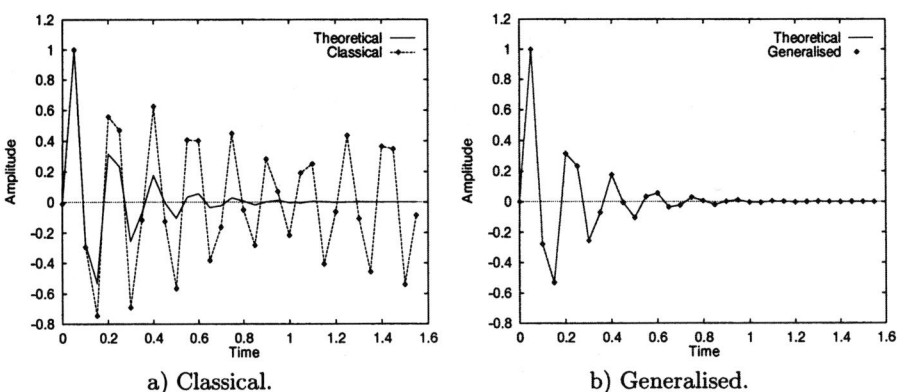

a) Classical.   b) Generalised.

Fig. 11. Classical and Generalised System Identification with 31-Bit Gold Sequence.

possible using the generalised approach.

5.3 Example-2 : Kasami Sequences

Kasami sequences are similar to Gold sequences in structure and method of generation. Sets can be synthesised to provide a group of sequences with favourable CCF
properties; again, their ACFs are not ideally impulsive. Figures 12(a) and 12(b)
show a typical 63-bit Kasami sequence [1] and its reciprocal sequence obtained
using the method of section (5.1). Its ACF and the convolution with its reciprocal
sequence are given in figure 13. Once again the ACF of the Kasami sequence is of
the quasi-impulsive type, whilest the convolution with its corresponding reciprocal
sequence is an ideal impulse.

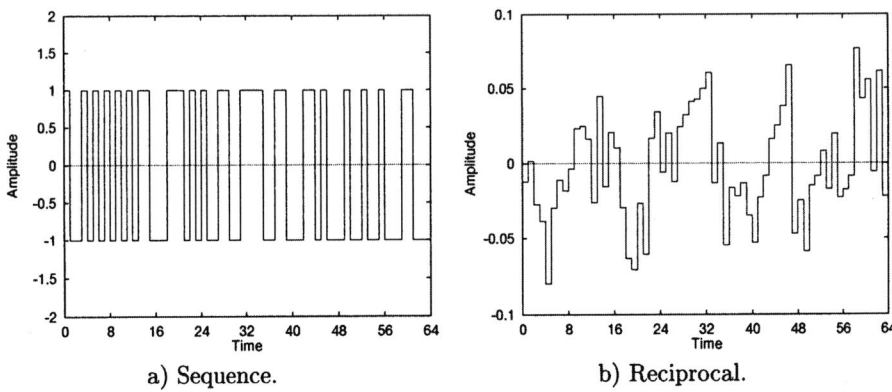

a) Sequence.   b) Reciprocal.

Fig. 12. 63-Bit Kasami Sequence and Its Reciprocal Sequence.

The Kasami sequence was first used in the identification of a known dynamic
system with an impulse response of $h(t) = \exp[-2t]\sin(15t)$ using the classical

13

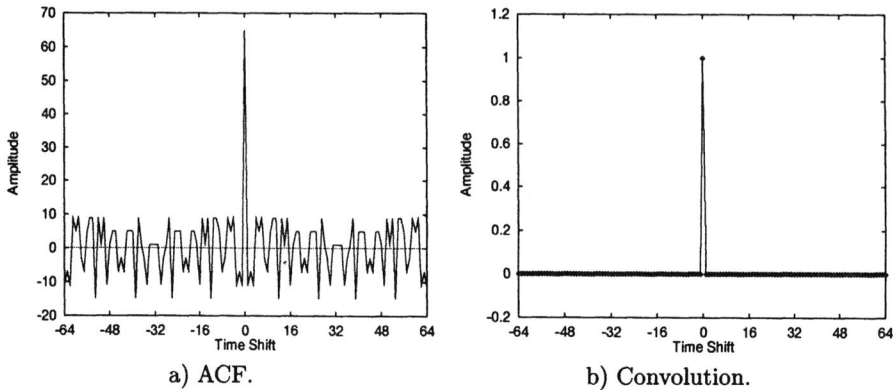

Fig. 13. ACF of the Kasami Sequence and the Convolution with its Reciprocal Sequence.

method; the results are shown in figure 14(a). The identification of the same system was then repeated using the generalised technique; the results of this identification are given in figure 14(b). Again, the non-impulsive nature of the test signal resulting in large estimation errors is evident from the identification using the classical method, whilst the generalised method yields a very accurate estimate of the impulse response.

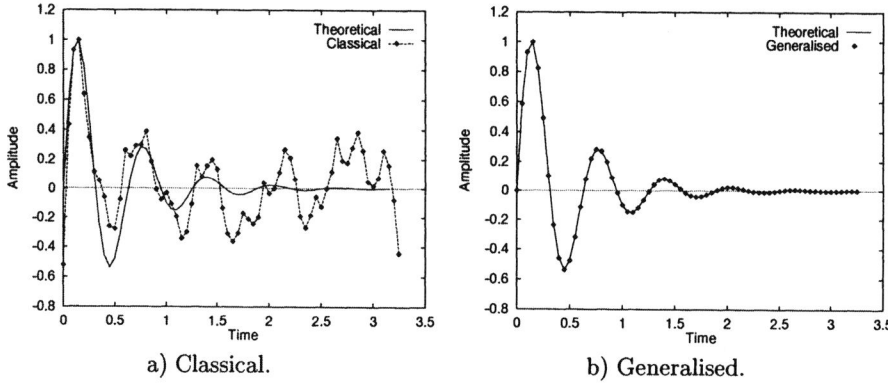

Fig. 14. Classical and Generalised System Identification with 63-bit Kasami Sequence.

5.4 Example-3 : Trajectory-Derived (TD) Sequences

TD sequences are a class of non-binary (quasi-analogue) PR sequences [16]. Figure 15 shows a 96-digit trajectory-derived sequence, together with its periodic ACF. Using the procedure outlined above, the reciprocal sequence was computed for this multi-level sequence and is shown in figure 16.

The trajectory-derived sequence was used in the identification of an LTI system with an impulse response of $h(t) = 0.272t \cdot \exp[-0.1t]$; the results of this estimation using both the classical and generalised technique are given in figures 17(a) and (b), respectively. It is clear from these results that, using the new method, a sequence with a highly imperfect ACF can produce an accurate estimate the impulse response of an unknown LTI system.

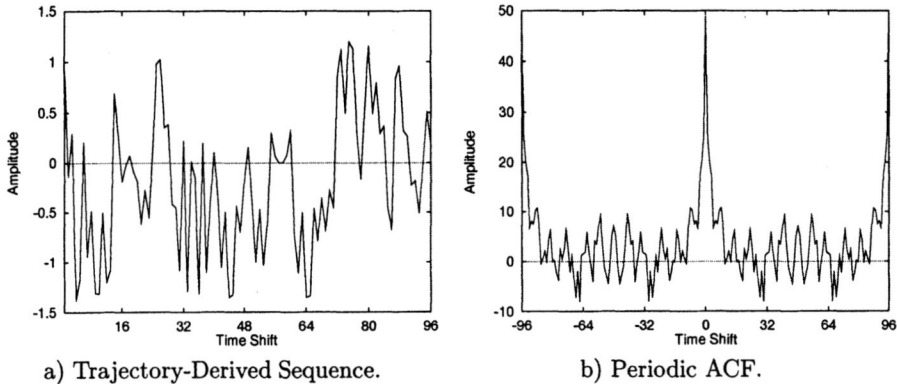

a) Trajectory-Derived Sequence.　　　b) Periodic ACF.

Fig. 15. Trajectory-derived Sequence and its ACF.

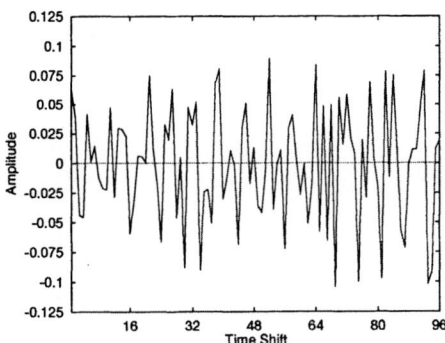

Fig. 16. Reciprocal Sequence of the Trajectory-Derived Sequence.

6　Concluding Remarks

In this paper, the authors have addressed the problem of generalised system identification via impulse response estimation using periodic test signals with non-impulsive ACFs. The induced error in the impulse response estimates due to the imperfection in the ACF of the test sequence has been fully characterised, and a method for its compensation derived. A new formulation of the identification problem has been proposed and its usefulness demonstrated using a Gold, Kasami

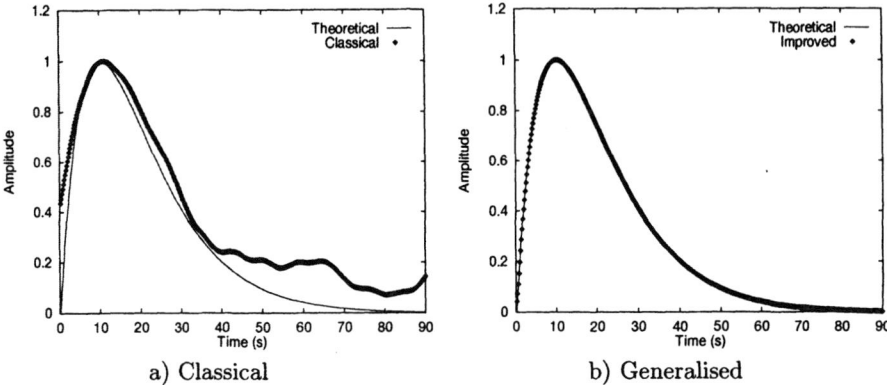

a) Classical   b) Generalised

Fig. 17. Classical and Generalised System Identification with 96-Digit TD sequences.

and TD sequences. The classical correlation-based method is shown to be a special case of this new generalisation. Future research will concentrate on the extension of the generalised technique to incorporate the use of aperiodic test signals.

Acknowledgement

The support of the Defence and Evaluation Research Agency (Malvern) for this work is gratefully acknowledged.

References

1. P Fan and M Darnell, *Sequence Design for Communications Applications*, Research Studies Press. John Wiley & Sons, ISBN: 0 86380 201 X, 1996.
2. M Grayson, *Improved Synchronisation Techniques for Time-Varying Dispersive Radio Channels*, Ph.D. thesis, University of Hull, 1993.
3. K Godfrey, *Perturbation Signals for System Identification*, Prentice Hall, ISBN: 0-13-656414-3, 1993.
4. J Ruprecht and M Rupf, "On the search for good aperiodic binary invertible sequences," *IEEE Tran on Info Theory*, vol. 42, no. 5, pp. 1604–1611, September 1996.
5. J Massey, "The invertibility principle for maximum likelihood estimation," in *4th International Symposium on Communication Theory and Applications, Ambleside*, 13-18, July 1997, pp. 273–274.
6. J Destefano, A Stubberud, and I Williams, *Feedback Control Systems*, Shaum's Outline Series, ISBN: 07 084369 4, 1987.
7. Oppenheim, Willsky, and Young, *Signal And Systems*, Prentice-Hall International Editions, ISBN: 0-13-811175-8, 1983.
8. F Stremler, *Introduction to Communication Systems*, Addison-Wesley, ISBN: 0 201 07259 9, 1982.
9. A Al-Dabbagh, "General purpose response measurement system," Final year project report, University of Hull, Dept of Electronic Engineering, Hull, HU6 7RX, UK, 1990.

10. P Lynn, *An Introduction to the Analysis and Processing of Signals*, McMillan, ISBN: 0-333-48887-3, 3rd edition, 1994.
11. Y W Lee, *Statistical Theory of Communication*, Wiley, 1960.
12. L F H Habil, *Correlation Techniques, Foundations and Applications of Correlation Analysis in Modern Communications, Measurement and Control*, Liffe Books Lts, 1967.
13. A Al-Dabbagh, M Darnell, A Noble, and S Farquhar, "Accurate system identification using inputs with imperfect autocorrelation properties," *Electronic Letters*, vol. 33, no. 17, pp. 1450–1451, Aug 1997.
14. R Gold, "Optimal binary sequences for spread spectrum multiplexing," *IEEE Tran on Info Theory*, vol. IT, no. 13, pp. 619–621, October 1967.
15. R Gold, "Maxmimal recursive sequences with 3-valued recursive crosscorrelation functions," *IEEE Tran on Info Theory*, vol. IT, no. 14, pp. 154–156, 1967.
16. A Al-Dabbagh and M Darnell, "New pseudo-random sequences derived from reflections in enclosures," in *First International Symposium on Communication Theory and Applications, Crieff*, September 1991.

Zero-Error Codes for Correlated Information Sources

A. Kh. Al Jabri and S. Al-Issa

EE Dept., College of Engineerin
King Saud University
P.O. Box 800, Riyadh 11421
Saudi Arabia
Email : f45e011@ksu.edu.sa

Abstract. Slepian and Wolf [2] gave a characterization for the compression region of distributed correlated sources. Their result is for codes with a decoding probability of error approaching zero as the code length is increased. Of interest in many applications is to find codes for which the *probability of error is exactly zero*. For this latter case, block codes using the zero-error information between the sources have been proposed by Witsenhausen [3]. Better codes, however, can be obtained by further exploitation of the statistical dependency impeded in the correlation information. In this paper variable-length zero-error codes are proposed that are generally more efficient than Witsenhausen codes. A method for their construction is presented and an example demonstrating such construction with the achieved rate region are given.

1 Introduction

In most practical situations, data produced by sources have some form of redundancy that can be removed without affecting their information contents. Such removal is necessary for efficient data transmission and/or storage. The amount of redundancy in the data from the source is determined by the Shannon entropy [1]. In multi-user environments, however, information may be collected at several places and brought to a common site for processing or for other use. If the data from the sources are correlated, then further reduction in the total number of transmitted bits required to represent the sources is possible; a fact that was shown by Slepian and Wolf (SW) [2]. A simple example of such situation with the achievable rate region is shown in Figure 1 and will be considered for investigation in this paper.

Consider the two discrete memoryless correlated sources shown in Figure 1a with outputs represented by the random variables X and Y taking values on the alphabet $\mathcal{X} = \{x_1, x_2, \ldots, x_{n_x}\}$ and $\mathcal{Y} = \{y_1, y_2, \ldots, y_{n_y}\}$, respectively, and each pair (x, y), $x \in \mathcal{X}$, $y \in \mathcal{Y}$, of symbols is generated according to the joint probability mass function (pmf) $P_{XY}(x, y)$. Let $\|P_{XY}\| = [P_{XY}(x, y), x = 1, 2, \ldots, n_x, y = 1, 2, \ldots, n_y]$ denotes the matrix representation of this pmf.

An obvious encoding method is to encode each source independently. The achievable rates R_X and R_Y in this case satisfy

$$R_X \geq H(X) \quad \text{and} \quad R_Y \geq H(Y),$$

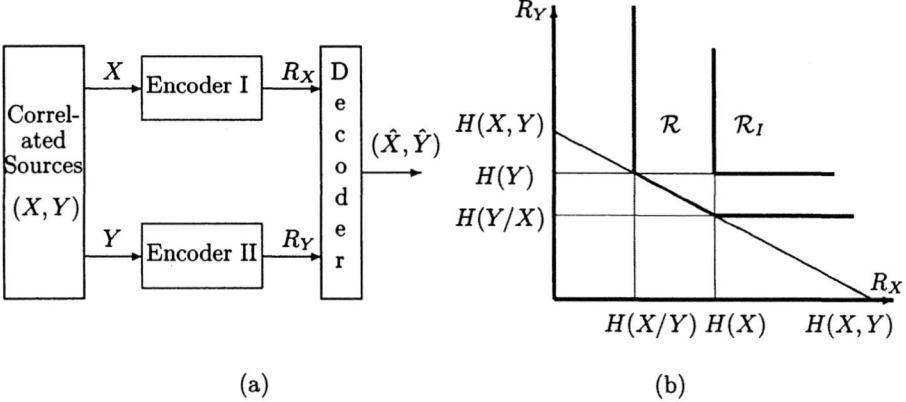

Fig. 1. Correlated source coding configuration and achievable rate region.

where $H(X)$ and $H(Y)$ are the entropies of X and Y, respectively. This determines the two dimensional region $\mathcal{R}_I$ shown in Figure 1b. On the other hand, using the correlation information Slepian and Wolf showed that correlated sources X and Y can be separately described at rates R_X and R_Y and recovered with arbitrarily low probability of error, P_e, by a common decoder if and only if

$$R_X \geq H(X/Y),$$
$$R_Y \geq H(Y/X),$$
$$R_X + R_Y \geq H(X,Y),$$

where $H(X,Y)$ is the joint entropy of X and Y [2].

This paper is concerned with codes for which P_e *is exactly zero*. For a single source, zero-error codes with rates as closed as desired to the source entropy can be constructed using Huffman codes with the implication that zero-error codes do also exist for correlated sources that achieve $\mathcal{R}_I$. Beyond this region, the only known result reported in the literature, to the best of knowledge, is that of Witsenhausen [3]. Finding the region for which the codes P_e is exactly zero is still, however, an open problem. In Witsenhausen work, the minimum number of transmission symbols needed to transmit information with zero-error about say Y in the presence of full description of X at the decoder is related to the chromatic number of the adjacency graph of the transition channel between Y and X [3]. If M is the number of such symbols, then the achieved rate, R_Y, will be

$$R_Y = \frac{1}{n} \log_2 M \qquad \text{bits/source symbol.}$$

In this paper we show the possibility of constructing better codes. The idea is based on the observation that the correlation information can provide in certain cases, in addition to the zero-error information, statistical information that can be used for constructing a class of variable-length zero-error codes that are generally more efficient than block codes.

2   Codes Construction

In the proposed coding method, the data from one of the sources is first encoded by one of the encoders (called here the primary encoder) then the result is delivered to the decoder. The other encoder, on the other hand, (called here the secondary encoder) exploits its zero-error information and their statistical structure obtained through $\|P_{XY}\|$ to encode its own data and pass it to the decoder. From these information, the decoder should be able to recover the data from both sources. It is interesting to note that the selection of which encoder to be the primary and which one to be the secondary is important in the codes construction and efficiency. This is due, in general, to the non symmetry in the conditional statistics viewed from one side to the other.

   This paper introduces preliminary results of a class of zero-error codes that, in general, achieve rates below that obtained by independent encoding of the sources and also more efficient than other known codes. The correlation information, represented by $\|P_{X,Y}\|$, can be viewed as a communication channel connecting the alphabet of the two sources. Such a channel may pass zero-error information and their statistics from one site to the other. By allowing one source to encode its data in an optimal way, the other source, using the side information obtained through the correlation channel, needs only to pass necessary information to the decoder needed to resolve its ambiguity about the data from the second source. The amount of side information depends on the structure of the correlation channel. In general, the correlation channel may not be symmetric. In fact, there are situations where one direction passes zero-error information while the other does not. The erasure channel is a simple example of such situation.

   For correlated sources the following algorithm can be used to construct zero-error variable-length codes. With out loss of generality, let X be the primary encoder and Y be secondary one. The proposed code construction method is as follows:

1. Choose the same block size, m, of input symbols to the encoders. This defines the extended sources X^m and Y^m with alphabet $\mathcal{X}^m$ and $\mathcal{Y}^m$, respectively.
2. Let $\{\underline{x}_1, \underline{x}_2, \ldots, \underline{x}_{n_x^m}\}$ be the set of symbols from X^m. Encode these blocks using Huffman code and let $\{\underline{h}_1, \underline{h}_2, \ldots, \underline{h}_{n_x^m}\}$ be the set of corresponding codewords.
3. Construct for every i, $i = 1, 2, \ldots, n_y^m$, the set $\mathcal{X}_i$ with elements from $\mathcal{X}^m$ such that
$$\mathcal{X}_i = \{\underline{x} : \underline{x} \in \mathcal{X}^m \text{ and } p(\underline{x}/\underline{y}_i) > 0\}.$$
4. Partition the set $\mathcal{Y}^m$ into a number of subsets, say $\mathcal{Y}_1, \mathcal{Y}_2, \ldots, \mathcal{Y}_{d(m)}$, for some number $d(m)$ such that the elements of every $\mathcal{Y}_i$, $i = 1, 2, \ldots, d(m)$ have disjoint corresponding subsets in $\mathcal{X}^m$ as defined in the previous step.

5. Let the symbol s_i represents $\mathcal{Y}_i$, $i = 1, 2, \ldots, d(m)$ with a corresponding probability $p(s_i)$. Encode the set of symbols $\{s_1, s_2, \ldots, s_{d(m)}\}$ using Huffman code and let $\{g_1, g_2, \ldots, g_{d(m)}\}$ be the set of corresponding codewords. This Huffman code and the one given in step 2 are the required codebooks. Notice here that it may be possible that there are more than one way to partition in Step 4. In such case one needs to select the partition that yields the minimum entropy. That is, if $\mathcal{P}$ is the set of all possible partitions, then one needs to find P that yields

$$\min_{P \in \mathcal{P}} \sum_{i=1}^{d(m)} p(s_i) \log_2 \frac{1}{p(s_i)}.$$

This ends the construction phase.

To decode, one needs to perform the following three steps:

1. Decode the symbols received from the primary encoder.
2. Decode the received symbols from the secondary encoder to obtain g_i and hence s_i for some i, $i = 1, 2, \ldots, d(m)$.
3. Since x_i, $i = 1, 2, \ldots, n_x^m$ and s_i are known to the decoder, it will be able to resolve the ambiguity about which symbols have been transmitted from the secondary source and, therefore, recover both encoded data.

The rate achieved depends on the particular partition rule. The achievable rate is then given by

$$R_X \geq H(X)$$

and

$$R_Y \geq \min_{P \in \mathcal{P}} H(Y'),$$

where Y' is a random variable over the set $\{s_1, s_2, \ldots, s_{d(m)}\}$ with a pmf as defined above.

3 Numerical Results

In this part we give an example showing the method of code construction and the performance of the constructed code. Consider the case of two sources for which the joint pmf matrix is given by

$$\|P_{XY}\| = \begin{pmatrix} 3/20 & 3/20 & 0 & 0 & 0 \\ 1/15 & 0 & 1/15 & 1/15 & 0 \\ 1/20 & 0 & 1/20 & 1/20 & 0 \\ 0 & 1/30 & 0 & 1/30 & 1/30 \\ 0 & 1/12 & 0 & 1/12 & 1/12 \end{pmatrix}$$

The marginal pmf of X and Y are given by

i	1	2	3	4	5
$p(x_i)$	3/10	1/5	3/20	1/10	1/4
$p(y_i)$	4/15	4/15	7/60	7/30	7/60

The partitions of the $\mathcal{Y}$ symbols for $m = 1$ are shown below

Symbols in $\mathcal{Y}$	Corresponding subset of $\mathcal{X}$
y_1	$\{x_1, x_2, x_3\}$
y_2	$\{x_1, x_4, x_5\}$
y_3	$\{x_2, x_3\}$
y_4	$\{x_2, x_3, x_4, x_5\}$
y_5	$\{x_4, x_5\}$

There are different ways to take the union among the partitions. These and the achievable rate are summarized below

Partition	Achievable rate
$\{y_1\}, \{y_2\}, \{y_3\}, \{y_4\}, \{y_4\}$	2.23012
$\{y_1\}, \{y_2\}, \{y_3, y_5\}, \{y_4\}$	1.99679
$\{y_1\}, \{y_2, y_3\}, \{y_4\}, \{y_5\}$	1.89028
$\{y_1, y_5\}, \{y_2\}, \{y_3\}, \{y_4\}$	1.89028
$\{y_1, y_5\}, \{y_2, y_3\}, \{y_4\}$	1.55044

The above shows the effect of partition selection on the achieved rate with the last raw yielding the minimum entropy and, hence, the required partition. If, on the other hand, the primary encoder is taken to be on the Y side, then there will be no zero-error codes below $H(X)$. The achievable rate region $\mathcal{R}_0$ by the proposed codes and the SW region are shown in Figure 2. The achieved rates by Witsenhausen code is defined by the region ($R_X \geq H(X)$ and $R_Y \geq \log_2(3)$) which is included in $\mathcal{R}_0$.

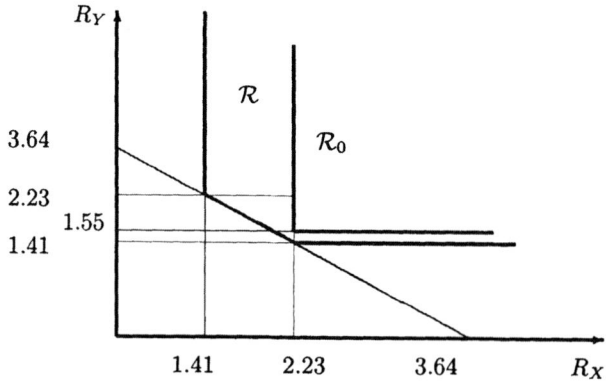

Fig. 2. Achieved rates for different codes ($m = 1$).

4 Conclusions

In this paper a method for generating a class of variable-length zero-error codes for correlated information sources have been proposed. The idea is based on the fact that the correlation between the sources can provide, in many situations, zero-error and statistical information that can be used for designing such codes. The proposed method of encoding yields codes that are generally more efficient than previously known block codes for such sources.

References

1. T. Cover and J. Thomas, Elements Of Information Theory, John Wiley, 1991
2. D. Slepian and J. Wolf, "Noiseless Coding of Correlated Information Sources", *IEEE Trans. Inform. Theory*, Vol. IT-19, pp. 471-480, July 1973.
3. H. S. Witsenhausen,"The zero-error side information problem and chromatic numbers", *IEEE Trans. Inform. Theory*, Vol. IT-22, pp. 592-593, 1976.

Trellis Decoding Techniques and Their Performance in the Adder Channel for Synchronous and Asynchronous CCMA Codes

P. Benachour, G. Markarian and B. Honary
Lancaster Communications Research Centre
SECAMS, Lancaster University, Lancaster, LA1 4YR
E-mail : benachou@lancs.ac.uk

Abstract

A novel technique that allows efficient trellis decoding for M-choose-T synchronous collaborative coding multiple access (CCMA) codes is investigated. The results show that a sufficient performance improvement is achieved without the need for additional information regarding the active users. The application of the proposed technique to asynchronous CCMA codes is also analysed. The modified decoding procedure shows a further improvement when compared to the conventional hard decoding technique.

1. Introduction

In the MA adder channel, the output symbol value is the arithmetic sum of the input symbol values [1,2]. The original model of such a channel was proposed by Kasami and Lin [2] and represents a uniquely decodable code pair of block length $n=2$, as is shown in Tab. 1.

Tab. 1. Conventional MA Coding Scheme With 2 Users

$User_2 \backslash User_1$	(00)	(11)
(00)	(00)	(11)
(01)	(01)	(12)
(10)	(10)	(21)

In this table, $User_1$ has only two codewords, $C_1=(00,11)$, and $User_2$ has three code words, $C_2=(00,01,10)$. Since all the sum codewords are distinct, the decoder can reconstruct the two messages without ambiguity. The overall rate of this MA coding scheme is $R=1.292$ which is better than time sharing [1,2]. Trellis design for the MA adder channel has recently been investigated and implemented [3]. The technique is based on the concept of the Shannon product of trellises [4] and is applicable to MA coding schemes with M (M $\geq$ 2) users. Based on this technique, trellis diagrams for some known 2-user schemes have been constructed; in addition, a DC-free 2-user binary code together with its trellis structure was introduced and investigated. It was shown that such a technique has allowed the achievement of about 2.2 dB energy gain for the overall coding scheme in comparison with the conventional hard decision case.

The results achieved so far use code constructions that are based on a MA model where all the M-users are simultaneously active and the individual users codewords are combined in the channel to form a composite codeword. However, a more practical channel model describes the situation where there are M potential users in the system of which at most T are active simultaneously [5]. However, the need to identify the set of active users using a signalling codeword and the use of hard decision decoding reduces the efficiency of such a scheme.

In this paper, we present a novel decoding technique for the M-choose-T system and we show that a sufficient performance improvement could be achieved without the need for additional information regarding the active users. Initially, the system is tested under the assumption of perfect block and bit synchronisation. For the T-user asynchronous case, the overall trellis structure is a modified version with one sub-trellis corresponding to the synchronised case and a further n sub-trellises were n is the maximum number of possible bit shifts. The modified decoding procedure for the asynchronous scheme shows a further improvement when compared to the conventional hard decoding technique.

2 Trellis Decoding in the M-choose-T Synchronous CCMA Scheme

The decoding technique is termed *look-ahead/look-back* and is based on the codes developed in [5]. Such a technique attempts to identify the set of active users without the need for a special signalling codeword; in fact no extra information is required on the set of active users at all. This approach is based on the assumption that a set of active users will be active during a finite period (frame) in which they will transmit their information. The system keeps track of several received composite codewords and makes a decision when it is confident that a set of users has been active for some time.

2.1 Decoding Procedure
The *look-ahead/look-back* decoding steps are as follows :
Step 1)   Implement *look-ahead* operation by performing maximum likelihood decoding (MLD) on all received composite codewords. At the same time, keep track of metric values accumulated over the transmission period.
Step2)   *look-back* to the metric values and make a decision on the set of active users by identifying the decoder with the least accumulated metric.

2.2 Metric Calculation and Accumulation
The overall metric accumulation process is based on calculating the Euclidean distance for each received composite codeword and updating the metric for the next codeword. Each codeword is decoded according to it's own minimum metric, but an overall metric is updated based on the accumulation of all metrics calculated during the transmission period. Since the scheme is implemented in an adder channel, the output symbol with added noise will be :

$$s = \sum_{i=1}^{T} x_i + N \qquad (1)$$

where x_i, $i=1, 2, ..., T$, is the *ith* user channel input symbol , s the composite channel output symbol and N the Gaussian noise random variable. The Euclidean distance between each received composite vector and all the possible composite codewords is expressed as :

$$\|d_j\|^2 = \sum_{i=1}^{n} \left[r_i - \left(\sum_{i=1}^{T} x_i + N \right) \right]^2 \qquad (2)$$

where r_i and n are the received data and the number of bits in the sequence (codeword) respectively and $\|d_j\|$ means that the distance is a metric where $j=1,2,...,$ $^M C_T$ refers to the number of decoders in such a system. The total metric value accumulated over a transmission period τ can be expressed as follows :

$$\sum_{t=0}^{\tau} \|d_j\|^2 = \sum_{t=0}^{\tau} \sum_{i=1}^{n} \left[r_i - \left(\sum_{i=1}^{T} x_i + N \right) \right]^2 \qquad (3)$$

At the end of transmission, the system makes a decision on which set of active users have combined in the channel by identifying the decoder which corresponds to the smallest accumulated metric over τ .

$$\sum_{t=0}^{\tau} \|d_j\|_{min^j}^2 = \min \left\{ \sum_{t=0}^{\tau} \|d_1\|_{min^1}^2, \sum_{t=0}^{\tau} \|d_2\|_{min^2}^2, ..., \sum_{t=0}^{\tau} \|d_{^M C_T}\|_{min^{^M C_T}}^2 \right\} \qquad (4)$$

3 Trellis Decoding in the T-user Asynchronous CCMA Scheme

3.1 Description of System

In this case, we use an existing 2-user uniquely decodable code in [5] and we shift one of the user's codeword . The original codes for such a scheme are presented below in Tab. 2.

Tab. 2. 2-User Synchronous Scheme

User$_2$\User$_1$	000	001	110
000	000	001	110
010	010	011	120
101	101	102	211

In this table, User$_1$ and User$_2$ have three codewords each, $C_1=(000,001,110)$ and $C_2= (000,010,101)$ and the overall rate is $R=1.057$. If User$_2$ codewords were shifted by 1 bit, the result for such shifted version is presented in Tab. 3 as follows.

26

Tab. 3. 2-User asynchronous Scheme

User$_2$\User$_1$	000	001	110
00X	000	001	110
(000 or 001)	001	002	111
10X	101	102	211
(101 or 100)	100	101	210
01X	010	011	120
(010 or 011)	011	012	121

Inspection of Tab. 3 reveals that the 2-user scheme is not uniquely decodable which makes the decoding process unreliable and delivering correct data to its corresponding destination will not be achievable in some cases . Although this can be seen as a drawback, we will use this effect for good cause. It can also be seen in Tab. 3 that User$_2$ constituent codes have doubled due to the shifting operation performed and therefore $C_2=$ *(000,001,101,100,010,011)* and consequently the overall rate has increased to *R=1.389*.

3.2 Trellis Construction and Representation

The original trellis structures for the User$_1$ and User$_2$ codes in Tab. 2 are illustrated in Fig. 1 and 2 respectively.

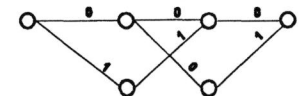

Fig. 1. Trellis Structure of User$_1$ Code

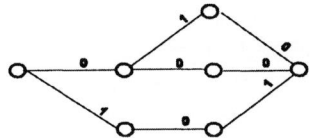

Fig. 2. Trellis Structure of User$_2$ Code

Tab. 3 can be seen to have three main rows, each having two sub rows : upper and lower. The upper sub-rows can be seen to represent the original synchronous scheme in Tab. 2 and the lower sub-rows represent the remainder from the effect of the shifting operation. Therefore, trellis construction for the User$_2$ case will be based on upper and lower sub-trellises. The modified version for such a code is shown in Fig. 3.

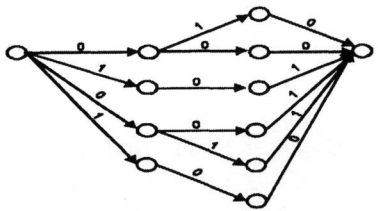

Fig. 3. Modified Trellis Structure of User$_2$ Code

27

In order to construct the composite trellis structure of this scheme, the Shannon Product of trellises is applied to the codes in Fig. 1 & 3. Since User$_2$ trellis has been modified as illustrated in Fig. 3, the composite trellis should also be made up of two sub-trellises. Consequently, the overall composite trellis is illustrated below in Fig. 4.

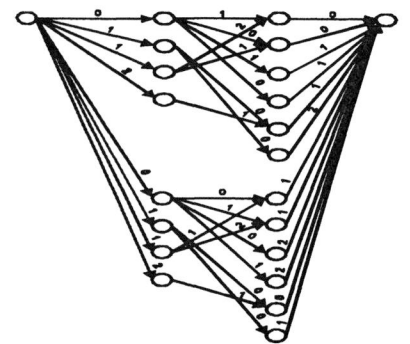

Fig. 4. Shannon Product Trellis Structure of
User$_1$ + User$_2$ Code

4 Simulation Tests and Results

4.1 Simulations for Active Users Identification

The system was tested under additive white Gaussian noise (AWGN) conditions and graphical results for each of the decoder's accumulated metrics for the example in section 2) are presented in Fig. 5. The results show that for higher ratio values of E/No, where E is the average energy per user, the scheme performs reliably and the system can identify the set of active users successfully by detecting the decoder with the least accumulated metric. In Fig. 5, curves 1, 2 and 3 correspond to decoders for active users 1&2, 1&3 and 2&3 respectively and the set of users activated for transmission is users 1&2.

4.2 Simulations for the Synchronous and Asynchronous CCMA Schemes

In Fig. 6, curves 1 and 2 represent the conventional and trellis decoding characteristics for the synchronous scheme. It can be seen that the soft decision case shows a performance improvement when compared to the conventional hard decoding technique. The modified decoder " curve 4 " for the asynchronous case also shows an improvement and outperforms the conventional decoding technique by achieving a coding gain of about 0.5 dB. Curve 3 represents the asynchronous performance when such a scheme is fed through the synchronised decoder in place of the modified version; the graphical results show that even though the modified technique " curve 4 "does not outperform the original synchronous decoder "curve 2", there is a substantial improvement when compared to curve 3.

5. Conclusion

The performances of the synchronous and asynchronous CCMA coding schemes have been investigated and tested. For the M-choose-T synchronous case, a sufficient performance improvement has been realised without the need for additional information on the set of active users. In the T-user asynchronous system, it has been demonstrated that it is possible to improve such a scheme by modifying the constituent codes in use with further increase in overall code rate at the expense of decoder complexity.

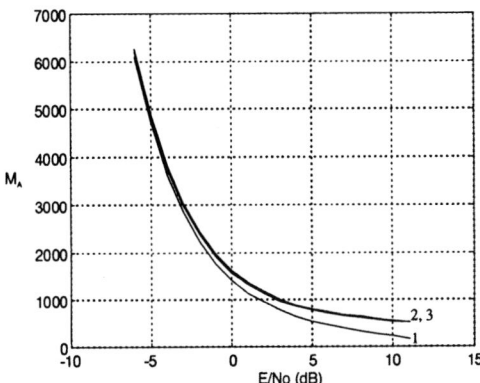

Fig. 5. Graphical plot of accumulated metrics versus E/No
M_A**: Metrics Accumulated**

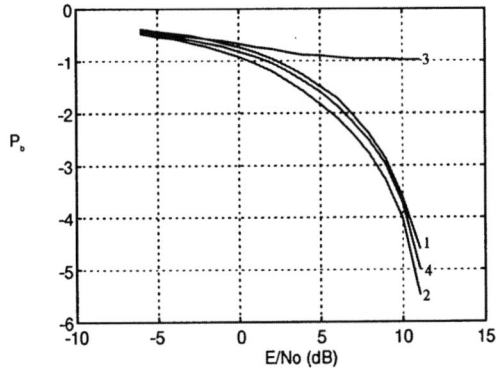

Fig. 6. Performance of the Synchronous and Asynchronous 2-user schemes
P_b **: Probability of Bit Error Rate**

6. References

[1]   P. G, Farrell "Survey of channel coding for multi-user systems", in *Skwirzynski, J.K.* (Ed.) : 'New concepts in multi-user communications' (Sijthoff and Noordhoff, 1981), pp. 133-159.

[2]   T.Kasami and S.Lin, "Coding for a multiple access channel", *IEEE Trans. on Inform. Theory*, Vol. IT-22, No.2, pp 129-137, March 1976.

[3]   G.Markarian, B.Honary, P.Benachour, ''Trellis decoding techniques for the

binary adder channel with M users'', *IEE Proceedings on Communications,* vol.144, 1997.

[4]   V.Sidorenko, G.Markarian, B.Honary, ''Minimal trellis design for linear codes based on the Shannon product '', *IEEE Trans. Inform. Theory,* vol. 42 pp.2048-2053, Nov.1996.

[5]   P.Mathys, ''A class of codes for T active users out of N, for a multiple access

communication system,'' *IEEE Trans. Inform. Theory,* vol. 36 pp.1206-1219, Nov.1990.

Key Agreement Protocols and Their Security Analysis
(Extended Abstract)

Simon Blake-Wilson[1]*, Don Johnson[2], and Alfred Menezes[3]

[1] Dept. of Mathematics, Royal Holloway, University of London, Egham,
Surrey TW20 0EX, United Kingdom. Email: phah015@rhbnc.ac.uk
[2] Certicom Corp., 200 Matheson Blvd West, Mississauga,
Ontario L5R 3L7, Canada. Email: djohnson@certicom.com
[3] Dept. of Discrete and Statistical Sciences, Auburn University,
Auburn, AL 36849-5307, U.S.A. Email: menezal@mail.auburn.edu

Abstract. This paper proposes new protocols for two goals: authenticated key agreement and authenticated key agreement with key confirmation in the asymmetric (public-key) setting. A formal model of distributed computing is provided, and a definition of the goals within this model supplied. The protocols proposed are then proven correct within this framework in the random oracle model. We emphasize the relevance of these theoretical results to the security of systems used in practice. Practical implementation of the protocols is discussed. Such implementations are currently under consideration for standardization [2, 3, 18].

1 Introduction

The *key agreement problem* is stated as follows: two entities wish to agree on keying information in secret over a distributed network. Since the seminal paper of Diffie and Hellman, solutions to the key agreement problem whose security is based on the *Diffie-Hellman problem* in finite groups have been used extensively.

Suppose now that entity i wishes to agree on secret keying information with entity j. Each party desires an assurance that no party other than i and j can possibly compute the keying information agreed. This is the *authenticated key agreement (AK) problem*. Clearly this problem is harder than the key agreement problem in which i does not care who (or what) he is agreeing on a key with, for in this problem i stipulates that the key be shared with j and no-one else.

Several techniques related to the Diffie-Hellman problem have been proposed to solve the AK problem [19, 17, 1]. However, no practical solutions have been provably demonstrated to achieve this goal, and this deficiency has lead in many cases to the use of flawed protocols (see [22, 16, 21]). The flaws have, on occasion, taken years to discover; at best, such protocols must be employed with the fear that a flaw will later be uncovered.

Since in the AK problem, i merely desires that only j *can possibly* compute the key, and not that j has *actually* computed the key, solutions are often said to provide *implicit (key) authentication*. If i wants to make sure in addition that

---

* The author is an EPSRC CASE student sponsored by Racal Airtech. Work performed while a visiting student at Auburn University funded by the Fulbright Commission.

j really has computed the agreed key, then *key confirmation* is incorporated into the key agreement protocol, leading to so-called *explicit authentication*. The resulting goal is called *authenticated key agreement with key confirmation (AKC)*. It is a thesis of this paper that key confirmation essentially adds the assurance that i really is communicating with j to the AK protocol. Thus the goal of key confirmation is similar to the goal of *entity authentication*, as defined in [6]. More precisely, the incorporation of entity authentication into the AK protocol provides i the additional assurance that j *can* compute the key, rather than the (slightly) stronger assurance that j has *actually* computed the key.

Practical solutions that employ asymmetric techniques to solve the AK and AKC problems are clearly of fundamental importance to the success of secure distributed computing. The motivation for this paper stems in part from the recent successes of the 'random oracle model' [7] in providing practical, provably good asymmetric schemes [7, 8, 10, 11, 23], and in part from the desire of various standards' bodies (in particular IEEE P1363 [18]) to lift asymmetric techniques in widespread use above the unsuccessful 'attack-response' design methodology. The goal of this paper is to make strides towards the provision of practical solutions for the AK and AKC problems which are provably good — firstly by providing clear, formal definitions of the goals of AK and AKC protocols, and secondly by furnishing practical, provably secure solutions in the random oracle model. The model of distributed computing adopted appears particularly powerful, and the definitions of security chosen particularly strong. The approach we take closely follows the approach of [6], where provable security is provided for entity authentication and authenticated key transport using symmetric techniques. Also relevant is the adaptation of techniques from [6] to the asymmetric setting found in [13].

Roughly speaking, the process of proving security comes in five stages: (1) specification of model; (2) definition of goals within this model; (3) statement of assumptions; (4) description of protocol; and (5) proof that the protocol meets its goals within the model. We believe that the goals of AK and AKC currently lack formal definition. It is one of our central objectives to provide such definitions.

We particularly wish to stress the important roles that appropriate assumptions, an appropriate model, and an appropriate definition of protocol security play in results of provable security—all protocols are provably secure in some model, under some definitions, or under some assumptions. Thus the emphasis in such work should be how appropriate the assumptions, definitions, and model which admit provable security are, rather than the mere statement that such-and-such a protocol attains provable security. It is a central thesis of this work, therefore, that the model of distributed computing we describe models the environment in which solutions to the AK and AKC problems are required, and that the definitions given for the AK and AKC problems are the 'right' ones.

§2 discusses the requirements of a secure key agreement protocol. §3 describes the model of distributed computing adopted. §4 discusses AKC protocols and introduces the protocols we propose. In §5 we turn our attention to the AK problem. Finally, practical issues are discussed in §6, and §7 makes concluding remarks. The full version of this paper, containing all definitions and proofs as well as extensive additional discussion of the results is available [14].

2   Properties of Key Agreement Protocols

There is a vast literature on key agreement protocols (see [21] for a survey). Unlike other primitives, such as encryption or digital signatures, it is not clear what constitutes an attack on a key agreement protocol. A number of distinct types of attacks have been proposed against previous schemes, as well as a number of less serious weaknesses. Besides resistance to passive and active attacks, a number of desirable *attributes* of key agreement protocols have also been identified:

1. *known session keys.* A protocol still achieves its goal in the face of an adversary who has learned some previous session keys.
2. *(perfect) forward secrecy.* If long-term secrets of one or more entities are compromised, the secrecy of previous session keys is not affected.
3. *unknown key-share.* i cannot be coerced into sharing a key with j without i's knowledge, i.e., when i believes the key is shared with some $l \neq j$.
4. *key-compromise impersonation.* Suppose i's secret value is disclosed. Clearly an adversary that knows this value can now impersonate i, since it is precisely this value that identifies i. However, it may be desirable that this loss does not enable an adversary to impersonate other entities to i.
5. *loss of information.* Compromise of information that would not ordinarily be available to an adversary does not affect the security of the protocol.
6. *message independence.* Individual flows of a protocol run between two honest entities are unrelated.

Each attribute may be desirable for either AK or AKC protocols, or both. For example, we argue in §5 that flaws in AKC protocols that exploit known session keys are a much more serious weakness than such flaws in AK protocols without key confirmation. Similarly, message independence is more desirable in AK protocols; conceptually AKC protocols inherently contain message dependence. A discussion of which of the attributes the protocols proposed in this paper possess can be found in the full paper [14].

3   Model of Distributed Environment

Before formal statements of the problems can be made, we need a formal model to work in. The model is a variant of the *Bellare-Rogaway model*, described in [6, 9]. Our description is necessarily terse; see [6] for further details.

3.1   Set-up

$\{0,1\}^*$ denotes the set of finite binary strings, and λ denotes the empty string. $I = \{1, \ldots, N_1\}$ is the set of entities in this environment (the adversary is not included as an entity). A real-valued function $\epsilon(k)$ is *negligible* if for every $c > 0$ there exists $k_c > 0$ such that $\epsilon(k) < k^{-c}$ for all $k > k_c$.

Definition 1. A *protocol* is a pair $P = (\Pi, \mathcal{G})$ of probabilistic polytime computable functions (polytime in their first input), where Π specifies how (honest) players behave; and $\mathcal{G}$ generates key pairs for each entity. The domain and range of these functions is as follows. Π takes as input the security parameter 1^k, $i \in I$ (identity of sender), $j \in I$ (identity of intended recipient), $K_{i,j}$ (i's key pair K_i together with j's public value), and *tran* (a transcript of the protocol run so far – the ordered set of messages transmitted and received by i so far in this run of the protocol). $\Pi(1^k, i, j, K_{i,j}, tran)$ outputs a triple (m, δ, κ), where $m \in \{0,1\}^* \cup \{*\}$ is the next message to be sent from i to j ($*$ indicates no message is sent), $\delta \in \{\texttt{Accept}, \texttt{Reject}, *\}$ is i's current decision ($*$ indicates no decision yet reached), and κ is the agreed key.

Our protocols are described in terms of arithmetic operations in the subgroup generated by an element α of prime order q in the multiplicative group $\mathbb{Z}_p^* = \{1, 2, \ldots, p-1\}$, where p is a prime. In each case, an entity's private value is an element S_i of $\mathbb{Z}_q^* = \{1, 2, \ldots, q-1\}$, and the corresponding public value is $P_i = \alpha^{S_i} \bmod p$ (the operator '$\bmod p$' will henceforth be omitted), so that i's key pair is $K_i = (S_i, P_i)$. The protocols can be described equally well in terms of the arithmetic operations in any finite group; of course, we would then have to convert our security assumptions on the Diffie-Hellman problem to that group.

$\mathcal{G}$ takes as input the security parameter 1^k and selects the triple of global parameters (p, q, α) to be used by all entities by running $\mathcal{G}_{DH}$, the parameter generator for a Diffie-Hellman scheme, on input 1^k. The operation of $\mathcal{G}_{DH}$ will be discussed in §4, when a Diffie-Hellman scheme is defined. $\mathcal{G}$ then picks a secret value S for each entity by making N_1 independent random samples from $\mathbb{Z}_q^*$, and calculates the public value $P = \alpha^S$ of each entity. $\mathcal{G}$ then forms a directory *public-info* containing the global parameters (p, q, α) and an entry corresponding to each entity — the entry corresponding to entity i consists of the pair (i, P_i) of i's identifier and i's public value. $\mathcal{G}$ outputs each entity's key pair along with the directory *public-info*.

$\mathcal{G}$ is a technical description of the key generation process. It is a formal model designed to capture the attributes of the techniques typically used to generate keys in a distributed environment. Of course, in real life, each entity will usually generate key pairs itself and then get them certified by a Certification Authority.

A generic execution of a protocol between two players is called a *run* of the protocol. While a protocol is formally specified by a pair of functions $P = (\Pi, \mathcal{G})$, in this paper it is informally specified by the description of a run between two arbitrary entities. Any particular run of a protocol is called a *session*. The word 'session' is often associated with anything specific to one particular execution of the protocol. For example, the keying information agreed in the course of a protocol run is referred to as a *session key*. The individual messages that form a protocol run are called *flows*.

3.2 Description of Model

Our adversary is afforded enormous power. She controls all communication between entities, and can at any time ask an entity to reveal its long-term secret key. Furthermore she may at any time initiate sessions between any two entities, engage in multiple sessions with the same entity at the same time, and in some cases ask an entity to enter a session with itself.

With such a powerful model it is not clear what it means for a protocol to be secure. Informally, we say that an AK protocol is secure if no adversary can learn anything about a session key held by an uncorrupted entity i (an entity whose long-term keying material she has not revealed), provided that i has computed that session key in the belief that it is shared with another entity j (who is also uncorrupted). Again informally, we will say that an AKC protocol is secure if the protocol distributes a key just like an AK protocol, and has the additional property that an accepting entity i is assured that it has been involved in a real-time communication with j. Therefore to make an entity accept in an AKC protocol, the adversary effectively has to act just like a wire.

We now formalize the above discussion. An *adversary*, E, is a probabilistic polytime Turing Machine taking as input the security parameter 1^k and the directory *public-info*. E has access to a collection of oracles: $\{ \Pi_{i,j}^s : i \in I, j \in I, s \in \{1, \ldots, N_2\}\}$. Oracle $\Pi_{i,j}^s$ behaves as entity i carrying out a protocol session in the belief that it is communicating with j for the sth time (i.e. the sth run of the protocol between i and j). Each $\Pi_{i,j}^s$ oracle maintains its own variable *tran* to store its view of the run so far. E is equipped with a polynomial number of $\Pi_{i,j}$ oracles, so that $N_2 = T_2(k)$ for some polynomial function T_2. Each $\Pi_{i,j}^s$ oracle takes as initial input the security parameter 1^k, the key pair K_i assigned to entity i by $\mathcal{G}$, a *tran* value of λ, and the directory *public-info*. E is allowed to make three types of queries of its oracles, as illustrated in the table below.

Query	Oracle reply	Oracle update
Send($\Pi_{i,j}^s, x$)	$\Pi^{m\delta}(1^k, i, j, K_{i,j}, tran.x)$	$tran \leftarrow tran.x.m$
Reveal($\Pi_{i,j}^s$)	$\Pi^{\kappa}(1^k, i, j, K_{i,j}, tran)$	—
Corrupt(i, K)	K_i	$K_i \leftarrow K$

In the table, $\Pi^{m\delta}$ denotes the first two arguments of $\Pi_{i,j}^s$'s output, and Π^κ denotes the third. The Send query represents E giving a particular oracle message x as input. E initiates a session with the query Send($\Pi_{i,j}^s, \lambda$), i.e. by sending the oracle it wishes to start the session the empty string λ. Reveal tells a particular oracle to reveal whatever session key it currently holds. Corrupt tells all $\Pi_{i,j}^s$ oracles, for any $j \in I$, $s \in \{1, 2, \ldots, N_2\}$, to reveal entity i's long-term secret value to E, and further to replace K_i with any valid key pair K of E's choice. In addition, all oracles' copies of i's public value in *public-info* are updated.

Our security definitions now take place in the context of the following experiment — the experiment of running a protocol $P = (\Pi, \mathcal{G})$ in the presence of an adversary E using security parameter k: (1) toss coins for $\mathcal{G}$, E, all oracles $\Pi_{i,j}^s$, and any public random oracles; (2) run $\mathcal{G}$ on input 1^k; (3) initialize all oracles; and (4) start E on input 1^k and *public-info*. Now when E asks oracle $\Pi_{i,j}^s$ a query, $\Pi_{i,j}^s$ calculates the answer using the description of Π. This definition of the experiment associated with a protocol implies that when we speak of the probability that a particular event occurs during the experiment, then this probability is assessed over all the coin tosses made in step 1 above.

The first step in defining the security of a protocol is to show that the protocol is 'well-defined'. To assist in this process we sometimes need to consider the following particularly friendly adversary. For any pair of oracles $\Pi_{i,j}^s$ and $\Pi_{j,i}^t$, the *benign adversary* on $\Pi_{i,j}^s$ and $\Pi_{j,i}^t$ is the deterministic adversary that always performs a single run of the protocol between $\Pi_{i,j}^s$ and $\Pi_{j,i}^t$, faithfully relaying flows between these two oracles.

An oracle $\Pi_{i,j}^s$ has *accepted* if $\Pi^\delta(1^k, i, j, K_{i,j}, tran) = \texttt{Accept}$, it is *opened* if there has been a $\texttt{Reveal}(\Pi_{i,j}^s)$ query, and it is *corrupted* if there has been a $\texttt{Corrupt}(i, \cdot)$ query.

4 AKC

The model described in §3 provides the necessary framework for our security proofs; however, before we can prove anything about any protocol, a formal definition of the goal of a secure AKC protocol must be given.

4.1 Definition of Security

As stated in the introduction, a central thesis to this paper is that the goal of an AKC protocol is essentially identical to the goal of an authenticated key transport protocol [6, 13]. The intent of an AKC protocol is therefore to assure two specified entities that they are involved in a real-time communication with each other. Further, the protocol must provide the two entities with a key distributed uniformly at random from $\{0,1\}^k$. No adversary should be able to learn any information about the agreed key held by an uncorrupted entity i, provided the entity j that i believes it is communicating with is also uncorrupted.

PROTOCOL SECURITY. *Matching conversations* provide the necessary formalism to define the assurance provided to entity i during an AKC protocol that it has been involved in a real-time communication with entity j. The idea of matching conversations was first formulated in [12], refined in [17], and later formalized in [6]. Roughly speaking, it captures when the adversary E merely acts like a wire. Let $\texttt{No-Matching}^E(k)$ denote the event that, when protocol P is run against adversary E, there exists an oracle $\Pi_{i,j}^s$ which accepted but there is no oracle $\Pi_{j,i}^t$ which has engaged in a matching conversation to $\Pi_{i,j}^s$. Further, we require $i, j \notin \mathcal{C}$ (where $\mathcal{C}$ denotes the set of entities corrupted by the adversary E during the experiment).

The notion that no adversary can learn information about session keys is formalized along the lines of polynomial indistinguishability. Specifically at the end of its execution, the adversary should be unable to gain more than a negligible advantage when it tries to distinguish the actual key held by an uncorrupted entity from a key sampled at random from $\{0,1\}^k$. Therefore we make the following addendum to the experiment. Call a $\Pi_{i,j}^s$ oracle *fresh* if it has accepted, neither i nor j has been corrupted, it remains unopened, and there is no opened oracle $\Pi_{j,i}^t$ with which it has had a matching conversation. After the adversary has asked all the queries it wishes to make, E selects any fresh $\Pi_{i,j}^s$ oracle. E asks this oracle a single new query: $\texttt{Test}(\Pi_{i,j}^s)$. To answer the query, the oracle flips a fair coin $b \leftarrow \{0,1\}$, and returns the session key $\kappa_{i,j}^s$ if $b = 0$, or else a random key sampled from $\{0,1\}^k$ if $b = 1$. The adversary's job is now to guess b. To this end, E outputs a bit $\texttt{Guess}$. Let $\texttt{Good-Guess}^E(k)$ be the event that $\texttt{Guess} = b$. Then we define: $advantage^E(k) = |Pr[\texttt{Good-Guess}^E(k)] - \frac{1}{2}|$.

Definition 2. A protocol $P = (\Pi, \mathcal{G})$ is a *secure AKC protocol* if: (1) in the presence of the benign adversary on $\Pi^s_{i,j}$ and $\Pi^t_{j,i}$, both oracles always accept holding the same session key κ, and this key is distributed uniformly at random on $\{0, 1\}^k$; and if the following conditions hold for every adversary E: (2) if uncorrupted oracles $\Pi^s_{i,j}$ and $\Pi^t_{j,i}$ have matching conversations then both oracles accept and hold the same session key κ; (3) the probability of $\text{No-Matching}^E(k)$ is negligible; (4) $advantage^E(k)$ is negligible.

The first condition says that in the presence of a benign adversary, oracles always accept holding the same, randomly distributed key. The second says that in the presence of any adversary if two entities behave correctly, and the transmissions between them are not tampered with, then both accept and hold the same key. The third says that essentially the only way for any adversary to get an uncorrupted entity to accept in a run of the protocol with any other uncorrupted entity is by relaying communications like a wire. The fourth says that no adversary can learn any information about a session key held by a fresh oracle.

This definition is identical to the definition of an authenticated key transport protocol as defined in [6]. While it seems strange to suggest that the definitions of security for two goals that have traditionally been regarded as distinct should in fact be identical, the justification for this security definition for AKC is straightforward. It is intuitively what we require from an AKC protocol. Further, a review of the literature reveals that a number of the attacks proposed on previous protocols can be explained by the observation that the protocols concerned do not meet this definition.

4.2 Description of Primitives

The two primitives used in our protocols are are *message authentication codes* (MACs) and *Diffie-Hellman schemes* (DHSs).

MESSAGE AUTHENTICATION CODES. Provably secure message authentication codes have been discussed in [5, 4]. We restrict attention to MACs that have key space uniformly distributed on $\{0, 1\}^k$.

Definition 3 [4]. A *message authentication code* is a deterministic polytime algorithm $MAC_{(.)}(\cdot)$. To authenticate a message m, an entity with key κ' computes: $(m, a) = MAC_{\kappa'}(m)$. The *authenticated message* is the pair (m, a); a is called the *tag* on m. To verify (m, a) is indeed an authenticated message, any entity with key κ' checks that $MAC_{\kappa'}(m)$ does indeed equal (m, a).

An adversary F (of the MAC) is a probabilistic polytime algorithm which has access to an oracle that computes MACs under a randomly chosen key κ'. F's output is a pair (m, a) such that m was not queried of the MACing oracle.

Definition 4 [4]. A MAC is a *secure MAC* if for every adversary F of the MAC, the function $\epsilon(k)$ defined by $\epsilon(k) = Pr[\kappa' \leftarrow \{0, 1\}^k; (m, a) \leftarrow F : (m, a) = MAC_{\kappa'}(m)]$ is negligible.

Roughly speaking, this means a MAC is secure only if the probability of forging a valid tag on any message that has not yet been authenticated using a call to the MACing oracle is negligible. Thus we require a MAC to withstand an adaptive chosen-message attack.

DIFFIE-HELLMAN SCHEMES. The assumption that the Diffie-Hellman problem is hard is common in the cryptographic literature. In order to formalize what we mean by 'the Diffie-Hellman problem is hard', we first define a DHS.

Definition 5. A *Diffie-Hellman scheme* (DHS) is a pair of polytime algorithms, $(\mathcal{G}_{DH}, calc)$, the first being probabilistic. On input 1^k, $\mathcal{G}_{DH}$ generates a triple of global parameters (p, q, α). p and q are primes such that q divides $p - 1$, and α is an element of order q in $\mathbb{Z}_p^*$. $calc$ exponentiates in $\mathbb{Z}_p^*$ — it takes as input $((p, q, \alpha), g, x)$ where the triple (p, q, α) has been generated by $\mathcal{G}_{DH}$, g is in $\mathbb{Z}_p^*$, and x is an integer satisfying $0 \leq x \leq p - 2$. $calc$ outputs $g^x \bmod p$.

An adversary F (of the DHS) is a probabilistic polytime algorithm which takes as input a parameter set (p, q, α) generated using $\mathcal{G}_{DH}$, and a pair $(\alpha^{R_1}, \alpha^{R_2})$ for R_1 and R_2 chosen independently at random from $\mathbb{Z}_q^*$. F's output is an element g of $\mathbb{Z}_p^*$.

Definition 6. A *secure DHS* is one for $Pr[(p, q, \alpha) \leftarrow \mathcal{G}_{DH}(1^k); R_1, R_2 \leftarrow \mathbb{Z}_q^*; g \leftarrow F((p, q, \alpha), (\alpha^{R_1}, \alpha^{R_2})) : g = \alpha^{R_1 R_2}]$ is negligible for every adversary F.

The Diffie-Hellman problem is *hard* if there exists a secure Diffie-Hellman scheme. This formal definition corresponds precisely with our intuitive notion that the Diffie-Hellman problem is 'hard' (in subgroups of prime order) — i.e. it is extremely unlikely that anyone can guess $\alpha^{R_1 R_2}$ given only α^{R_1} and α^{R_2}.

The proofs in this paper assume that i does not enter protocol runs with itself. This condition can be removed provided the DHS being employed is also secure against an adversary that takes as input a pair $(\alpha^{R_1}, \alpha^{R_1})$ rather than $(\alpha^{R_1}, \alpha^{R_2})$. Call a secure DHS *-secure* if it remains secure against this modified opponent. Work done by Maurer and Wolf [20] suggests that secure DHSs and *-secure DHSs are equivalent.

RANDOM ORACLES. The security proofs of this paper take place in the 'random oracle model' [7]. All parties involved in the protocols are supplied with a 'black-box' random function $\mathcal{H}(\cdot) : \{0, 1\}^* \longrightarrow \{0, 1\}^k$. It is often convenient to think of $\mathcal{H}$ defined in terms of its coin tosses in the following way. When $\mathcal{H}$ is queried for the first time, say on string x, it returns the string of length k corresponding to its first k coin tosses as $\mathcal{H}(x)$. When queried with a second string, say x', first $\mathcal{H}$ compares x and x'. If $x' = x$, $\mathcal{H}$ again returns its first k coin tosses $\mathcal{H}(x)$. Otherwise $\mathcal{H}$ returns its second k tosses as $\mathcal{H}(x')$. And so on.

Of course, a random oracle is a theoretical construct designed to facilitate security analysis. In instantiations, $\mathcal{H}$ will be modeled by a hash function H (or some more complex 'key derivation function' [18]). We emphasize that the security proofs take place in the random oracle model, and that instantiating a random oracle using a specific function is a heuristic step, known as 'the random oracle paradigm'. See [7] for further discussion of the 'random oracle paradigm'; here we merely echo the assertion of those authors that 'it is important to neither over-estimate nor under-estimate what the random-oracle paradigm buys you in terms of security guarantees'. Also [11] contains an excellent discussion of the implications of 'provable security in the random oracle model'.

4.3 Protocol 1

We can now describe the first AKC protocol proposed. It is represented graphically in Figure 1. Use $\in_R$ to denote an element chosen independently at random, and commas to denote a unique encoding through concatenation (or any other unique encoding). $\mathcal{H}_1$ and $\mathcal{H}_2$ represent independent random oracles. When entity i wishes to initiate a run of P with entity j, i selects $R_i \in_R \mathbb{Z}_q^*$ and sends α^{R_i} to j. On receipt of this string, j checks that $2 \leq \alpha^{R_i} \leq p-1$ and $(\alpha^{R_i})^q = 1$, then chooses $R_j \in_R \mathbb{Z}_q^*$, and computes α^{R_j} and $\kappa' = \mathcal{H}_1(\alpha^{S_i S_j})$. Finally, j uses κ' to compute $MAC_{\kappa'}(2, j, i, \alpha^{R_j}, \alpha^{R_i})$, and sends this authenticated message to i. (Recall that $MAC_{\kappa'}(m)$ represents the pair (m, a), not just the tag a.) On receipt of this string, i checks that the form of this message is correct, and that $2 \leq \alpha^{R_j} \leq p-1$ and $(\alpha^{R_j})^q = 1$. i then computes κ' and verifies the authenticated message it received. If so, i accepts, and sends back to j $MAC_{\kappa'}(3, i, j, \alpha^{R_i}, \alpha^{R_j})$. Upon receipt of this string, j checks the form of the message, verifies the authenticated message, and accepts. Both parties compute the agreed session key as $\kappa = \mathcal{H}_2(\alpha^{R_i R_j})$. If at any stage, a check or verification performed by i or j fails, then that party terminates the protocol run, and rejects.

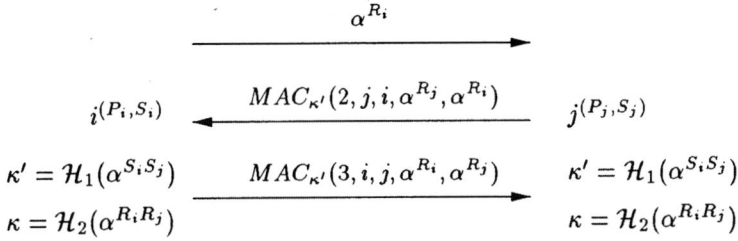

$$i^{(P_i, S_i)} \qquad\qquad j^{(P_j, S_j)}$$

$$\kappa' = \mathcal{H}_1(\alpha^{S_i S_j}) \qquad\qquad \kappa' = \mathcal{H}_1(\alpha^{S_i S_j})$$

$$\kappa = \mathcal{H}_2(\alpha^{R_i R_j}) \qquad\qquad \kappa = \mathcal{H}_2(\alpha^{R_i R_j})$$

Fig. 1. Protocol 1

Theorem 7. *Protocol 1 is a secure AKC protocol provided the DHS and MAC are secure and $\mathcal{H}_1$ and $\mathcal{H}_2$ are independent random oracles.*

COMMENTS. In practice, entity i may wish to append its identity to the first flow of Protocol 1. Doing so in no way affects the security proof. We omit this identity because certain applications may desire to identify the entities involved at the packet level rather than the message level — in this instance, identifying i again is therefore superfluous.

Note that entities use two distinct keys in Protocol 1 — one key for confirmation, and a different key as the session key for subsequent use. This separation appears important. In particular, the common practice of using the same key both for confirmation and as the session key is clearly dangerous if this means the same key is used by more than one primitive.

It is easy to show that the probability that a $\Pi_{i,j}^s$ oracle with $i, j \notin \mathcal{C}$ has a matching conversation with more than one $\Pi_{j,i}$ oracle in a run of Protocol 1 is negligible (the same is true of all the protocols described in this paper).

Protocol 1 is different from most proposed AKC protocols in the manner that entities employ their long-term secret values and session-specific secret values. Most proposed protocols use both long-term secrets and short-term secrets in the formation of all keys. In Protocol 1, long-term secrets and short-term secrets are used in quite independent ways. Long-term secrets are used only to form a session-independent confirmation key and short-term secrets only to form the agreed session key. Conceptually this approach has both advantages and disadvantages over more traditional techniques. On the plus side, the use of long-term keys and short-term keys is distinct, serving to clarify the effects of a key compromise — compromise of a long-term secret is fatal to the security of future sessions, and must be remedied immediately, whereas compromise of a short-term secret effects only that particular session. On the negative side, both entities must maintain a long-term shared secret key κ' in Protocol 1.

Separation of an AK phase of this AKC protocol appears impossible.

4.4 Protocol 2

Protocol 2 is an AKC protocol designed to deal with some of the disadvantages of Protocol 1. It is represented graphically in Figure 2. The actions performed by entities i and j are similar to those of Protocol 1, except that the entities use both their short-term and long-term values in the computation of both the keys they employ. Specifically, the entities use $\kappa' = \mathcal{H}_1(\alpha^{R_i R_j}, \alpha^{S_i S_j})$ as their MAC key for this session, and $\kappa = \mathcal{H}_2(\alpha^{R_i R_j}, \alpha^{S_i S_j})$ as the agreed session key.

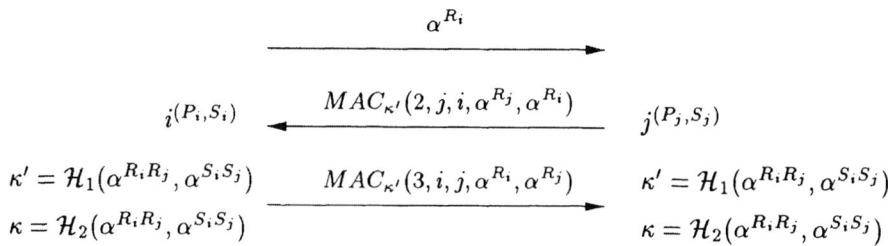

Fig. 2. Protocol 2

Theorem 8. *Protocol 2 is a secure AKC protocol provided the DHS and MAC are secure and $\mathcal{H}_1$ and $\mathcal{H}_2$ are independent random oracles.*

COMMENTS. Unlike Protocol 1, both long-term secrets and both short-term secrets are used in Protocol 2 to form each key. While this makes the effect of a compromise of one of these values less clear, it also means that there is no long-term shared key used to MAC messages in every session between i and j. However, the two entities do still share a long-term secret value $\alpha^{S_i S_j}$. This value must therefore be carefully guarded against compromise, along with S_i and S_j themselves. Conceptually it is possible to separate the AK phase and the key confirmation phase in Protocol 2. This will be the subject of §5.

5 AK

DEFINITION OF SECURITY. In the past, defining the goal of an AK protocol has proved difficult. The clarity of Definition 2 provided for AKC protocols allows us to separate out a definition of security for AK protocols. Informally, we require an AK protocol to distribute a key to two specified entities in such a way that no adversary can learn any information about the agreed key. This is translated into the formal language of our model as follows.

Definition 9. A protocol P is a *secure AK* protocol if: (1) in the presence of the benign adversary on $\Pi_{i,j}^s$ and $\Pi_{j,i}^t$, both oracles always accept holding the same session key κ, and this key is distributed uniformly at random on $\{0,1\}^k$; and if the following conditions hold for every adversary E: (2) if uncorrupted oracles $\Pi_{i,j}^s$ and $\Pi_{j,i}^t$ have matching conversations then both oracles accept and hold the same session key κ; (3) *advantage*$^E(k)$ is negligible.

Conditions 1 and 2 say that a secure AK protocol does indeed distribute a key of the correct form. Condition 3 says that no adversary can learn any information about the key held by a fresh oracle.

PROTOCOL 3. Our first attempt at specifying a secure AK protocol tries to separate an AK phase from Protocol 2. Figure 3 contains a graphical representation of the actions taken by i and j in a run of Protocol 3.

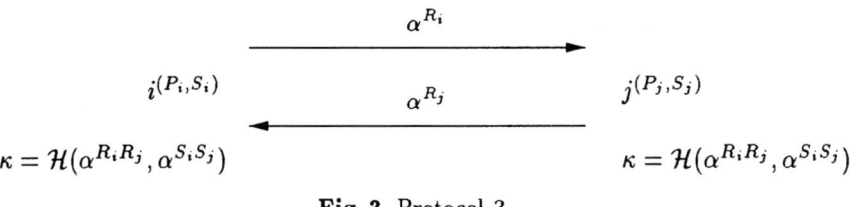

Fig. 3. Protocol 3

Theorem 10. *Protocol 3 is a secure AK protocol as long as E makes no* **Reveal** *queries, and provided the DHS and MAC are secure and $\mathcal{H}$ is a random oracle.*

COMMENTS. To see that Protocol 3 is not a secure AK protocol if an adversary can reveal unconfirmed session keys, notice the following attack. E begins two runs of the protocol, one with $\Pi_{i,j}^s$, and one with $\Pi_{i,j}^u$. Suppose $\Pi_{i,j}^s$ sends α^{R_i}, and $\Pi_{i,j}^u$ sends $\alpha^{R_i'}$. E now forwards α^{R_i} to $\Pi_{i,j}^u$, and $\alpha^{R_i'}$ to $\Pi_{i,j}^s$. E can now discover the session key $\kappa = \mathcal{H}(\alpha^{R_i R_i'}, \alpha^{S_i S_j})$ held by $\Pi_{i,j}^s$ by revealing the (same) key held by $\Pi_{i,j}^u$.

Theorem 11 really says that care must be taken when separating authenticated key agreement from key confirmation. Protocol 3 above is not a secure AK protocol in the full model of distributed computing we've been adopting, but can nonetheless be turned into a secure AKC protocol, as in Protocol 2. At issue here is whether it is realistic to expect that an adversary can learn keys that have not been confirmed. Indeed, studying the list of suggested reasons for session key compromise in [16], it can be seen that the majority of the scenarios discussed lead to the disclosure of *confirmed* keys.

Therefore, although in this paper we have tried to separate the goals of AK and AKC, the principle that Theorem 10 suggests is that no key agreed in an

AK protocol should be used without key confirmation. The only reason we have endeavored to separate authenticated key agreement from key confirmation is to allow flexibility in how a particular implementation chooses to achieve key confirmation. For example, architectural considerations may require key agreement and key confirmation to be separated — some systems may provide key confirmation during a 'real-time' telephone conversation subsequent to agreeing a session key over a computer network, while others may instead prefer to carry out confirmation implicitly by using the key to encrypt later communications.

The reason that we have specified the use of a subgroup of prime order by the DHSs in this paper is to avoid various known session key attacks on AK protocols that exploit the fact that a key may be forced to lie in a small subgroup of $\mathbb{Z}_p^*$. Note however that this condition is not necessary for the security proofs to work — from the point of view of the security proofs, we could equally well have made assumptions about DHSs defined in $\mathbb{Z}_p^*$ rather than a subgroup of $\mathbb{Z}_p^*$.

Theorem 10 testifies to the strength of our definition for security of an AK protocol. Notice in particular that, as is the case with Protocol 3, many previous AK protocols (e.g., those of [19]) do not contain asymmetry in the formation of the agreed key to distinguish which entity involved is the protocol's initiator, and which is the protocol's responder. Such protocols certainly will not meet the security requirements of Definition 9.

PROTOCOL 4. We speculate that the following protocol meets the full rigor required by Definition 9. Again, instead of describing the actions of i and j verbally, we illustrate these actions in Figure 4.

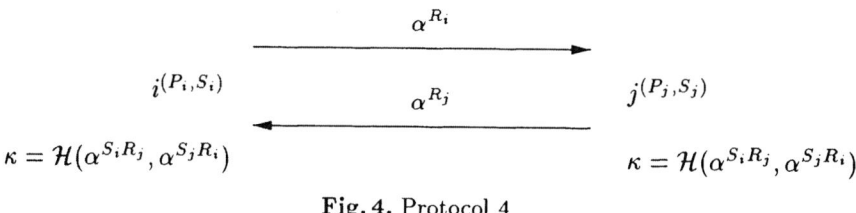

Fig. 4. Protocol 4

Conjecture 11. *Protocol 4 is a secure AK protocol provided the DHS and MAC are secure and $\mathcal{H}$ is a random oracle.*

COMMENTS. While at first glance, Protocol 4 may look almost identical to the well-known MTI protocol [19], where the shared value computed is $\alpha^{S_i R_j + S_j R_i}$, notice the following important distinction. Entity i calculates a different key in Protocol 4 depending on whether i believes it is the initiator or responder. In the first case, i computes $\kappa = \mathcal{H}(\alpha^{S_i R_j}, \alpha^{S_j R_i})$, and in the second case $\kappa = \mathcal{H}(\alpha^{S_j R_i}, \alpha^{S_i R_j})$. As we remarked above, such asymmetry is essential in a secure AK protocol under Definition 9. Of course, such asymmetry is not always desirable — a particular environment may require that i calculate the same key no matter whether i is the initiator or responder. In such a case, Definition 9 would require (slight) modifications.

If indeed it can be shown that Protocol 4 is a secure AK protocol, then we imagine it can be turned into a secure AKC protocol in the same spirit as Protocol 2.

6 Practicalities

6.1 'Real-World' Implications

What are the implications of these theoretical results to the 'real world'?

Until the recent advent of 'practice-oriented provable security', systems which offered any degree of provable security were impractical due to the large computational overheads incurred by their operation. As in [7, 8, 10, 23], this is not the case here. All the protocols in this paper are examples of the 'unified model' of key agreement, which it is our task to present. Practical implementations of the unified model are as efficient as any implementations used in practice; indeed the unified model is currently under consideration for standardization [2, 3, 18].

However, while the results of this paper ensure *theoretical* correctness of the protocols, the theoretical proofs take place in the random oracle model. Therefore the security of a *practical* implementation of any of the protocols relies on the ability of a hash function to instantiate a random oracle. The potential for such an instantiation to introduce weaknesses has led to criticism of the random oracle paradigm. Let us address some common concerns.

Firstly, as with all proofs in the random oracle model, our results guarantee security against *generic* attacks — attacks which do not exploit any special properties of the hash function instantiating the random oracle. It is precisely such generic attacks that have caused the downfall of many previous key agreement protocols. Let us therefore emphasize that such attacks are prohibited by our results within the model of distributed computing employed.

Secondly, let's consider the typical cost of generic attacks on key agreement protocols. In the case of signature and encryption schemes which employ hash functions, generic attacks usually carry a high computational overhead, so it is unclear whether a non-generic attack that exploits the structure of the hash function used will involve as much work as a generic attack. In contrast, generic attacks on key agreement protocols typically involve almost no computational burden – non-generic attacks on the other hand will require the typically greater computational expense of exploiting a weakness in the hash function being used.

Thirdly, let's examine the strength of the requirements made on the instantiation of the random oracle in implementations of the protocols. All the protocols essentially employ a random oracle mapping *fixed-length* inputs to *fixed-length* outputs. Therefore the security of the hash function is not stretched by the need to produce arbitrary length outputs. In most implementations a single application of the hash function chosen will produce sufficiently many output bits. Furthermore, the use of the hash function is likely to be infrequent, since the hash function needs to be used only once or twice each time a new key is agreed. This is in contrast to the use of hash functions in practical implementations of provably secure signature schemes or encryption schemes, where the frequent need to use the hash function makes its efficiency vital. In this instance, an implementation may choose to use a less efficient, but (supposedly) more secure hash function construct. As a concrete suggestion, hash functions and MACs whose construction employs an underlying block cipher could be chosen. Using such instantiations, the security of practical implementations of the protocols can be made to rest on, say, the security of DES and the Diffie-Hellman problem, both cryptographic schemes that have withstood 20 years of extensive investigation.

Finally, we must discuss the justification for using a hash function in a key agreement protocol (aside from its ability to facilitate provable security in the random oracle model). Traditionally, the most significant bits of a Diffie-Hellman number have been used as the agreed session key. While recent work [15] has shown that some of the most significant bits of a Diffie-Hellman number are as hard to compute as the entire number, it is not clear what this exact number of hard bits is. Moreover, it is not known whether these most significant bits are pseudorandom. Thus it seems sensible that a hash function should be used to 'distill' pseudorandomness from the whole Diffie-Hellman number. Hashing in this way also renders known key attacks less damaging — for while it may be unclear whether revealing previous Diffie-Hellman numbers calculated using S_i and S_j enables information about S_i and S_j or other session keys calculated using S_i and S_j to be inferred, the one-way property of hash functions adds to any confidence one may have that disclosure of previous session keys gives nothing away when the agreed value has been hashed to form the session key.

In summary, when employing the proposed protocols in practice, an implementor is assured that no subtle flaws exist in the form of the protocols, and further that an attack on their implementation is likely to incur the heavy computational burden associated with breaking one of the underlying cryptographic primitives. No currently employed protocol can give such assurances, and as discussed in §1, a large number of flaws have been found in the protocols previously proposed. Thus, not only do our protocols provably achieve the goals of AK and AKC in the random oracle model, but in addition, practical implementations of the protocols that employ a hash function to instantiate the random oracle offer superior security assurances compared to any currently in use.

6.2 Implementation Issues

This subsection discusses some practical issues, such as efficiency, that may arise when implementing the protocols.

One issue is how to instantiate the random oracles. SHA-1 should provide sufficient security for most applications. It can be used in various ways to provide instantiations of independent random oracles. For example, an implementation of Protocol 1 may choose to use: $\mathcal{H}_1(x) := \text{SHA-1}(01, x)$ and $\mathcal{H}_2(x) := \text{SHA-1}(10, x)$. A particularly efficient instantiation of the random oracles used in Protocol 2 is possible using SHA-1 or RIPEMD-160. Suppose 80-bit session keys and MAC keys are required. Then the first 80 bits of $\text{SHA-1}(\alpha^{R_i R_j}, \alpha^{S_i S_j})$ can be used as κ' and the second 80 bits used as κ. Of course, such efficient implementations may not offer the highest conceivable security assurance of any instantiation.

It is easy to make bandwidth savings in implementations of the AKC protocols. Instead of sending the full authenticated messages (m, a) in flows 2 or 3, in both cases the entity can omit much of m, leaving the remainder of the message to be inferred by its recipient.

In some applications, it may not be desirable to carry out a protocol run each time a new session key is desired. For example, in Protocol 2, entities may wish to compute the agreed key as: $\mathcal{H}_2(\alpha^{R_iR_j}, \alpha^{S_iS_j}, counter)$. Then instead of running the whole protocol each time a new key is desired, most of the time the counter is simply incremented. Entities need then only resort to using the protocol itself every now and then to gain some extra confidence in the 'freshness' of the session keys they're using.

In Protocols 1, 2, and 3, performance and security reasons may make it desirable to use a larger (and presumably more secure) group for the static Diffie-Hellman number $(\alpha_1^{S_iS_j})$ than for the ephemeral Diffie-Hellman number $(\alpha_2^{R_iR_j})$ calculation. The larger group is desirable because the static number will be used more often. The static numbers may be cached to provide a speed up in session key calculation.

Finally, note that a practical instantiation of $\mathcal{G}$ using certificates should check knowledge of the secret value before issuing a certificate on the corresponding public value. We believe that this is a sensible precaution in any implementation of a Certification Hierarchy.

7 Conclusions and Further Work

This paper has proposed formal definitions of secure AK and AKC protocols within a formal model of distributed computing. The 'unified model' of key agreement has been introduced, and several variants of this model have been demonstrated to provide provably secure AK and AKC protocols in the random oracle model. Strong evidence has been supplied that practical implementations of the protocols also offer superior security assurances than those currently in use, while maintaining similar computational overheads.

The definitions we have suggested for secure AK and AKC protocols are new, and the first question to ask is: are these the correct definitions for AK and AKC? We have supplied justification for the definitions we've chosen; further debate of the appropriateness of these definitions is clearly required.

A number of other questions are suggested by our results. Is the model of distributed computing adopted ideal? What impact do security proofs have on protocols? Can these methods be applied to protocols with different security goals? For which other goals would implementors like to see proven secure solutions? At a more concrete level: is it possible to remove, or at least minimize, the random oracle assumptions on which the security proofs rely? Do the reductions in the proofs yield meaningful measures of exact security [10]?

8 Acknowledgements

The authors are grateful to Karl Brincat, Mike Burmester, and Peter Wild for comments on an early draft of this work, and to Phil Rogaway for comments and an enlightening conversation at PKS'97. The authors also wish to thank Rick Ankney for his contributions to the description of the unified model in X9.42.

References

1. N. Alexandris, M. Burmester, V. Chrissikopoulos, and D. Peppes, "Key agreement protocols: two efficient models for provable security", *Proc. IFIP SEC '96*, 227–236.

2. ANSI X9.42-1996, *Agreement of Symmetric Algorithm Keys Using Diffie-Hellman*, September 1996, working draft.

3. ANSI X9.63-1997, *Elliptic Curve Key Agreement and Key Transport Protocols*, October 1997, working draft.

4. M. Bellare, R. Canetti, and H. Krawczyk, "Keying hash functions for message authentication", *Crypto '96*, 1–15.

5. M. Bellare, J. Kilian, and P. Rogaway, "The security of cipher block chaining", *Crypto '94*, 341–358.

6. M. Bellare and P. Rogaway, "Entity authentication and key distribution", *Crypto '93*, 232–249. A full version of this paper is available at http://www-cse.ucsd.edu/users/mihir

7. M. Bellare and P. Rogaway, "Random oracles are practical: a paradigm for designing efficient protocols", *1st ACM Conference on Computer and Communications Security*, 1993, 62–73.

8. M. Bellare and P. Rogaway, "Optimal asymmetric encryption", *Eurocrypt '94*, 92–111.

9. M. Bellare and P. Rogaway, "Provably secure session key distribution—the three party case", *Proc. 27th ACM Symp. Theory of Computing*, 1995, 57–66.

10. M. Bellare and P. Rogaway, "The exact security of digital signatures – how to sign with RSA and Rabin", *Eurocrypt '96*, 399–416.

11. M. Bellare and P. Rogaway, "Minimizing the use of random oracles in authenticated encryption schemes", *Proceedings of PKS'97*, 1997.

12. R. Bird, I. Gopal, A. Herzberg, P. Janson, S. Kutten, R. Molva, and M. Yung, "Systematic design of two-party authentication protocols", *Crypto '91*, 44–61.

13. S. Blake-Wilson and A.J. Menezes, "Entity authentication and authenticated key transport protocols employing asymmetric techniques", to appear in *Security Protocols Workshop '97*, 1997.

14. S. Blake-Wilson, D. Johnson, and A.J. Menezes, "Key agreement protocols and their security analysis", full version of the current paper, available from the first author, 1997.

15. D. Boneh and R. Venkatesan, "Hardness of computing the most significant bits of secret keys in Diffie-Hellman and related schemes", *Crypto '96*, 129–142.

16. M. Burmester, "On the risk of opening distributed keys", *Crypto '94*, 308–317.

17. W. Diffie, P.C. van Oorschot, and M.J. Wiener, "Authentication and authenticated key exchanges", *Designs, Codes, and Cryptography*, 2 (1992), 107–125.

18. IEEE P1363, *Standard for Public-Key Cryptography*, July 1997, working draft.

19. T. Matsumoto, Y. Takashima, and H. Imai, "On seeking smart public-key-distribution systems", *The Transactions of the IECE of Japan*, E69 (1986), 99–106.

20. U.M. Maurer and S. Wolf, "Diffie-Hellman oracles", *Crypto '96*, 268–282.

21. A.J. Menezes, P.C. van Oorschot, and S.A. Vanstone, *Handbook of Applied Cryptography*, Chapter 12, CRC Press, 1996.

22. J.H. Moore, "Protocol failure in cryptosystems", in *Contemporary Cryptology: the Science of Information Integrity*, G.J. Simmons, editor, IEEE Press, 1992, 541–558.

23. D. Pointcheval and J. Stern, "Security proofs for signature schemes", *Eurocrypt '96*, 387–398.

Low Density Parity Check Codes Based on Sparse Matrices with No Small Cycles

J. Bond

Science Applications International Corp

4015 Hancock Street

San Diego, CA 92110

bond_jw@nosc.mil

S. Hui

Department of Mathematical Sciences

San Diego State University

San Diego, CA 92182

hui@saturn.sdsu.edu

H. Schmidt

Technology Service Corporation

962 Wayne Avenue, Suite 800

Silver Spring, MD 20910

hschmidt@tscwo.com

September 19, 1997

Abstract

In this paper we give a systematic construction of matrices with constant row weights and column weights and arbitrarily large girths. This resolves a problem raised by D. MacKay. The matrices are used in the generator matrices of linear codes. We give the experiment performance results for codes whose associated matrices have girth 8. We also give a randomized construction of matrices with constant row sums and column sums and few 4-cycles. The codes generated using the matrices are used to encode bit streams for a Gaussian channel and decoded using a decoding algorithm that combines features of the algorithms given by MacKay and Cheng and McEliece. The experimental performance results for codes generated using the random matrices are compared to those of the systematically constructed codes. The results show that the codes generated using the random codes with smaller block sizes perform as well as the systematic codes with bigger block sizes. The performance of the systematic codes, for specified weights, can be used to tailor the random codes. MATLAB routines for the construction for the girth 8 case and a special girth 4 case are included.

1 Introduction

A linear code is completely determined by a generator matrix G, or equivalently by a parity check matrix H, which satisfies $HG = 0 \mod 2$. We say that a matrix is *uniform* if the row sums and column sums are constant, respectively. For a matrix containing just 0's and 1's, this is equivalent to the requirement that each row contains a fixed number of 1's and each column contains a possibly different fixed number of 1's. The generation of low density parity check codes using sparse matrices with a uniform structure has attracted much interest recently (see, for example, [1], [4], [5]). Low density parity check codes were first studied by Gallager [2], [3] and rediscovered by MacKay and others. See [4] for a nice discussion of why Gallager's work on low density codes was forgotten for a long time. When decoded with algorithms derived from the Pearl's belief propagation algorithm (see [6]), the sparse matrix codes performed as well as some of the more computationally intensive codes. The results in these papers suggest that these codes may have applications in situations where the block size is not critical.

In [4], a uniform sparse matrix is indirectly used to obtain a generator matrix while in [1], a uniform sparse matrix is used directly in the generator matrix. In both of these papers, the uniform matrices are generated randomly. In [4], the matrix is constrained so that the bipartite graph associated with the matrix has no 4-cycles. The authors suggest that the small cycles may play an important role in the performance of these codes.

Recall that the bipartite graph associated with a $m \times n$ $\{0, 1\}$-matrix A has vertex classes $R = \{r_1, \ldots, r_m\}$ and $C = \{c_1, \ldots, c_n\}$ corresponding to the rows and columns of the matrix. The node r_i is adjacent to node c_j if $A_{ij} = 1$. We will not distinguish between a matrix and its associated bipartite graph when discussing the properties of the bipartite graph.

A problem proposed in [4] is the systematic construction of graphs with given weights and girths. We found an inductive method of constructing such graphs and we will give an algorithmic construction of uniform sparse matrices with arbitrarily large girths. We experimentally obtain the performance results for the codes whose associated matrices with small weights that do not contain 4 or 6-cycles.

We also present the construction of random matrices with few 4-cycles and the experimental performance results of the associated codes. In addition to being useful comparisons to the systematic codes, these codes have much smaller block size for the same row and column sums, and may therefore be more attractive in certain applications. Simulations show that these codes perform better than the systematic codes at high bit-energy-to-noise level while the systematic codes are better at low bit-energy-to-noise level.

2 The Systematic Construction of Uniform Matrices with No Small Cycles

In this section, we give the algorithmic construction of uniform sparse matrices with arbitrarily large girths. We present the algorithms for the construction of matrices whose bipartite graphs contain no 4 or 6-cycles and the general case separately. To avoid unnecessary technical details, we will only give the detailed proof for the no 4 or 6-cycle algorithm. The proof of the general case is similar.

An algorithm for generating uniform matrices with girth 5 for prime row sums is included in the Appendix.

2.1 The Girth 8 Case

Suppose we want to construct a matrix whose column sums are k and whose row sums are h with no 4 or 6-cycles. Without loss of generality, we can assume that $k \leq h$. Otherwise, just switch the k and h and take the transpose. Recall that the Kronecker product $A \otimes B$ of the $m \times n$ matrix $A = (a_{ij})$ and $p \times q$ matrix $B = (b_{ij})$ is the $mp \times nq$ matrix obtained by replacing every element a_{ij} of A by the block $a_{ij}B$. This operation is available in most commercially available software packages, such as MATLAB and Mathematica. Let $U_{m,n}$ be the $m \times n$ matrix with all 1's and let I_n be the $n \times n$ identity matrix.

Algorithm: (k,h)-Regular Graphs with Girth 8

1. **Initialize:** $M = U_{1,h}$

2. **For $j = 2$ to k:**

$$n = \# \text{ of columns of } M$$

$$M = \begin{pmatrix} I_h \otimes M \\ U_{1,h} \otimes I_n \end{pmatrix}$$

end

We include for the convenience of the reader a MATLAB version of the above algorithm.

MATLAB Routine

```
% Routine for (k,h)-regular matrix with girth 8

        function[M]=girth8(k,h)
        U=ones([1 h]);
        M=U;
        for j=2:k,
                [m n]=size(M);
                M=[kron(eye(h),M);kron(U,eye(n))];
        end
```

Figure 1 illustrates our construction for the case of $k = 3$ and $h = 4$. The left-hand plot is a representation of the 48 × 64 matrix with girth 8 where each dot represents 1 with all other entries 0. The right-hand plot is the associated bipartite graph with the first 4 row nodes on top and then alternating rows of column nodes and row nodes. Note the systematic structure of the matrix and its graph. The innermost box in the left plot contains the initial matrix of the construction and the middle box contains the matrix after one iteration. The construction is complete after two iterations since $k = 3$.

It is easy to see by induction that the matrix we constructed above has dimension $kh^{k-1} \times h^k$. It is an open question what the dimension of the smallest

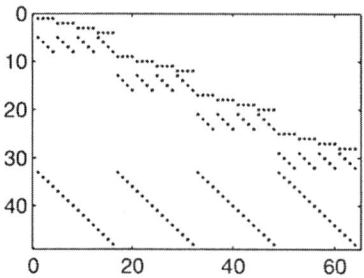 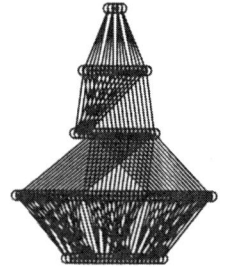

Figure 1: The matrix and its bipartite graph for $k = 3$, $h = 4$

(k, h) uniform matrix is. Note that the above algorithm does not depend on our assumption that $k \leq h$. We made that assumption since $h^k \leq k^h$ for $h \geq k \geq 3$. This follows easily from the fact that the function $\log x / x$ is decreasing for $x > e$.

We now prove that the above algorithm works. Clearly, the initial matrix $U_{1,h}$ has no cycles since the associated graph is a tree with one root and h branches. So the initial M, call it M_1, has no cycles. Suppose $j \geq 2$ and we have iterated to obtain M_{j-1} with row sum h and column sum $j - 1$ and with no 4 or 6-cycles. At the next iteration, we form a new matrix using two Kronecker products. The top block of the new matrix is $I_h \otimes M_{j-1}$. Note that this is a block matrix with h M_{j-1}'s down the main diagonal and zeros elsewhere:

$$
I_h \otimes M_{j-1} = \begin{pmatrix}
M_{j-1} & 0 & 0 & \cdot & \cdot & \cdot & 0 \\
0 & M_{j-1} & 0 & \cdot & \cdot & \cdot & 0 \\
0 & 0 & M_{j-1} & \cdot & \cdot & \cdot & 0 \\
\cdot & \cdot & \cdot & \cdot & \cdot & \cdot & \cdot \\
0 & 0 & 0 & \cdot & \cdot & \cdot & M_{j-1}
\end{pmatrix}.
$$

The blocks are independent and it is clear that the matrix is $(j - 1, h)$-regular. Since there is no interaction between the different blocks, the top block of M_j has no 4 or 6-cycles.

Let us now consider the bottom block of M_j, $U_{1,h} \otimes I_n$, where n is the number of columns of M_{j-1}. It has the form

$$U_{1,h} \otimes I_n = \left(\begin{array}{ccc} I_n & \cdots & I_n \end{array} \right).$$

It is helpful to consider $U_{1,h} \otimes I_n$ as being obtained from $U_{n,h}$ by putting each 1 in a different column while holding its relative row position. More explicitly, the transformation of a column takes the form

$$\begin{pmatrix} 1 \\ \vdots \\ 1 \end{pmatrix} \rightarrow \begin{pmatrix} 1 & 0 & 0 & \cdots & 0 \\ 0 & 1 & 0 & \cdots & 0 \\ & \cdot & \cdot & \cdots & \cdot \\ 0 & \cdot & \cdot & \cdots & 1 \end{pmatrix}.$$

Since n is the number of columns of M_{j-1} and there are n 1's in each column of $U_{n,h}$, the block obtained by stretching each column of $U_{n,h}$ matches each M_{j-1} block in the top half. Clearly, each row of the bottom block has weight h and contributes one additional 1 to each column. Therefore the weight of each column in M_j is one more than that of M_{j-1}. It follows that M_j is (j, h)-regular. Since there is only one nonzero entry in each column in $U_{1,h} \otimes I_n$, there are no cycles in the bottom half of M_{j-1}. The only cycles that involve the rows of $U_{1,h} \otimes I_n$ must also go through the top half. Note that $U_{n,h}$ has no 2-cycles and, since all cycles are even, no 3-cycles. In other words, to start at a node of the associated bipartite graph corresponding to a row or column of the $U_{n,h}$, it takes at least 4 "jumps" to get back to that node. In $U_{1,h} \otimes I_n$, each column node is connected to exactly one row node and it is not possible to move from one row node to another row node directly, which takes 2 jumps. To go from one row node to another row node of $U_{1,h} \otimes I_n$, the path must go through a row node of the top block. Thus, the number of jumps it takes to go from one row node to another row node of $U_{1,h} \otimes I_n$ is at least doubled. Therefore it takes at least 8 jumps to go from one row node back to itself. Combined with the fact that there are no 4 or 6-cycles in the top block, we conclude that there are no 4 or 6-cycles in M_j. By induction, we have proved that the algorithm gives a (k, h)-regular matrix with girth 8.

2.2 The General Case

For the general case, the construction is more complicated and the algorithm requires the construction of (k, h)-regular matrices with half the required girth.

Let $G(k, h, n)$ denote a (k, h)-regular matrix with girth at least n. For example, for n=8, we can use the previous algorithm to generate $G(k, h, 8)$. Let NzD denote the operation of sequentially converting each column of a matrix into a diagonal matrix and then deleting the columns whose entries are all zero. For example,

$$NzD\left(\begin{pmatrix} 1 & 0 \\ 2 & 3 \end{pmatrix}\right) = \begin{pmatrix} 1 & 0 & 0 \\ 0 & 2 & 3 \end{pmatrix},$$

is obtained by the process

$$\begin{pmatrix} 1 & 0 \\ 2 & 3 \end{pmatrix} \rightarrow \begin{pmatrix} 1 & 0 & 0 & 0 \\ 0 & 2 & 0 & 3 \end{pmatrix} \rightarrow \begin{pmatrix} 1 & 0 & 0 \\ 0 & 2 & 3 \end{pmatrix}.$$

Observe also that

$$NzD\left(U_{n,h}\right) = U_{1,h} \otimes I_n.$$

An algorithm for generating a (k,h)-regular matrix is the following.

Algorithm: (k,h)-Regular Graphs with Girth g

1. Initialize: $M = U_{1,h}$

2. For $j = 2$ to k:

$$n = \text{\# of columns of } M$$
$$G = G(n, h, g/2)$$
$$q = \text{\# of columns of } G$$

$$M = \begin{pmatrix} I_q \otimes M \\ NzD(G) \end{pmatrix}$$

end

Note that when $g = 8$, $G(n, h, 4) = U_{n,h}$ and $NzD\left(U_{n,h}\right) = U_{1,h} \otimes I_n$. We see that the algorithm for the girth 8 case is a special case of the general algorithm. The proof that the general algorithm works is very similar to the special case and will be omitted.

The dimensions of the matrices are far from optimal in many cases. For example, if we apply the algorithm for $g = 5$, we actually obtain a matrix with girth 8 and h^k columns. However, there is a way of constructing (k,h)-regular matrices with girth 5 for h prime and $k \leq h$ such that there are only h^2 columns. For completeness, we include this algorithm in the Appendix.

We illustrate the above construction with a simple example where we iterate through $k = 1, 2, 3$ with $h = 3$ and desired girth 10. For $k = 1$, which corresponds to the initialization step, we have $M_1 = U_{1,3}$. There are 3 columns in this matrix and, since $h = 3$ and $g = 10$, we look for $G(3,3,5)$, a $(3,3)$-regular matrix with girth 5. One such matrix is the following 9×9 matrix:

$$G(3,3,5) = \begin{pmatrix} 1 & 0 & 0 & 1 & 0 & 0 & 1 & 0 & 0 \\ 0 & 1 & 0 & 0 & 1 & 0 & 0 & 1 & 0 \\ 0 & 0 & 1 & 0 & 0 & 1 & 0 & 0 & 1 \\ 1 & 0 & 0 & 0 & 1 & 0 & 0 & 0 & 1 \\ 0 & 1 & 0 & 0 & 0 & 1 & 1 & 0 & 0 \\ 0 & 0 & 1 & 1 & 0 & 0 & 0 & 1 & 0 \\ 1 & 0 & 0 & 0 & 0 & 1 & 0 & 1 & 0 \\ 0 & 1 & 0 & 1 & 0 & 0 & 0 & 0 & 1 \\ 0 & 0 & 1 & 0 & 1 & 0 & 1 & 0 & 0 \end{pmatrix},$$

which is generated using the algorithm, for h prime, given in the Appendix. To obtain M_2, we let the top block be $I_9 \otimes M_1$, or just 9 $U_{1,3}$ down the main diagonal, which gives a 9×27 block. The bottom block is obtained by applying NzD to $G(3,3,5)$. For example, the first column undergoes the expansion

$$\begin{pmatrix} 1 \\ 0 \\ 0 \\ 1 \\ 0 \\ 0 \\ 1 \\ 0 \\ 0 \end{pmatrix} \rightarrow \begin{pmatrix} 1 & 0 & 0 \\ 0 & 0 & 0 \\ 0 & 0 & 0 \\ 0 & 1 & 0 \\ 0 & 0 & 0 \\ 0 & 0 & 0 \\ 0 & 0 & 1 \\ 0 & 0 & 0 \\ 0 & 0 & 0 \end{pmatrix}.$$

Thus M_2 is 18×27. The matrix M_2 is $(2,3)$-regular with girth 10. See Figure 2 for an illustration of M_2 and its associated bipartite graph. In the graph, the first 9 row nodes are on top.

For M_3, we observe that there are 27 columns in M_2 and we look for a $(27,3)$-regular matrix with girth at least 5. At present, the best we have is the construction that actually gives a $(27,3)$-regular matrix G with girth 8 and it is 19683×2187.

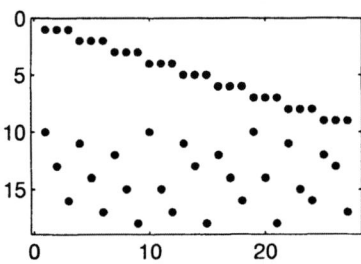 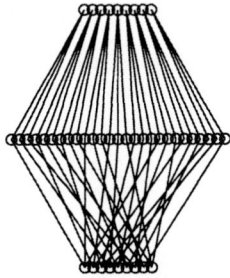

Figure 2: The matrix and its bipartite graph for $k = 2$, $h = 3$, girth=10.

The top block is $I_{2187} \otimes M_2$, which is 39366×59049. The bottom block is $NzD(G)$, which is 19683×59049. Thus M_3, a $(3,3)$-regular matrix with girth 10, is 59049×59049. We do not know what the smallest $(3,3)$-regular matrix with girth 10 is.

3 Construction of the Random Matrices

In this section, we describe how the random sparse matrices are generated. Let k and h be given and let the desired dimension of the matrix be $m \times n$ with $mh = nk$. The matrices generated are uniform but they may have 4-cycles, no more than $h(h-1)/2$ and usually less than 10. Our approach is based on the approach in [1] and modified as suggested in [4] to remove most of the 4-cycles.

The rows of the matrix are selected iteratively as follows. For each row, except for the last row, 1's are put in h distinct randomly selectly columns, from a total of n columns, and compared with the previous rows to ensure that there are no 4-cycles. If 4-cycles are found, the row is discarded and another independent attempt to generate the row is made. The process described thus far guarantees that the row sums are h. To ensure that the column sums are k, we use an adaptive random number generator for the column selection. At each row, each column is weighted in the random draw by the unused sum of that column. More explicitly, if the sum of a column constructed up to that point is c, then the weight for that column in the random generator is $k - c$. It follows that the column sums are no more than k. Using this method, it much more likely that each column sum is exactly k. Assume that $m - 1$ rows have been constructed, the last row is completely determined and 4-cycles may arise.

It is worth nothing that the above procedure is probabilistic in nature and need not converge. However, when m and n are large relative to k and h, the procedure converged in all attempts.

4 Experimental Results

In this section, we present the experimental performance results obtained by simulations. In the first subsection, we present the the performance results for the systematic codes and the comparisons with the random codes. In the second subsection, we discuss the possible roles played by the weights and the block sizes for the random codes.

In our experiments, we encode a bit stream with the systematic or random codes, add Gaussian noise to the bits, and then decode. We use a modified form of the decoding algorithm that given in [1]. This algorithm is derived from Pearl's belief propagation algorithm. See [1], [4], [6] for details.

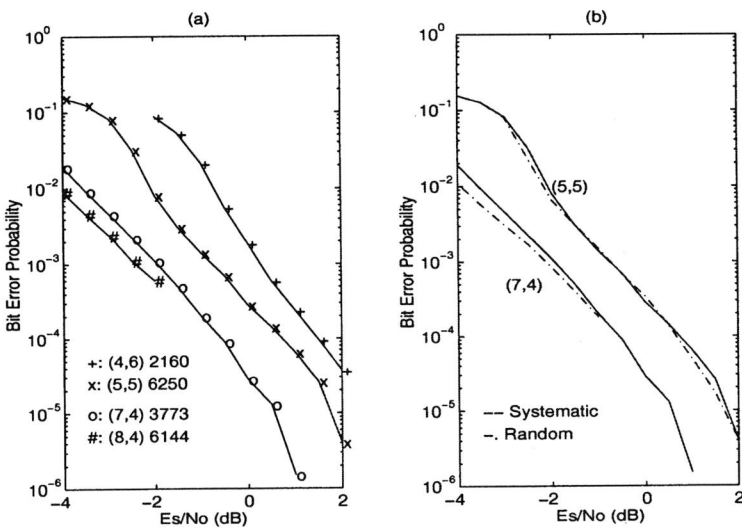

Figure 3: (a) Performances of the Systematic Codes (b) Comparison of Systematic and Random Codes for (5,5) & (7,4)

We generated matrices with girth 8 for small k's and h's using our systematic construction and using the random method constructed matrices of various sizes for the same k's and h's. This allows a meaningful comparison of the performances of the random and the systematic constructions. The random matrices have very few 4-cycles and the systematic matrices have no 4 or 6-cycles.

4.1   The Systematic Codes

The performance curves for the systematic codes with girth 8 and (k, h) equal to $(5,5)$ and $(7,4)$ for E_s/N_0 in the range of -4 dB to 2 dB are given in Figure 3(a). Partial results for the $(8,4)$ and $(4,6)$ systematic codes are also included in the same figure. The curves are essentially flat with the same slope in the E_s/N_0 range under consideration. The performance improves when the rate is decreased.

In Figure 3(b), the performance curves for the $(5,5)$ and $(7,4)$ systematic and random codes are presented. The curves for the systematic codes are the same as those in Figure 3(a). It is clear from the figure that for the same block size and weights, the systematic codes and the random codes have about the same levels of performance. The systematic codes have no 4 or 6-cycles while the random codes have less than ten 4-cycles each. We conclude that the 6-cycles in the random codes did not affect their performances relative to the systematic codes. This suggests that when the block size is sufficiently large, an average number of 6-cycles in a code with few 4-cycles will not affect performance. As we will see in the next section, the performance of the codes deteriorates with decreasing block size.

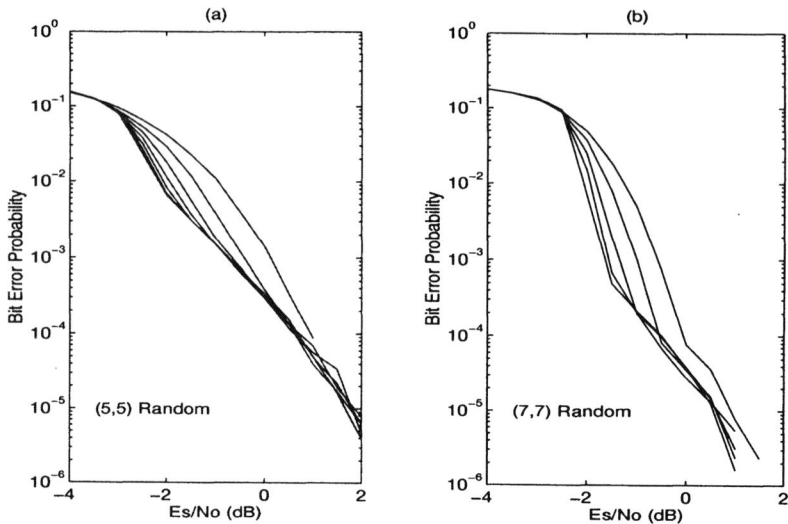

Figure 4: Affect of Increasing Block Size (a) (5,5) Random Codes (b) (7,7) Random Codes

4.2 The Roles of the Weights and Block Sizes in Random Codes

In this section, we present the results on the changing performances due to changes in the block sizes and the weights for rate 1/2 codes. For rate 1/2 codes, the row sum h is equal to the column sum k.

Experiments on the block size changes were conducted for (k, h) equal to $(5, 5)$ and $(7, 7)$. The block size for the $(5, 5)$ code was increased from 100 to 6250 by doubling, except from 3200 to 6250, and for the $(7, 7)$ code from 400 to 6250 by doubling as before. The performance curves for the $(5, 5)$ codes are in Figure 4(a) and for the $(7, 7)$ codes in Figure 4(b). The results show that the performance improves with increasing block sizes. For that reason, we did not label the curves individually. Note that the performance curves for the $(5, 5)$ codes for block sizes 800 to 6250 are virtually indistinguishable. For the particular random $(5, 5)$ codes we used, the number of 6-cycles ranges between 600 and 800 and is generally decreasing with increasing block size. We conjecture that past a certain "average number of 6-cycles per bit" threshold, the number of 6-cycles is not a determining factor for performance. The performance curves for the $(7, 7)$ codes are more separated than the $(5, 5)$ curves but the curves again "converge". We suspect that the performance will not improve significantly with bigger block sizes and that the performance of the $(5, 5)$ systematic code is in some sense an upper bound of achievable performance for $(5, 5)$ random codes with no or few 4-cycles.

Figure 5(a) and Figure 5(b) contain the performance curves for $k = 5, 7,$ and 10 and block sizes 1600 and 3200, respectively. The $(10, 10)$ code is the best for high E_s/N_0 while $(5, 5)$ code is best for low E_s/N_0. This is consistent with the results in [1].

56

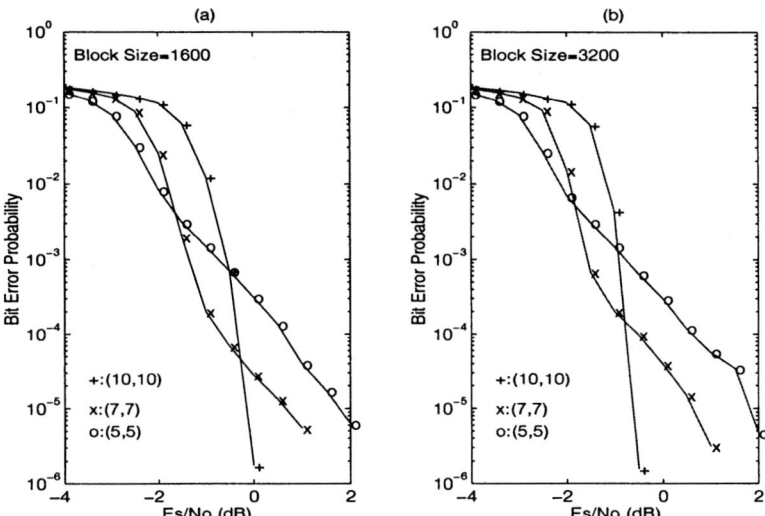

Figure 5: Affect of Increasing K on Rate $\frac{1}{2}$ Random Codes (a) Block Size=1600
(b) Block Size=3200

Based on the above observations, we can use the systematic codes with girth 8 to provide a guide to the available performance for various weights and then use the random codes with smaller block sizes for the actual implementation. This allows us to tailor the random codes to meet performance requirements.

5   Conclusion

In this paper, we gave a systematic construction of uniform matrices with arbitrarily large girths and a randomized construction of uniform matrices with few 4-cycles. The performances of the systematic codes and the random codes are compared and the possible roles played by the weights and the 6-cycles in these codes are considered. Even though there is no satisfactory theoretical explanation for the roles played by these factors, the experiments indicate that the minimum of the weights determine the slope of the performance curves while the 6-cycles determine performance at low bit-energy-to-noise levels. The slope of the performance curves for the random codes can be tailored by the selection of weights and block sizes.

6   Appendix

In this section, we give the algorithm that generates (k,h)-regular matrices with girth 6 for h prime. The matrix generated has dimension $kh \times h^2$. The algorithm fails in general for h not prime. We require the primality of h because the algorithm depends on the fact that $\mathbb{Z}_h$, the integers modulo h, is a field when h is prime. The proof that this algorithm works is not difficult but will be omitted.

Let $h = p$ be prime. Let $E_0 = I_p$ and for $j = 1, \ldots, p-1$, let E_j be the right cyclic shift of the columns of E_{j-1}. For example, when p=3,

$$E_1 = \begin{pmatrix} 0 & 1 & 0 \\ 0 & 0 & 1 \\ 1 & 0 & 0 \end{pmatrix}, \quad E_2 = \begin{pmatrix} 0 & 0 & 1 \\ 1 & 0 & 0 \\ 0 & 1 & 0 \end{pmatrix}.$$

A (k, h)-regular matrix with girth 6 is given by the $kp \times hp$ consisting of kh blocks of $p \times p$ matrices B_{ij}, $i = 1, \ldots, k$ and $j = 1, \ldots, h$ such that

$$B_{ij} = E_{(i-1)(j-1) \bmod p}.$$

For example, when $k = 2$ and $h = p = 3$, we have the matrix

$$
\begin{aligned}
M &= \begin{pmatrix} B_{11} & B_{12} & B_{13} \\ B_{21} & B_{22} & B_{23} \end{pmatrix} \\[4pt]
&= \begin{pmatrix} I & I & I \\ I & E_1 & E_2 \end{pmatrix} \\[4pt]
&= \begin{pmatrix}
1 & 0 & 0 & 1 & 0 & 0 & 1 & 0 & 0 \\
0 & 1 & 0 & 0 & 1 & 0 & 0 & 1 & 0 \\
0 & 0 & 1 & 0 & 0 & 1 & 0 & 0 & 1 \\
1 & 0 & 0 & 0 & 1 & 0 & 0 & 0 & 1 \\
0 & 1 & 0 & 0 & 0 & 1 & 1 & 0 & 0 \\
0 & 0 & 1 & 1 & 0 & 0 & 0 & 1 & 0
\end{pmatrix}.
\end{aligned}
$$

The bipartite graph associated with this matrix is given is Figure 6.

The matrices constructed with the above algorithm can be used as the initial matrix for constructing matrices with larger k's and h's. For example, it is quite easy to construct the $(p+1, p+1)$-regular matrix with girth 5 and dimension $p(p-1)+1 \times p(p-1)+1$ using the (p, p)-regular matrix. We include a MATLAB version of the algorithm.

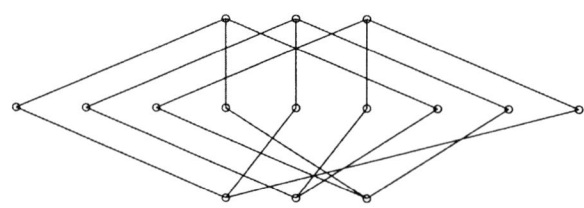

Figure 6: Bipartite graph for $k = 2$, $h = 3$, and girth 6.

MATLAB Routine

```
function[M]=girth4(k,h)
    E=sparse(h,h^2);
    E(1,:)=sparse(ones([h 1]),(1:h+1:h^2)',ones([h 1]));
    for j=2:h,
            E(j,1:h^2-h)=E(j-1,h+1:h^2);
            E(j,h^2-h+1:h^2)=E(j-1,1:h);
    end
    M=[kron(spones(ones([1 h])),speye(h));E];
    for i=3:k,
            F=sparse(h,h^2);
            for j=1:h,
                    q=rem((i-1)*(j-1),h);
                    F(:,(j-1)*h+1:j*h)=E(:,q*h+1:(q+1)*h);
            end
    M=[M;F];
    end
```

References

[1] J. F. Cheng and R. J. McEliece, Some high rate near capacity codes for the Gaussian channel, Preprint.

[2] R. G. Gallager, Low density parity check codes, IRE Transactions on Information Theory, Jan. 1962.

[3] R. G. Gallager, *Low Density Parity Check Codes*, MIT Press, 1963.

[4] D. J. C. MacKay, Good error-correcting codes based on very sparse matrices, Preprint.

[5] D. J. C. MacKay and C. P. Hesketh, Sensitivity of low density parity check codes to decoding assumptions, Preprint.

[6] J. Pearl, *Probabilistic Reasoning in Intelligent Systems: Networks of Plausible Inference*, Morgan Kaufmann Publishers, 1988.

On the SAFER Cryptosystem

Karl Brincat[1] and Alko Meijer[2]

[1] Mathematics Department, RHUL, Egham Hill, Egham TW20 0EX, UK.
[2] Mathematics Department, UND, King George V Avenue, Durban 4001, RSA.

Abstract. An abstraction and some desirable properties of the PHT layer as defined in SAFER are identified. These properties lead to the construction of 3071 other structures which could be used as alternative PHT layers. Results of preliminary investigations carried out on these structures to ascertain their suitability as alternatives are briefly discussed, together with other possible uses and open problems.

1 Introduction

The SAFER cryptosystem, introduced in [3] is an iterated block cipher whose design principles were motivated by the principles of diffusion and confusion as set out in [7]. The confusion is attained mainly by the use of s-boxes in the form of functions over GF(257). Diffusion is accomplished by the use of "an unorthodox linear transform ... the Pseudo-Hadamard Transform". Other novel features introduced in the definition of the system include the use of key biases to eliminate the possibility of "weak keys".

The suitability of the choice of functions for confusion was investigated by Serge Vaudenay in [8]. In this paper he also points out that the Pseudo-Hadamard Transform (PHT) is not a multipermutation and that there are no multipermutations of the required size which are linear over $\mathbb{Z}_{256}$. The choice of functions for the non-linear layer of SAFER was shown to be optimal from a certain point of view, establishing that a particular linear attack was not possible for this choice of functions while it is possible for other choices.

The main potential attacks on SAFER were presented in [2] and [6]. Under certain analysis, it was shown how to exploit properties of the PHT layer to the detriment of the cryptosystem. Murphy in [6] reveals weaknesses in the structure of the PHT layer under a module-theoretic approach. Knudsen and Berson in [2] make use of certain structural properties of the PHT layer as given in [4] to launch Truncated Differential attacks on SAFER with 5 rounds which are better than exhaustive search.

The main thrust of this paper is to investigate the PHT layer as used in SAFER. First, an abstraction of the "good" diffusion properties of the layer is made. The main contribution of this paper is the use of this abstraction to construct 3072 structures, each of which yields a different PHT level. One of the 3072 structures corresponds to that used in SAFER. A preliminary investigation, using certain criteria, was carried out on the other 3071 items with the scope to see whether there are any suitable alternatives to the one used in SAFER. The results of these initial investigations are promising. A non-trivial set of possible alternatives satisfying certain empirical conditions has been determined. Further work is to be done to establish which, if any, of the 3071 alternatives can be used instead of the PHT structure defined in SAFER.

The rest of this paper is organised as follows. In section 2 a brief description of SAFER and some related work is given. In section 3 the main contribution of this paper is presented by exhibiting the set of objects that satisfy the "good" properties of the PHT layer. In section 4 some results of the preliminary investigations undertaken to examine the alternative structures are briefly discussed. Finally in section 5 conclusions and other issues are presented.

2 Description of SAFER and previous work

SAFER is an r round cipher, where the original recommendation for r was a minimum of 6 and a maximum of 10 for SAFER K-64, [3]. SAFER operates on bytes rather than bits and uses a mix of operations over different algebraic structures to get good diffusion and confusion after a sufficient number of rounds. The general encryption scheme for all the SAFER variants is the same, the differences being in the length of the keys employed (64, 128 or 40 bits) and the actual sub-key generation for encryption (and decryption). After Lars Knudsen pointed out some weaknesses of SAFER if used for hash functions [1], he gave a new key-scheduling scheme which prevents the attacks in his own paper and also attacks by Murphy [6]. This resulted in another sub-class of variants, SAFER SK-64, SAFER SK-128 and recently SAFER SK-40, employing the new key-schedule rules. In the present paper the actual key-schedule used is not important and therefore is not described. Interested readers are referred to [3] and [4] for the original definitions and [1] for the new key-schedule. The description of SAFER SK-40 [5] is available from Professor James Massey.

Each of the r rounds in SAFER is similar in that it is made up of four main layers. The key of the cryptosystem is used to generate $2r + 1$ subkeys $K_1, K_2, \ldots, K_{2r}, K_{2r+1}$, where subkeys K_{2i-1} and K_{2i} are used in round i and subkey K_{2r+1} is used in a final output transformation after round r. The four main layers are as follows:

1. *Mixed XOR/Addition layer.* Here bytes 1, 4, 5 and 8 of the round input are XORed with the respective bytes (1, 4, 5 and 8) of subkey K_{2i-1}, while bytes 2, 3, 6 and 7 of the input are added byte-wise modulo 256 to the respective bytes of K_{2i-1}.

2. *Non-linear layer.* This layer makes use of two functions — $45^{(x)}$ and its inverse $\log_{45}(x)$. The element 45 is a primitive in the field of 257 elements, $GF(257)$, and hence the transformation $45^x \bmod 257$ defines a permutation on the set of non-zero residues modulo 257. A byte x can be canonically associated with an integer in the range $[0 : 255]$. The transformation $45^{(x)}$ is defined as $45^{(x)} = 45^x \bmod 257$, for all bytes x considered as integers, except for $x = 128$, where in this case $45^{(128)} = 0$. The function $45^{(x)}$ is used on bytes 1, 4, 5 and 8, while the function $\log_{45}(x)$ is used on bytes 2, 3, 6 and 7 of the output from the previous layer.

3. *Mixed Addition/XOR layer*: Bytes 1, 4, 5 and 8 from the output of the non-linear layer are added byte-wise to the respective bytes of subkey K_{2i} and bytes 2, 3, 6 and 7 of the output of the previous layer are XORed with the respective bytes of K_{2i}

4. *Three-level "linear" layer (PHT)*: This layer consists of three levels, each consisting of four applications of the linear transform 2-PHT (2-point pseudo-Hadamard transform). The 2-PHT takes two input bytes (a_1, a_2) and outputs (b_1, b_2) where $b_1 = 2a_1 + a_2$ and $b_2 = a_1 + a_2$. All operations are done modulo 256. The outputs of levels 1 and 2 are permuted, using the decimation-by-2 permutation, to give the inputs to levels 2 and 3 respectively. The byte-pairs $(2j-1, 2j)$ for $j = 1, 2, 3, 4$ are operated on by the 2-PHT transforms in each level. The overall effect of this layer is to take the 8-byte vector v, output of the Mixed Addition/XOR layer and produce the 8-byte vector vM, where M is the following matrix and all operations are carried out in the ring $\mathbb{Z}_{256}$:

$$M = \begin{pmatrix} 8 & 4 & 4 & 2 & 4 & 2 & 2 & 1 \\ 4 & 2 & 4 & 2 & 2 & 1 & 2 & 1 \\ 4 & 2 & 2 & 1 & 4 & 2 & 2 & 1 \\ 2 & 1 & 2 & 1 & 2 & 1 & 2 & 1 \\ 4 & 4 & 2 & 2 & 2 & 2 & 1 & 1 \\ 2 & 2 & 2 & 2 & 1 & 1 & 1 & 1 \\ 2 & 2 & 1 & 1 & 2 & 2 & 1 & 1 \\ 1 & 1 & 1 & 1 & 1 & 1 & 1 & 1 \end{pmatrix}.$$

The output of this layer is the round output.

The output transformation is an application of the Mixed XOR/Addition layer with the output of round r and the last subkey, K_{2r+1}. This produces the cryptogram of the message which was the input of round 1. Decryption is essentially the inverse operations described for encryption. It is not obtained simply by reversing the key-schedule, but slightly different operations (the inverse operations) have to be employed. The decryption process is not important to this paper and is therefore not described. The interested reader is referred to [3] for more details.

The PHT layer plays an important role in both attacks exhibited to date by Sean Murphy [6] and by Lars Knudsen and Thomas Berson [2]. In [6] Murphy analyses the structure of PHT as a module homomorphism $\alpha : V \longrightarrow V$ where V is a module of rank 8 over the ring of integers modulo 256. V represents the message and cryptogram spaces and also the space of all the 8-byte vectors obtained in the intermediate steps during the encryption process, including the space corresponding to the input and output of the PHT layer. This interpretation of the structure of SAFER yields several interesting features of the cryptosystem. As an example of the importance of this module-theoretic approach, Murphy shows that V can be decomposed into a direct sum of subspaces, one of which is α-invariant. Several possible exploitations of this structure are listed in [6]. The net conclusion is that there are simple statistics of the output which are independent of a quarter of the key bytes, implying that SAFER does not satisfy fully the design principle of *confusion* as stated by Shannon in [7].

Properties of the PHT layer play an important role in the truncated differential attacks described by Knudsen and Berson in [2]. The attacks described are independent of the actual s-boxes used in SAFER — the s-boxes used are the functions $45^{(x)}$ and $\log_{45}(x)$ described above. Differential attacks on a cryptosystem make use of probabilities of getting certain output differences given certain input differences to obtain information about the (secret) key used for encryption. In truncated differentials only certain positions in messages and cryptograms and the differences they exhibit are taken into consideration. The PHT layer for SAFER has a number of output configurations which prove useful to the launching of successful attacks on the system with a reduced number of rounds which is better than exhaustive search. More details concerning the probabilities associated with the s-boxes in SAFER can be found summarised in table 3 of [4]. Analysis regarding the properties of these s-boxes when compared to other possible choices is made in [8]. The interested reader is referred to [2] for more details regarding attacks using truncated differentials on SAFER.

3 A closer look at the PHT layer

The linear stage of each round of SAFER in encryption consists of three layers of four 2-PHT boxes each. Each 2-PHT box can be represented by a matrix

$$B' = \begin{pmatrix} 2 & 1 \\ 1 & 1 \end{pmatrix}.$$

Note that since none of the entries in B' is zero, each one of the two outputs is dependent on each of the two inputs. Denote by B the 8×8 diagonal block matrix

$$B = \begin{pmatrix} B' & 0 & 0 & 0 \\ 0 & B' & 0 & 0 \\ 0 & 0 & B' & 0 \\ 0 & 0 & 0 & B' \end{pmatrix}$$

which is the matrix representation of the transformation applied to an 8-byte input $(a_1, a_2, \ldots, a_8)$ by the four 2-PHT boxes, all arithmetic done modulo 256. In the first two levels the output bytes are permuted before entering as input into the next level of 2-PHT boxes, the matrix representing this permutation being

$$P = \begin{pmatrix} 1 & 0 & 0 & 0 & 0 & 0 & 0 & 0 \\ 0 & 0 & 0 & 0 & 1 & 0 & 0 & 0 \\ 0 & 1 & 0 & 0 & 0 & 0 & 0 & 0 \\ 0 & 0 & 0 & 0 & 0 & 1 & 0 & 0 \\ 0 & 0 & 1 & 0 & 0 & 0 & 0 & 0 \\ 0 & 0 & 0 & 0 & 0 & 0 & 1 & 0 \\ 0 & 0 & 0 & 1 & 0 & 0 & 0 & 0 \\ 0 & 0 & 0 & 0 & 0 & 0 & 0 & 1 \end{pmatrix}.$$

The overall effect of the 3-level linear stage is then represented by the matrix

$$M = BPBPB = (BP)^2 B.$$

The matrix P appears to have been selected to provide maximal diffusion over the three levels: each byte of the output at the linear stage depends on each byte of the input. However, as pointed out by Murphy [6], this good diffusion does not imply good confusion, since in [6] it is shown by module theoretic means that "there are simple statistics of the output ... that are independent of a quarter of the key bytes". The structure of the relevant submodules (for example, over $\mathbb{Z}$ or over $\mathbb{Z}_{256}$) depends, given B, on the permutation matrix P.

The set of permutations which yield this desired maximal diffusion is to be determined, that is, the permutation matrices ensuring that every byte of the output of the linear stage will depend on every byte of the input. Define the *wire sequence* of a permutation in the linear stage of SAFER to be $(a_1, a_2, \ldots, a_8)$ if the a_i^{th} byte of the output of level r ($r = 1, 2$) provides the ith byte of the input to level $r + 1$. Similarly, define the *permutation sequence* as the inverse of this sequence: if $a_i = j$, let $I(j) = i$, in which case the permutation sequence will be denoted by $(I(1), \ldots, I(8))$. Thus, for example, in SAFER the wire sequence is $(1, 3, 5, 7, 2, 4, 6, 8)$ and the permutation sequence is, accordingly, $(1, 5, 2, 6, 3, 7, 4, 8)$. The latter sequence, which may be considered shorthand for the permutation $\begin{pmatrix} 1\ 2\ 3\ 4\ 5\ 6\ 7\ 8 \\ 1\ 5\ 2\ 6\ 3\ 7\ 4\ 8 \end{pmatrix}$ can, in the usual way, be represented by the permutation matrix P. Similarly, the wire sequence can be represented by $P^{-1} = P^T$. For clarity of exposition it is better to concentrate on the wire sequence, specifically on the wire sequences which will ensure that every output byte of the 3-level 2-PHT linear stage depends on every input byte; it is clear how the corresponding permutation matrices are determined by the wire sequences.

Note that in SAFER the connection between the first and second levels of the 2-PHT boxes is identical to the connection between the second and third levels. Number the boxes at each level $1, 2, 3, 4$, with box 1 on the extreme left, and let $A(i)$ denote the parents of box i (in levels 2 and 3) at the previous level. Then A is a function from $\{1, 2, 3, 4\}$ to the family of 2-subsets of $\{1, 2, 3, 4\}$ satisfying:

1. $A^2(i) = \{1, 2, 3, 4\}$, since any box at the bottom level is effected by every box at the top level; and
2. For every i, $|\{j : i \in A(j)\}| = 2$, since every box has exactly two outputs.

It is shown that there are exactly twelve possible choices for A. Equivalently, it is shown that for any partition of $\{1, 2, 3, 4\}$ into 2-subsets, there are four possible assignments of the values of A.

For example, consider the partition $\{\{1, 2\}, \{3, 4\}\}$. If $A(1) = \{1, 2\}$, then $A^2(1) = A(\{1, 2\}) = A(1) \bigcup A(2) = \{1, 2\} \bigcup A(2) = \{1, 2, 3, 4\}$, so that $A(2)$ is now fixed as $A(2) = \{3, 4\}$. Now, $A(3)$ must be either $A(3) = \{1, 2\}$ or $A(3) = \{3, 4\}$, which implies in the same way $A(4) = \{3, 4\}$ or $A(4) = \{1, 2\}$, respectively. Note that $A(3) = \{2, 3\}$, say, is impossible, since then $A^2(3) = \{2, 3, 4\} \neq \{1, 2, 3, 4\}$. Thus the images of 3 and 4 have to be elements of the same set $\{\{1, 2\}, \{3, 4\}\}$ of 2-subsets as $A(1)$ and $A(2)$. With $A(1) = \{1, 2\}$ there are therefore two possibilities for the function A. Similarly, had $A(1)$ been chosen as $\{3, 4\}$, two possibilities for A would have arisen.

Thus any partition of $\{1, 2, 3, 4\}$ into 2-subsets yields four possible assignments of $A(1), \ldots, A(4)$. Since there are three such partitions, there is a total of twelve functions A with the required two properties listed above:

$$
\begin{array}{ccccc}
 & A(1) & A(2) & A(3) & A(4) \\
1. & \{1,2\} & \{3,4\} & \{1,2\} & \{3,4\} \\
2. & \{1,2\} & \{3,4\} & \{3,4\} & \{1,2\} \\
3. & \{3,4\} & \{1,2\} & \{1,2\} & \{3,4\} \\
4. & \{3,4\} & \{1,2\} & \{3,4\} & \{1,2\} \\
5. & \{1,3\} & \{1,3\} & \{2,4\} & \{2,4\} \\
6. & \{1,3\} & \{2,4\} & \{2,4\} & \{1,3\} \\
7. & \{2,4\} & \{1,3\} & \{1,3\} & \{2,4\} \\
8. & \{2,4\} & \{2,4\} & \{1,3\} & \{1,3\} \\
9. & \{1,4\} & \{1,4\} & \{2,3\} & \{2,3\} \\
10. & \{1,4\} & \{2,3\} & \{1,4\} & \{2,3\} \\
11. & \{2,3\} & \{1,4\} & \{2,3\} & \{1,4\} \\
12. & \{2,3\} & \{2,3\} & \{1,4\} & \{1,4\} \\
\end{array}
$$

To obtain all possible wire sequences corresponding to a given function A, note that the outputs from any 2-PHT box can be arranged in two ways (if $A(i) = \{j,k\}$, then either the first or the second output byte from box j, for example, can be fed into box i at the next level), and similarly for the inputs into any box. Thus, for any box, there are 2^2 ways of arranging inputs and outputs, making a total of 2^8 for the set of four boxes. Hence, any one of the twelve functions A listed above can be implemented in 2^8 possible ways, making a total of $12 \times 2^8 = 3072$ possible wire sequences.

A wire sequence $(a_1, a_2, \ldots, a_8)$ is *lexicographically ordered* if $a_{2i-1} < a_{2i}$, $i \in \{1,2,3,4\}$. The following are the sixteen lexicographically ordered wire sequences which can be obtained from the first function A in the above list:

$$
\begin{array}{cc}
(1,3,5,7,2,4,6,8) & (2,3,5,7,1,4,6,8) \\
(1,4,5,7,2,3,6,8) & (2,4,5,7,1,3,6,8) \\
(1,3,6,7,2,4,5,8) & (2,3,6,7,1,4,5,8) \\
(1,4,6,7,2,3,5,8) & (2,4,6,7,1,3,5,8) \\
(2,4,6,7,1,3,5,8) & (1,4,6,8,2,3,5,7) \\
(2,3,5,8,1,4,6,7) & (1,4,5,8,2,3,6,7) \\
(2,4,5,8,1,3,6,7) & (1,3,6,8,2,4,5,7) \\
(2,3,6,8,1,4,5,7) & (2,4,6,8,1,3,5,7) \\
\end{array}
$$

Note that the first of these is the wire sequence as used in SAFER which corresponds to the decimation-by-2 permutation. The next section contains some considerations of how the alternative permutation matrices which give the required diffusion in the linear stage of SAFER may effect the performance of the resulting SAFER-variant cryptosystem.

4 Results of initial investigations

Each one of the 3072 wire sequences derived in the previous section gives rise to a permutation P which can be used in the 3-level 2-PHT layer of any SAFER cryptosystem. As was pointed out earlier, one of these permutations is the one used in the definition of SAFER [3]. The next stage is to find out what happens to the performance of the cryptosystem when the other 3071 permutation matrices are used instead of the decimation-by-2 matrix used in the original definition. Specifically, is there any advantage in making use of other permutation matrices in the cryptosystem? To answer this question requires further investigation. Preliminary findings show that there is scope in further investigation since some of the matrices exhibit properties which could prove advantageous to the security of the resulting cryptosystem.

For example, in Murphy's paper [6], the property of the matrix M as used in the definition of SAFER which lead to the described attack was the α-invariant submodules obtainable from the use of the module homomorphism α corresponding to the matrix M. The example quoted in [6] made use of submodules constructed from V, the $\mathbb{Z}_{256}$-module of rank 8 corresponding to the 8-byte message space. These submodules were constructed by using the factorization over the integers of the characteristic polynomial of the matrix M. This was an example: other factorizations, over $\mathbb{Z}_{256}$ for example, may also have proved fruitful.

It is also the case that certain submodules of V do not correspond to any factorization of the characteristic polynomial of M over any suitable algebraic structure. Finding such non-trivial submodules is however more difficult. Because of this, the 3072 characteristic polynomials corresponding to the different M matrices obtained from the different permutations P were checked to see if any of them were: a) irreducible over the integers $\mathbb{Z}$; and b) irreducible over $\mathbb{Z}_2$. The reasoning behind this was to determine whether there are any matrices whose characteristic polynomials are irreducible over the stated structures and therefore do not lend themselves to easy determination of submodules and invariant submodules in particular. Of course, determination of such matrices does not guarantee that no submodules exist — only an indication that it is more difficult to determine them. An exhaustive search showed that there are 1680 permutations whose corresponding characteristic polynomials are irreducible over the integers, and of these 240 correspond to matrices whose characteristic polynomials are irreducible over $\mathbb{Z}_2$ (denote this set of matrices by $\mathcal{P}$). The latter were found by reducing the characteristic polynomial modulo 2 and then checking for irreducibility over $\mathbb{Z}_2$. Since these polynomials are irreducible modulo 2, they are also irreducible modulo 256 (any factorization modulo 256 is a factorization modulo 2). However, this purging may have removed matrices whose characteristic polynomials are reducible modulo 2 but not modulo 256, since a factorization modulo 2 *does not* imply a factorization modulo 256. Also, this purging does not consider matrices whose characteristic polynomials are reducible over some structure and yet the related submodules are *not* invariant under the corresponding homomorphism. In any case, the point of the investigation was to determine whether matrices satisfying the conditions above exist, and the answer is affirmative.

The next stage of the investigation concentrated on the 240 matrices in $\mathcal{P}$ found earlier which had characteristic polynomials irreducible over $\mathbb{Z}_2$. There are two heuristic reasons for this. First, the elements of $\mathcal{P}$ have properties which are "good" by the reasoning given above. Secondly, this second stage of preliminary investigations was more laborious and time consuming, and, as such, consideration of a smaller set rather than all 3071 other matrices was desirable. Since $\mathcal{P}$ comprises a smaller "good" set, it seemed reasonable to concentrate on it. The objective of this stage of the investigation was to determine whether any of these alternative matrices are more resistant to Truncated Differential attacks as described by Knudsen and Berson in [2].

The attacks described in [2] are independent of the s-boxes used in the cryptosystem and are dependent only on properties of the PHT matrices. The properties used are essentially the behaviour of these matrices under certain inputs of "low" weight (weights ranging from 1 to 4). Two tables (tables 2 and 3 in Appendix A of [2]) are given, listing truncated differentials for one round of SAFER and their corresponding probabilities. These tables were constructed using tables 4 to 10 in Appendix A of [4]. Note that there are 87 elements in total listed in tables 4, 5, 6 and 7. The number of elements in the equivalent tables for the matrices in $\mathcal{P}$ is less than 87, and for some matrices may be as low as 58. At least for these tables, since the number of elements in them is less for the elements of $\mathcal{P}$, the number of combinations of the entries which yield some of the elements equivalent to those listed in tables 2 and 3 in [2] is less. This could imply, for example, that there are less entries in the equivalent tables, or that the probabilities for some of the one-round truncated differentials are less. For example, this lower probability could translate into a smaller probability of success for truncated differential attacks on SAFER cryptosystem variants using matrices in $\mathcal{P}$. The preliminary investigations in this case are encouraging since it seems possible to find PHT matrices which may lead to systems more resistant to truncated differential attacks.

It is possible that some of the other matrices not in $\mathcal{P}$, albeit they have characteristic polynomials which are reducible over the integers or some other ring, may be more resistant to truncated differential attacks than the matrix already in use. Indeed, it is possible that some of these matrices are more resistant than those in $\mathcal{P}$. Matrices outside $\mathcal{P}$ are still important to consider. Recall that using the new key schedule presented in [1] prevents the attacks described in [6]. However, the truncated differential attacks are only made more difficult, rather than prevented altogether, in the SK versions of SAFER which make use of the new key schedule. Thus from a practical point of view, it is better to try to find an alternative PHT matrix which is more resistant to truncated differential attacks as described in [2].

Other properties of the PHT matrices should also be considered before choosing them as alternatives. A designer has to check the alternative candidates to see if they are resistant to differential cryptanalysis at least to the same number of rounds as the matrix in the original definition as described in [4]. The fact that for the elements of $\mathcal{P}$ the number of items in the corresponding tables $4 - 7$ is less indicates that some of these matrices could be at least as good as the PHT matrix in the original definition with regards to resistance to attacks described in [4].

5 Conclusions

In this paper, properties of the permutation in the PHT level defined in [3] which provide the required diffusion have been formalized using the concepts in section 3. A method of how to determine the set of 3072 permutations satisfying these properties has also been described. Each of these permutations corresponds to a PHT matrix which offers the same kind of "good" diffusion as that offered by the PHT matrix obtained by the decimation-by-2 permutation matrix used in SAFER.

Preliminary investigations have been promising. A subset $\mathcal{P}$ of the 3072 possible PHT matrices has certain "desirable" properties, as described in section 4, for which under certain conditions it is difficult to launch attacks similar to those described in [6]. The matrices in $\mathcal{P}$ also have properties which indicate a potentially greater resistance to truncated differential attacks as described in [2]. Further research is to be undertaken to determine whether any of these alternative matrices could (or should) be used instead of the one proposed in the original definition. For example, does there exist a matrix which gives a 5-round variant of SAFER more resistant to truncated differential attacks than the present system?

The method used to describe part of the list of permutation matrices in section 3 lends itself naturally to a tree-structure algorithm which could be used to generate a PHT matrix out of the whole set of possibilities from a small primitive set. The primitive set used could consist of the twelve A functions listed in section 3. Such a tree structure could be used to select a key-dependent PHT matrix in some variant of the SAFER system as defined up to now. Note that 12 bits are sufficient to uniquely determine one of the 3072 possibilities. This is not a large overhead to the existing size of the keys. A key-dependent PHT matrix could also make successful cryptanalysis more difficult for an attacker, since then the PHT used in the cryptosystem is not known — recall that knowledge of the matrix is important in the attacks on SAFER described to date. At the same time, the generation of the matrix at the start of encryption is not too costly an overhead for the users of the system.

The existence of 3071 other possibilities for the PHT matrix besides the one used in the original definition allows for the search of an "optimal" matrix. Further work is required to determine what exactly the conditions for optimality are and which, if any, of the possible matrices satisfy them. It is possible to envisage variants of SAFER where it is desirable to find an "optimal" PHT matrix. Such a variant could, for example, make use of other algebraic structures than those employed in SAFER to date. One example (suggested by Gideon Kuhn, University of Pretoria) is a variant using $GF(2^8)$, the finite field with 256 elements, instead of $\mathbb{Z}_{256}$ in the PHT linear layer. Use of this field instead of the ring is preferable if the cryptosystem is to be hard-wired in hardware, since in this case operations in $GF(2^8)$ are easier to implement than those in $\mathbb{Z}_{256}$. Certainly in this case it would be wise to find the "best" PHT matrix since a hardware implementation is more difficult to undo than a software implementation if found lacking.

Acknowledgements. Thanks to Peter Wild and Sean Murphy for helpful comments.

References

1. Lars R. Knudsen: A key-schedule weakness in SAFER K-64. D. Coppersmith editor, Advances in Cryptology – CRYPTO'95, Lecture Notes in Computer Science, 963 (1995) 274–286.
2. Lars R. Knudsen, Thomas A. Berson: Truncated Differentials of SAFER. D. Gollman editor, Fast Software Encryption, Lecture Notes in Computer Science, 1039 (1996) 15–26.
3. J.L. Massey: SAFER K-64: A Byte-Oriented Block-Ciphering Algorithm. Fast Software Encryption, Lecture Notes in Computer Science, 809 (1994) 1–17.
4. J.L. Massey: SAFER K-64: One Year Later. B. Preneel editor, Fast Software Encryption, Lecture Notes in Computer Science, 1008 (1995) 212–241.
5. J.L. Massey: Announcement of 40-bit key scedule for SAFER. E-mail announcement by author, 22 October 1995.
6. Sean Murphy: An Analysis of SAFER. Journal of Cryptology (to appear).
7. C.E. Shannon: Communication theory of secrecy systems. Bell Systems Tech. Jnl., 28 (1949) 656–715.
8. Serge Vaudenay: On the need for Multipermutations: Cryptanalysis of MD4 and SAFER. Fast Software Encryption, Lecture Notes in Computer Science, 1008 (1995) 286–297.

An Adaptive Approach to T of N User Multi-access Communications Channel Using an Orthogonal Coded Multiplexer

K.Brown and M.Darnell
Institute of Integrated Information Systems
School of Electronic & Electrical Engineering
University of Leeds
Leeds, LS2 9JT
UK

Abstract

A practical approach is described for adaptively multiplexing in real-time T of N users in a multi-channel communications system using a majoritive logic multiplexer and quasi-othogonal binary coding. The adaptive multiplexer employs the users' class of service and channel noise constraints to optimize T and fidelity criteria.

1. Introduction

Several approaches to the T of N user coding problem have been addressed in literature. One multi-access approach, the T-adder channel, was presented by Kasami and Lin in 1975 [1]. This approach allows high throughput by several users without sub-division in frequency or time; the approach was further developed in key papers by Gallager [2], and Mathys [3]. In addition, several novel and diverse encoding schemes for the adder channel have been described, including [4,5]. However, the Achilles heel of the T-adder channel was its lack of optimum decoding. This was addressed by a maximum likelihood decoding scheme developed by Honary et al [6]. However, these sophisticated approaches, when complexity is considered, may offer no significant advantages over the most simple algebraically added multiple access channel systems as T approaches N in the steady state.

Another approach to the multiple access channel was introduced initially in 1970 by Gordon and Barrett [7]. This approach, termed "Correlation - Recovered Adaptive Majority Multiplexing", applies a majority selector to N orthogonally coded channels and reconstructs the original channel contents based on the coding gain of the individual channels. The design of such a coding scheme was based on a claim by the authors, and an associated proof by Titsworth, asserting that, in the binary case, the majority function is the optimal algorithm for combining orthogonally coded binary channels.

Neither the collaboratively coded or correlation recovered multi-access channel have found advantage over other multiplexing methods for incorporation into communications systems. However, a simplistic approach for adaptively coding and multiplexing the T of N channels potentially exists within them both. Assuming a low bandwidth reverse channel, the multiple access channel can be optimized for fidelity or throughput for individual or multiple channels.

2. Channel Description

As part of a multi-cell, multi-user, communications system, a simplistic approach to adaptively multiplex multiple channels using the correlation-recovered approach can be found. In the channel multiplexing scheme in figure 1, a 3-user transmitting local area cell is linked by a simulated Code Time Division Multiple-Access (CTDMA) channel under flat Rayleigh fading. The scheme employs a matched, 2-step, multiple-access approach, analogous to one described by Fan and Darnell [8]. The channels within the local cell are multiplexed by applying a majority logic adder to the outputs of each data source. The data source of each channel is encoded using a set of binary codes that exhibit quasi-orthogonal behaviour when summed coherently by the majority adder and aperiodically correlated with the original encoding sequences.

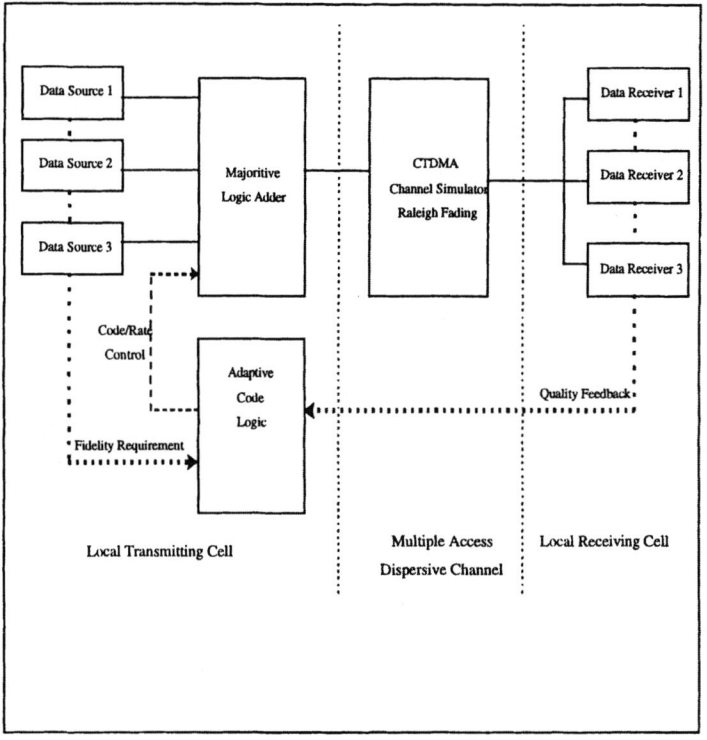

Figure 1. Channel Diagram

Gordon and Barrett [7] found that, for n channels using code words meeting the following equality,

$$D_c = S^*\{A(S^*D_c^T A)^T\},$$

where D_c is the received column vector, S is the vector sign operator, ^{T} is the transpose operator and A is the set of code words, and where either

(a) the code words form an algebraic group and $L=2^N$ or

(b) the code words are the first n phase shifts of a pseudorandom sequence of length $L=2^N-1$,

the information rate approaches 1 as T approaches N. As T decreases from N, fewer code words sum to form the transmitted code word and the summed noise is reduced for each of the T channels. This property is referred to as *reassurance* and, for an ideal family of coding sequences, a $2e + 1$ length code word will correct e errors when $T=1$. These properties allow an adaptive multiplexer to be constructed with low computational complexity at the user transmitter and a corresponding demultiplexer at the receiver.

3. Description of Model

Majoritive addition requires an odd number of inputs to combine without errors. The magnitude of these errors diminishes as N becomes large. Hence, the smallest multiplexer model exhibiting ideal adaptive performance is an $N=3$ system, where T is either 3 or 1. Because the objective in this case is not to achieve the ideal information rate, which approaches 1 when $T=N$, code sets with non-ideal majority added properties can be used to investigate various aspects of system performance. A computer program has been written to perform a direct search, using the constraints identified above, to generate encoding sequences with ideal properties.

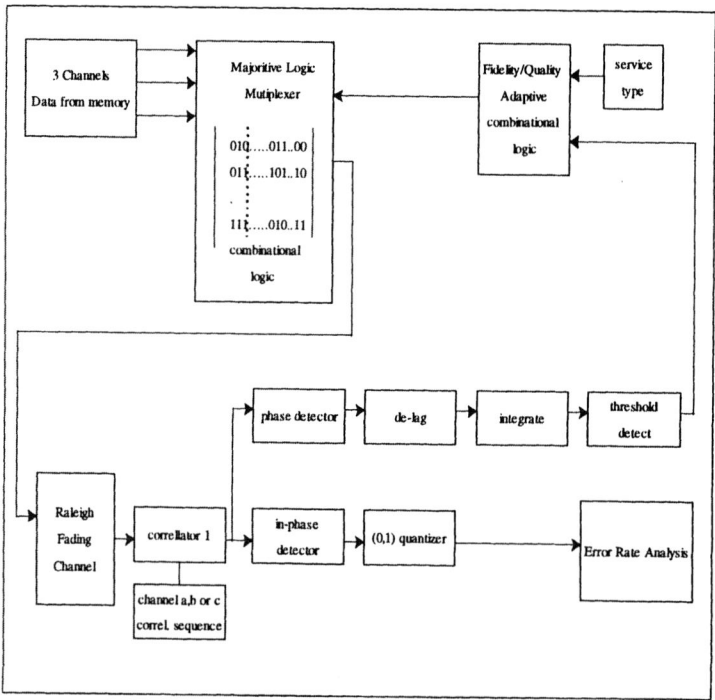

Figure 2. Computer Simulation Diagram

To construct an adaptive multiplexer model, a string within the multiple access channel model is derived for computer modelling, as shown in figure 2. The user code sequences are constructed in a matrix which contains all codewords and is addressed by the binary data streams of each user; this matrix is implemented in memory as a truth table. The resulting output is a unique bipolar code word representing the majority-added composite of all user code sequences. The code word is then corrupted by a simulated Raleigh fading channel and additive white Gaussian noise. A single channel of the multiplexed signal is then recovered by a simulated receiver comprising a correlator and binary quantizer. The receiver output stream is then buffered in time and compared with the appropriate input channel data for error rate analysis.

Adaptation in response to the simulated channel is based on the observation of noise threshold crossing from the correlator output. The output noise tracking is integrated by a FIR filter to provide averaging and a fixed delay; this then provides the first input to the adaptation combinational logic. The second input to the adaptation logic is the class of service or fidelity constant. The output of the adaptation logic varies the multiplexer logic to optimize T, based on the class of service requirement and channel noise observation.

The recovered raw in-phase correlator output is shown for a fully loaded cell, $T=3$, in figure 3. The adapted cell, $T=1$, raw correlator output is shown in figure 4. A higher average correlator output for T-1 demonstrates the coding gain achieved through adaptation.

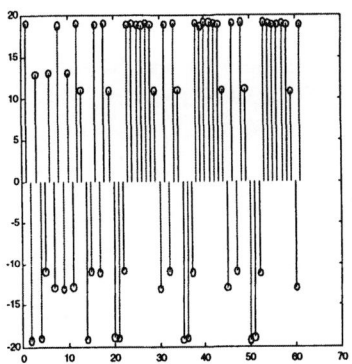 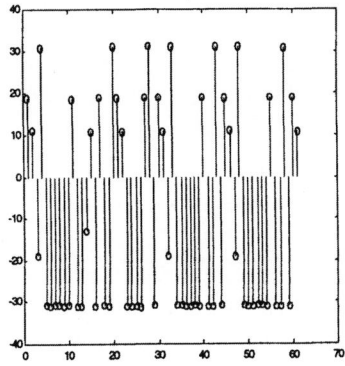

Figure 3.
In-Phase Correlator Output:
Channel 1 $T=N=3$

Figure 4.
In-Phase Correlator Output:
Channel 1 $T=1$

4. Concluding Remarks

In the case of this simulation of a $N=3$ adaptive multiplexer, the adaptation is between only two states, where $T=3$ and $T=1$. Hence the adaptive logic is simply a two input Boolean operation. As N increases, the adaptive logic complexity, C, increases as 2^T. However, the channel noise estimation complexity for the method used in this simulation remains approximately the same for all T. Thus for all practical instances, the adaptive component adds little complexity to system operation even as T becomes large. Similarly, the performance of the multiplexer is based on the selection of optimum code words and code word length. Each transmitter in a cell is only required to select from two sequences for any value of T. In addition, the paired receiver is only required synchronize and perform one correlation per received bit, regardless of the value of T. Thus the overall system is of small computational complexity compared with other methods of achieving high efficiency for the T of N user channel problem. It can also be shown that when N is large, an adaptive multiple-access multiplexer can be constructed to adapt to binary multimedia type data with multiple levels of user priority and channel noise conditions in real time and with low complexity.

5. References:

[1] T.Kasami and S.Lin, "Coding for a Multiple Access Channel", IEEE
    Transactions on Information Theory, Vol.IT-22, No.2, , March 1976, pp   129 -
137.

[2] R.G.Gallager, "A Perspective on Multiaccess Channels", IEEE   Transactions on Information Theory, Vol. IT-31, No.2, March 1985, pp   124 - 142.

[3] P.Mathys, "A Class of Codes for a T Active User Out of N Multiple -
    Access Communication System", IEEE Transactions on Information
    Theory, Vol. IT-36, No. 6, November, 1990, pp 1206 - 1219.

[4] P.Z.Fan, M.Darnell and B.Honary," Superimposed codes for multiaccess
    binary adder channel", Proceedings of 1994 IEEE Int. Workshop on
    Information Theory, (Moscow), July 3-8 1994, pp.30-32.

[5] S.Soysa, S, F.H. Ali, and S.A.G.Chandler, "Comparison of T-User M-PSK CV-
CCMA and Multi-level Modulation in a Rayleigh Fading Channel",

    Proceedings of ISCTA, Ambleside, July, 1997, pp 419 - 422.

[6] B.Honary, L.Kaya, G.S. Markarian, and M.Darnell, "Maximum-likelihood
    Decoding of Array Codes with Trellis Structure", IEE Proceedings-I,
    Vol.140, No. 5, October 1993, pp 340 - 346.

[7] J.A.Gordon, and R.Barrett "Correlation-recovered Adaptive Majority
    Multiplexing", Proc. IEE, Vol. 118, No. 3/4, March/April 1971, pp 417 -   422.

[8] P.Z.Fan, and M.Darnell, "Hybrid CCMA/SSMA Coding Scheme",
    Electron. Letters, Vol. 30, no. 25, December, 1994, pp 2105 - 2106.

The Breaking of the Lorenz Cipher: An Introduction to the Theory Behind the Operational Role of "Colossus" at BP

F.L. Carter B.Sc.

Bletchley Park Trust

1 Some Preliminary Information

The Lorenz Cipher Machine provided a means of communication which was used exclusively by the German Army High Command during the second world war. At Bletchley Park the cipher traffic generated by this machine, was called "Fish", and the different radio communication links used were given appropriate names e.g. "Tunny", "Bream", "Tarpon" etc. These cipher messages consisted of sequences of characters based on the International Teleprinter code. The transmissions were not made in Morse, but used a carrier wave modulated by six audio frequency tones, three for each of the two basic symbols used; these two basic symbols were referred to as "cross" and "dot" at Bletchley.

In the teleprinter code, each character is represented by a sequence of five of these basic symbols, so that in all, the system provides for a total of thirty two distinct characters. Twenty six of these are used to represent the standard letters of the alphabet, and five for the "control characters".

A $XX \bullet \bullet \bullet$	B $X \bullet \bullet XX$	C $XXXX \bullet$	D $X \bullet \bullet X \bullet$
E $X \bullet \bullet \bullet \bullet$	F $X \bullet XX \bullet$	G $XX \bullet XX$	H $X \bullet X \bullet X$
I $XXX \bullet \bullet$	J $XX \bullet X \bullet$	K $XXXX \bullet$	L $XX \bullet \bullet X$
M $X \bullet XXX$	N $X \bullet XX \bullet$	O $X \bullet \bullet XX$	P $XXX \bullet X$
Q $XXX \bullet X$	R $XX \bullet X \bullet$	S $X \bullet X \bullet \bullet$	T $X \bullet \bullet \bullet X$
U $XXX \bullet \bullet$	V XXXXX	W $XX \bullet \bullet X$	X $X \bullet XXX$
Y $X \bullet X \bullet X$	Z $X \bullet \bullet \bullet X$	3 $X \bullet \bullet X \bullet$	4 $XX \bullet \bullet \bullet$
8 XXXXX	+ $XX \bullet XX$	9 $X \bullet X \bullet \bullet$	/ $X \bullet \bullet \bullet \bullet$
3='carriage return'		4='line feed'	
8='letter shift'		+='figure shift'	
9='space'		/='null'	

Table 1. The International Teleprinter Code.

When the characters of a plain text message character were processed through the Lorenz machine, it generated a corresponding sequence of pseudo-random characters which were "added" to the plain text characters to produce the characters of the cipher. This sequence of random characters was called the cipher "key". The addition process was carried out on the five pairs of corresponding basic symbols representing each of the characters, using the following rules:-

$$X + X = \bullet \quad , \quad \bullet + \bullet = \bullet$$
$$X + \bullet = X \quad \text{and} \quad \bullet + X = X$$

This structure is isomorphic to addition in arithmetic modulo 2, and has the required property of being "self inverting", which means that if the same key character sequence is added to the cipher character sequence, the result is the original sequence of plain-text. Hence the cipher (Z) can be expressed as:-

$$Z = P + (key)$$

where P is the plain text, and

$$P = Z + (key)$$

Cipher communication was carried out using two Lorenz machines set to produce the same key character sequence, one operated to generate cipher from the plain-text, and the other to recover the plain-text from the cipher. This process was automatic, and if a message was prepared as a punched five hole paper tape, transmission would take place at high speed. The key character sequence used could be altered by making specific adjustments to the machines, and in the early days, the information about this would be directly transmitted across an additional communication link so that the operators at each end could set up their machines to produce the required identical key character sequence. Initially this information consisted of twelve distinct letters, known at BP as the "indicators", and it was assumed that messages with the same set of indicators, had the same key.

This much was known in 1941, and the early work done on breaking the ciphers were based on the reception of two such messages, giving what was known as a "depth of two"; however this did not often happen. Consider such a case:-

$$Z_1 = P_1 + (key) \text{ and } Z_2 = P_2 + (key)$$

Then

$$Z_1 + Z_2 = P_1 + P_2 + (key) + (key)$$
$$= P_1 + P_2$$
$$= U$$

After a considerable effort based on the experience of likely letter sequences in P_1 or P_2, BP succeeded in resolving some short sequences of U into its components P_1 and P_2. From these, corresponding sequences of pure key could be determined. These were useful if any further messages with the same indicators were received, and additionally gave some limited information about the logical structure of the cipher machine.

$$Z_1 = P_1 + (key) \text{ hence } Z_1 + P_1 = P_1 + P_1 + (key)$$
$$\text{or} \quad Z_1 + P_1 = (key)$$

On 30th August 1941, as a result of a serious mistake by two German operators, a 4000 character message was transmitted twice using the same key. Apart from some small differences at the beginning of one of these transmissions, and a few random typing errors, the plain text in each was the same. Hence when the two cipher messages were added together the result was the sum of two virtually identical messages. From this, the single message was recovered, which in turn lead to a 4000 character sequence of pure key. This considerable task was carried out by cryptanalyst J. Tiltman. From this long sequence of key characters, the

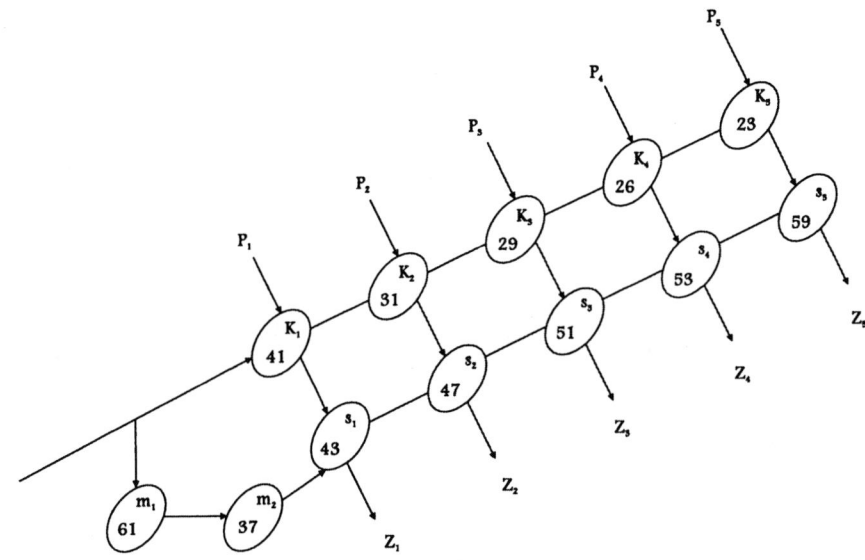

Fig. 1. The Structure of the Lorenz Machine

complete logical structure of the German Lorenz machine was determined by a systematic search for periodic patterns in the five individual pulse streams of the sequence. This was a very remarkable feat, largely due to the work of W.T. Tutte, a young Cambridge mathematician seconded to Bletchley. It is worth noting that although this important success can be attributed to the skill and enterprise of BP, it was the unusually weak characteristics of the cipher key that happened to be used on this particular day which enabled Tutte's method to succeed. It appears that the Germans realised that this second mistake had been made, as no subsequent key ever had this weakness. It is remarkable that this combination of two unlikely events lead to a development which probably influenced the subsequent course of the war.

The structure of the Lorenz machine is shown in the diagram. The machine automatically generated the random key characters by means of five pairs of wheels, one pair for each of the five component impulses required for each character. The wheels were coupled together by a system of gears and each wheel had a pattern of teeth on it, which could be changed. The key character generated each time depended on the tooth patterns and on the wheel positions, for each of the five pairs. The wheels were initially set to some pre-determined starting positions, and turned from one position to another after each character was typed, so that the key characters changed in a near random manner. The first wheel in each pair was known as a "Chi-wheel" and the second as an "Psi-wheel", from the Greek letters chi and psi, but in this paper the letters "K" and "S" are substituted. The K-wheels moved on one position after every letter typed in, but the S-wheels movement was more complex, and sometimes this set did not move, being under the control of two further so called "motor wheels".

The discovery of the structure of the machine represented a landmark in the struggle to break the Lorenz cipher. It was now clear to BP that the indicators represented the start positions for the twelve wheels of the machine. A device known as "Tunny" which simulated the operational function of the Lorenz machine was later constructed at BP. With this device the problem of decrypting Lorenz generated ciphers was theoretically reduced (!) to the task of determining the character generating patterns and the starting positions of the twelve wheels of the machine used for each message. If these could be found and set up on "Tunny", then the plain text of a cipher message could be obtained, in the same manner as if a genuine Lorenz machine was being used. The wheel settings were changed for every message, while the wheel patterns were also changed at regular intervals of time. Initially this was done each month, but later on a daily basis. Another level of difficulty arose when the Germans ceased to send transmit the wheel indicators in a direct form, thus making the identification of "depths" impossible.

The next major step was the introduction of statistical techniques for finding the wheel settings and the development of machines for their implementation. The Cambridge mathematician Max Newman was a key figure in this work, backed up by the engineering skill of a team from the Post Office Research Station headed by Dr T. Flowers. This co-operation between mathematicians and engineers ultimately lead to the construction of Colossus, a high speed electronic machine with many of the attributes now associated with a digital computer. This machine, first brought into operational use in the spring of 1944, was highly successful, and combined with the codebreaking skills of BP, lead to the regular decoding of Fish traffic.

2   The Theory of Wheel Setting

The first task undertaken using this new approach was to determine the settings or start positions of the five K-wheels for each intercepted message. The "key" characters generated by the Lorenz machine are the result of combining together the characters generated separately by the K-wheels and the S-wheels. The effect of the S-wheels is dependent on the motion of the "motor" wheels, and is represented by the symbol S', hence the expression Z = P + key can be extended to the form:-

$$Z = P + K + S'$$

It follows that if

$$D = Z + K \tag{1}$$

the also

$$P = D + S' \tag{2}$$

where the symbol D represents what was known as "pseudo plain-text". These two equations provided a basis for dividing the cipher breaking pr ocess into two stages:-

1. To obtain D from Z by means of machine based processes derived from Equation (1).
2. To deduce P from D using Equation (2), this task being carried out in another section at BP.

The decrypts obtained by hand methods showed that the German plain text messages generally contained a particular characteristic feature, namely the frequent presence of groups of two or more repeated characters, leading to a proportion of repeats significantly greater than would be expected by chance. In order to be able to identify these repeated characters by machine, a process for forming the "first difference" of the sequence of characters was introduced. This was done by adding together (modulo 2) adjacent characters, the new sequence so formed being referred to as "ΔP". Any repeated pair of characters in P resulted in a "/" in the sequence ΔP, that is by "●●●●●●", this leading to a "●" in each of the five individual pulse streams of ΔP.

Hence for most messages the five pulse streams making up ΔP would each consist of a sequence of "X"s and "●"s with the proportion of "●"s greater than that to be expected by chance. It was the recognition of this fact that ultimately provided a fast method for finding the wheel settings. Using the delta forms of equations (1) and (2), equation (1) and (2) become

$$\Delta D = \Delta Z + \Delta K \tag{3}$$

and

$$\Delta P = \Delta D + \Delta S' \tag{4}$$

respectively; a further form being

$$\Delta D = \Delta P + \Delta S' \tag{5}$$

Working with the five individual pulse streams the general form of equation (i) can be expressed as : -

$$\Delta D_i = \Delta Z_i + \Delta K_i$$

where $i = 1, 2, ...5$. The determination of the ΔD_i pulse stream, requires the correct setting position for the wheel K_i. One possible statistical approach for finding this, would be to examine, for all settings, the statistical properties of the pulse values derived from the expression $(\Delta Z_i + \Delta K_i)$, in the expectation that, as a consequence of a high proportion of "$\bullet$"s in ΔP_i, a larger number of "$\bullet$"s might be produced when the wheel is at the correct setting, so distinguishing it from all the others. In the case of K_1 for example, the ΔZ_1 cipher sequence would have to be added, term by term, to the corresponding ΔK_1 sequence and this would have to be done for all 41 possible starting positions. This would be a massive task if done by hand, but perhaps not altogether a hopeless one. The Germans, however, had by now prevented an attack of this form on a single impulse channel by applying a particular rule to the motor and S wheel patterns used, to ensure that the individual ΔD patterns were random at all settings and contained no hint of the non-random nature of the individual ΔP patterns;

Let $\Pr[\Delta P_i = " \bullet "] = p$, where it is anticipated that $p > \frac{1}{2}$. If the probability that the motor wheels step (move) the $S - wheels = a$, and $\Pr[\Delta S_i = "X"] = q$ then $\Pr[\Delta S_i' = "X"] = a.q$ By the application of equation (5), then $\Pr[\Delta D_i = " \bullet "] = \Pr[\Delta P_i = ""].\Pr[\Delta S_i' = " \bullet "] + \Pr[\Delta P_i = "X"].\Pr[\Delta S_i' = "X"]$, hence, $\Pr[\Delta D_i = " \bullet "] = p.(1 - a.q) + (1 - p).a.q = p(1 - 2aq) + a.q$.

The Germans imposed the rule $a.q = \frac{1}{2}$ on the patterns used, and introducing this value, $\Pr[\Delta D_i = " \bullet "] = \frac{1}{2}$ Hence the output from one impulse channel had purely random characteristics, no matter what the value of p happened to be. This was a somewhat discouraging outcome, however an analysis of the situation when two impulse channels are taken together, with this rule imposed, lead to a result of great value:-

Consider two impulse channels i and j, let $\Pr[\Delta S_i = "X"] = qi$ and $\Pr[\Delta S_j = "X"] = qj$. The application of the German rule to these two results leads to the conclusion that qi and qj must be equal, and replacing both by the symbol "q" it follows that

$$\Pr[\Delta S_i + \Delta S_j = " \bullet "] = \Pr[\Delta S_i = " \bullet "].\Pr[\Delta S_j = " \bullet "]$$
$$+ \Pr[\Delta S_i = "X"].\Pr[\Delta S_j = "X"]$$
$$= (1 - q)(1 - q) + q2$$

Next consider the value of $\Pr[\Delta S_i' + \Delta S_j' = " \bullet "]$, for this event either the S-wheels do not move (*probability* $= 1 - a$), or they do move and $(\Delta S_i + \Delta S_j = " \bullet ")$. Hence

$$\Pr[\Delta S_i' + \Delta S_j' = " \bullet "] = (1 - a) + a(1 - q)(1 - q) + q^2$$
$$= 1 - 2aq + 2aq^2$$
$$= q(\text{since } aq = \frac{1}{2}) \tag{6}$$

Let the proportion of repeated characters in the $plain-text = b$. Then the event $[\varDelta P_i + \varDelta P_j = " \bullet "]$ can happen in two ways:-

1. $\varDelta P_i = " \bullet "$ and $\varDelta P_j = " \bullet "$ as the result of a repeated character, with $[probability = b]$.
2. No repeat, but $\varDelta P_i + \varDelta P_j = " \bullet "$ due to chance, $[probability = (1-b)/2]$.

Hence

$$\Pr[\varDelta P_i + \varDelta P_j = " \bullet "] = \frac{1}{2}(1+b) \qquad (7)$$

Combining the results (6) and (7) above, gives:-

$$\Pr[\varDelta D_i + \varDelta D_j = " \bullet "] = \Pr[\varDelta P_i + \varDelta P_j = " \bullet "].Pr[\varDelta S_i' + \varDelta S_j = " \bullet "]$$
$$+ Pr[\varDelta P_i + \varDelta P_j = "X"].Pr[\varDelta S_i' + \varDelta S_j' = "X"]$$

hence,

$$Pr[\varDelta D_i + \varDelta D_j = " \bullet "] = \frac{1}{2}(1+b).q + \frac{1}{2}(1-b).(1-q)$$

$$= \frac{1}{2}1 + b(2q-1)$$

Previous experience of the motor wheels gave $a = \frac{52}{74}$ as a realistic approximation, and hence $q = 0.71$, leading to:- $\Pr[\varDelta D_i + \varDelta D_j = " \bullet "] = \frac{1}{2}(1+0.42b)$. This showed that, provided the proportion of repeats in the plain-text was high, then the sum of the two delta pulses did not form a random sequence.

From this result an algorithm for determining the correct settings for the pair of wheels K_1 and K_2 was developed. Assume that a cipher message Z consists of n characters, and that the corresponding n characters of D have been derived from this by adding to the characters of Z the *correct* sequence of K characters. Working with the impulse sequences 1 and 2, suppose a record is kept of the number of times the evaluation of the expression $[\varDelta D_1 + \varDelta D_2]$ results in a "$\bullet$". The previous analysis indicates that the expected total score will be $\frac{1}{2}(1+0.42b).n$. In contrast if the same procedure is carried out on Z again but with an incorrect K sequence (i.e.. with a wrong pair of wheel settings) the result would be a sequence of random characters for which the expected total score will be $\frac{n}{2}$. So a higher score is to be expected when the process is carried out at the correct wheel settings than at any other, because the correct K sequence is produced only when the correct wheel settings are used.

Since $\varDelta D_1 = \varDelta Z_1 + \varDelta K_1$ and $\varDelta D_2 = \varDelta Z_2 + \varDelta K_2$ it follows that the total score can be obtained by counting the "$\bullet$'s" generated by the expression:-

$$\varDelta Z_1 + \varDelta K_1 + \varDelta Z_2 + \varDelta K_2$$

This must be done for the entire length of code Z and for every possible pair of wheel settings for K_1 and K_2. The pair of settings giving the highest score is likely to be the correct one. An algorithmic solution to the problem! The penalty for this is of course the very large number of logical operations required.

Colossus was designed to perform this task at a very high speed. Working with K_1 and K_2 and a code message of say 3000 characters, $41 \times 31 \times 3000$ sets of such operations will be needed. The machine read about 5000 characters per second from a continuous loop of punched paper tape holding the cipher message, and without this high speed a statistical approach would not have been realistic. The wheel patterns were generated internally by electronic means. An earlier machine (Robinson) using two tapes, one of which held the wheel patterns, had not been a practical success, mainly due to problems in keeping the tapes in step at high running speeds. It had however demonstrated that in principle the method was feasible. Reducing the length of the code message obviously reduces the run time but at the expense of a loss in resolution. The important question about the minimum length of message required to resolve the settings in this way, will be addressed later.

The success of the method described above, depends on the above average occurrence of $[\Delta P_1 + \Delta P_2 = " \bullet "]$, resulting in the probability of the event $[\Delta D_1 + \Delta D_2 = " \bullet "]$ having a value significantly greater than the random expectation of one half. So far this has been attributed a high proportion of repeated characters in the plain-text, but it is important to realise that a frequent occurrence of other particular plain-text character pairs can also make significant contributions to the count of $[\Delta D_1 + \Delta D_2 = " \bullet "]$. In general the characteristics of ΔD are, to some degree, related to those of ΔP. To quote from a 1944 document :- "it is a rough approximation to the truth to state, that a ΔD sequence is a watered down version of the corresponding ΔP sequence". This important statement, which applies only to whole characters, can be justified in the following way:-

Since $\Delta D = \Delta P + \Delta S'$, it follows that whenever $\Delta S' = " \bullet "$, then $\Delta D = \Delta P$. This will happen when the S wheels do not move or when $DS = " \bullet "$. From the earlier analysis given above it can be seen that the probability of this is not less than $(1 - a) = \frac{22}{74}$, that is about 0.3 hence, very approximately, for about one third of the characters of ΔP the corresponding $\Delta S'$ character will be a $" \bullet "$. Hence the sequence of ΔD characters can be regarded as being the same as those of ΔP, but with approximately two thirds of them masked by obscuring $\Delta S'$ characters.

In the early days, knowledge about the statistical properties of the German ΔP traffic was limited to the fact that the five pulse streams contained significantly more $" \bullet "$s than would be expected by chance, and the first machine based attacks on the wheel settings used algorithms based on this assumption, of the form described above. Later, as more information was gained, it became clear that the ΔP traffic had other significant properties that could be used as the basis for a variety of other algorithms. The techniques were refined to the point that differences between the traffic over the different German communication links were identified and appropriate algorithms selected accordingly, to make the most effective attack on the settings. As an example , tables 2 and 3 contain information about the frequency of characters in the ΔP and ΔD traffic passing over one particular communication link:

The reasons for some of the high counts are due to the frequent occurrence

/ 156	R 76	A 102	D 79
9 75	C 73	U 196	F 134
H 72	V 51	Q 66	X 76
T 58	G 100	W 67	B 24
0 133	L 66	+ 361	Z 82
M 104	P 88	8 157	Y 94
N 61	I 55	K 68	S 93
3 150	4 63	J 139	E 81

Table 2. Characteristics of ΔP (3200 characters)

of the following pairs of characters in the plain text:-

1. The pairs 89 or 98 (8 = "letter shift", 9 = "space") giving "+".
2. The pair $+M$ (+ = "figure shift", then M = "full stop") giving "U".
3. The pair 9+ (9 = "space", + = "figure shift") giving "8".
4. Any pair of identical characters (a repeat) giving " /".

/ ●●●●● 128	R ●X●X● 92	A XX●●● 96	D X●●X● 89
9 ●●X●● 110	C ●XXX● 90	U XXX●● 124	F X●XX● 100
H ●●X●X 102	V ●XXXX 94	Q XXX●X 101	X X X●XXX 87
T ●●●●X 99	G ●X●XX 100	W XX●●X 89	B X●●XX 82
0 ●●●XX 104	L ●X●●X 92	+ XX●XX 143	Z X●●●X 89
M ●●XXX 100	P ●XX●X 96	8 XXXXX 112	Y X●X●X 97
N ●●XX● 100	I ●XX●● 96	K XXXX● 89	S X●X●● 104
3 ●●●X● 113	4 ●X●●● 90	J XX●X● 103	E X●●●● 89
(the teleprinter codes are given for future reference)			

Table 3. Characteristics of ΔD (3200 characters)

A comparison of these two tables confirms that high scores in ΔD tend to correspond with high scores in ΔP. At BP an exhaustive investigation of possible algorithms was undertaken, and over a period of time, assessments were made of their value in relation to the different communication links being intercepted. The most useful algorithm was the one described earlier, and this was represented in the "shorthand" notation used at the time as "$1 + 2 \bullet /$" or sometimes as "$1p2 \bullet /$". In all over fifty different algorithms could be selected for "runs" on Colossus, and a few examples of these are given, using the BP notation of the period, togther with the comments made at the time about their utility:-

1. $1 + 2 \bullet /$ "Useful break-in on all links. Gets weaker as the proportion of pure German increases."
2. $4 + 5 \bullet /$ "Moderate on all links, good on the same ones as $1 + 2 \bullet /$".
3. $4 = /1 = 2$ "Short run after break-in, finds K_4 setting, given those for K_1 and K_2".
4. $5 = /1 = 2 = 4$ "Short run after break-in, finds K_5 setting, given those for K_1, K_2 and K_4".

An understanding of this very abbreviated notation will be helpful, and some examples are given where algorithms represented in this way are described in full:-

1. "$1 \bullet /$":- A count is made of those positions in the code where $\Delta D_1 = " \bullet "$, for all the possible settings of K_1, (needs 41 passes through the code, known as a "short" run). Illustrative example only, never used.
2. "$1 + 2 \bullet /$":- A count is made of those positions in at which $\Delta D_1 + \Delta D_2 = " \bullet "$, for all possible settings of K_1 and K_2, (needs $41 \times 31 = 1271$ passes through the code and is an example of a "long" run).
3. "$4x5 \bullet /$":- A count is made of those positions in the code where $\Delta D_4 = "X"$ and $\Delta D_5 = \bullet$, for all possible settings of K_4 and K_5, (needs 598 passes through the code, another "long" run).
4. "$1 + /3 \bullet$":- A count is made which assumes that the correct setting for K_3 is known, and that K_1 is to be tried at all possible settings , each starting with K_3 at its correct setting. The score for each being a count of the number of positions in the code where $\Delta D_1 + \Delta D_3 = " \bullet "$, (needs 41 passes, another "short" run).
5. "$4 = /1 = 2$" A count is made which assumes that the correct settings for K_1 and K_2 are known and that K_4 is to be tried at all possible positions, each starting with K_1 and K_2 at their correct settings. The score for each, being a count of the number of positions in the code where $\Delta K_4 = \Delta K_1$ and $\Delta K_1 = \Delta K_2$, (needs 26 passes through the code, a "short" run).

The effectiveness of an algorithm depends on the statistical properties of the ΔD stream with which it is used. An estimate of this can be made, for a given communication channel, by working with a representative sample of ΔD traffic for that channel, and determining the value of the expression :- $(s-m)/m$, where s is the score obtained from the algorithm with the correct wheel settings, and m is the expected random mean score (i.e. the mean score for all the other settings). This value was called the "proportional bulge" (δ) at BP . Two examples are shown based on the ΔD traffic table given above:-

(r = the number of cipher characters used as a basis for the scores)

1. "$1 + 2 \bullet /$":- Let p be the random probability of the event "$1 + 2 \bullet /$", then $p = 1/2$, $(r = 3200)$. The random scores form a binomial distribution with a $mean = 1600$. From the ΔD table, $s = 1713$, hence $\delta = (1713 - 1600)/1600 = 0.071$.

2. "$4 = 5 = /1 = 2$":- Let p be the random probability of "$4 = 5 = /1 = 2$", then $p = 1/4$, ($r = 1713$). The random scores form a binomial distribution with a *mean* $= 428.25$. From the ΔD table $s = 493$, hence $\delta = (493 - 428)/428 = 0.151$.

The table gives a summary for some of the algorithms, when used with a ΔD stream with the same characteristics as the given example:-

		s	m	δd
(1)	$1 + 2 \bullet /$	1713	1600	0.071
(2)	$4 + 5 \bullet /$	1659	1600	0.037
(3)	$4 = /1 = 2$	886	856	0.035
(4)	$5 = /1 = 2$	493	856	0.046
(5)	$1 + 3 \bullet /$	1632	1600	0.020
(6)	$4 = 5 = /1 = 2$	493	428	0.151

Table 4. Table of ΔD stream.

The proportional bulge is a measure of the ability of the algorithm to discriminate between the score obtained when the wheel setting(s) are correct and all the other random scores. One way of expressing this, is that it determines the minimum length of code on which a run can be expected to succeed:-

$$\delta = (s - m)/m$$

where $m = p.r$, hence the deviation $(s - m) = \delta.p.r$. The standard deviation of the random scores $\sigma = \sqrt{p.(1 - p).r}$.

In order for the deviation from the random mean to be significant it must exceed some selected multiple of the standard deviation. Let this multiple be k. Then

$$\delta.p.r > k.\sqrt{p(1 - p).r}$$

and after squaring and simplification the following result is obtained:- $r > k2(1 - p)/p.\delta^2$. If $p = 1/2$ then this reduces to $r > k^2/\delta^2$ and if $p = 1/4$ to $r > 3k^2/\delta^2$.

It was the practice to take the value $k = 4$ for long runs, and $k = 3$ for short runs. Hence for a long run with $p = 1/2$ the minimum length of code $r > 16/\delta^2$. For the "$1 + 2 \bullet /$" algorithm the minimum length $= 3174$ characters, but for the "$4 + 5 \bullet /$" algorithm the minimum length $= 11{,}687$ characters. This demonstrates the high level of discrimination given by the "$1 + 2 \bullet /$" algorithm and explains why it was invariably chosen for the first run in a wheel setting sequence. Indeed if it failed to provide a useful result , then experience showed that it rarely happened that anything else would succeed. Often in these circumstances work would be abandoned on that particular message.

3 Some Statistical Considerations

In any wheel setting run, the objective is to find the setting(s) that give the highest score. It is likely that these settings are the correct ones, but there are occasions where by chance a random score will exceed that given by the correct settings. Whenever two or more high scores occur which are fairly close, there is a decision problem about which one to select as an indicator for the correct settings. In most instances the matter can be resolved by making other independent runs involving the same wheel(s). If the same setting for a wheel shows up with high scores on these runs , then this is evidence in its favour, which may be sufficient to enable a clear decision to be made. Runs which provided this additional evidence were called "pick-ups".

For any objective assessment, this evidence must be expressed in numerical form, and this was done by using the standard deviation of the random scores expected from each run. The deviation of the highest score from the random mean score, was expressed as a multiple of the standard deviation (σ) of the random scores (the "sigma-age" of the score). The question now arises about what value this multiple should take in order to justify the decision to assume that the corresponding settings are correct. To quote from a document of the period :-" *The value required depends on how certain we wish to be, i.e. on the balance to be struck between how long we can afford to take , and how often we can afford to be wrong.*" The answer to the last question was taken to be that not more than 1 in 20 messages (presumably the characters of D) marked as "certain", were expected to be wrong.

The "certain" settings for the wheels were identified in the following way:-

1. For a single long run:- a pair of settings with a "sigma-age" of 4.5 or more.
2. For a single short run:- a setting with a "sigma-age" of 3.8 or more.

For combining two runs, ("pick-ups") a set of tables was provided for the Colossus operators, which gave detailed numerical information upon which decisions could be based. So far, these tables, or the theory of their derivation, have not come to light. However it is known that the following rules, based on the sum of the "sigma-ages" of the two runs, were offered as approximations for the condition that a common setting to be considered "certain":-

* Two long runs:- *sum* > 7;
* one long and one short run:- *sum* > 6;
* two short runs:- *sum* > 5.

The results of the runs made on Colossus were printed out automatically by a specially modified electric typewriter, an advanced innovation at the time. One small problem was that, the total number of scores generated in a run was often quite large and yet it was only the highest scores that were of interest. For example in a long run for K_1 and K_2 the number of scores generated would be $41 \times 31 = 1271$. To eliminate the great majority of the results which were of no interest Colossus was programmed to only print out settings and scores when the latter exceeded a selected value. This value was called the "set-score" and it was usually made equal to:-

* $m + 5.s/2$, for "long" runs, and
* $m + s$ for "short" runs. (s = expected S.D. of the random scores).

Computer programs have been written to simulate the runs made on Colossus, using the algorithms that were originally employed. The message used was one of the few examples of genuine German cipher available, and had a length of 4056 characters. It is worth pointing out that one consequence of using a high level interpretative language on a PC., are run times longer than those made on Colossus over fifty years ago!

4 Finding the S and Motor Wheel Settings

Once the five K-wheel settings were known, then the full sequence of D characters could be found using the equation $D = Z + K$. As the movement of the S-wheels is intermittent, some of the characters of the ΔP stream must appear in the corresponding ΔD stream. This property of ΔD, enabled the settings for the S-wheels to be found, employing the more traditional "hand" methods (art?) of codebreaking.

An important factor in this was the recognition of the frequent presence in the plain-text , of some particular groupings of letters which were known as "cribs". For example the character group "+M89" ("full-stop" then "space"), occurred frequently in P, resulting in "$UA+$" in ΔP. The presence of all, or part of a crib, could often be detected in the ΔD stream, indicating that the corresponding characters of the plain-text were likely to be those of the crib. To quote from a document of the period:-"*Actually he relies on his skill, looks at D and deltas it in his head.*" In this way some of the letters of P were found, and these in turn provided information about some of the S' characters using the equation $S' = D + P$ (derived from $D = P + S'$). This knowledge about S' could be used to find the settings for the S-wheels, assuming that their patterns were already known.

The settings for the motor wheels could be determined by a special run on Colossus. working with the DD sequence already obtained. Methods also existed for finding the S-wheel settings with Colossus, but it would seem that at the time, the practice was to find these by hand, thus freeing the machine resource for the task of finding the K-wheel settings, as this was not amenable to hand methods. The hand work required people with consummate skills in logical thinking and an sound knowledge of "Military German". An additional requirement was the ability to carry out the "delta" operation mentally, with speed and accuracy.

5 Finding the Wheel Patterns

Since a knowledge of the wheel patterns is required before work on the settings can begin, it may appear somewhat illogical to have considered the problem of the settings first, but there are good reasons for taking this path. The initial role of Colossus was restricted to finding the K-wheel settings, as at that time the task of finding the patterns still relied on the interception of "depths" or partial "depths". The subsequent discovery of a method for finding the patterns which required only one cipher message, albeit a long one, was made by W.T. Tutte, the second important contribution he made during his time at BP. The method, which could be implemented on Colossus, was based on the techniques already established for finding settings.

It is proposed to offer only an outline description of this method. The starting point is a result given earlier, and on which the first wheel setting algorithm was based :-

$$\Pr[\Delta Z_1 + \Delta K_1 + \Delta Z_2 + \Delta K_2 = " \bullet "] > \frac{1}{2}$$

This is equivalent to :-

$$\Pr[(\Delta Z_1 + \Delta Z_2) = (\Delta K_1 + \Delta K_2)] > \frac{1}{2}$$

which in turn can be expressed in the following way:- "If the value of $(\Delta Z_1 + \Delta Z_2)$ is determined from a character of code, and the result is assumed to be an estimate for the corresponding value of $(\Delta K_1 + \Delta K_2)$, then there is a greater chance of this estimate being correct than there is of it being wrong. " A single estimate made in this way is hardly a reliable one to use, but if a sufficient number are made and combined in an appropriate way, then the result is much stronger. The procedure, is to systematically evaluate, from a long sequence of code, all the possible values of $\Delta Z_1 + \Delta Z_2$, and to enter each result in the appropriate place in a rectangular table containing 41x31 locations. The final outcome will be a table in which each location will contain several estimates of $\Delta Z_1 + \Delta Z_2$ ("X" or "$\bullet$"). From these the majority score can be determined and used as a basis for finding the best single estimate. The final result can be used as a good approximation for the table of the corresponding values of $\Delta K_1 + \Delta K_2$.

If now a first approximation for the set of differences of the pattern for one wheel is assumed, then this table can be used to derive a corresponding pattern for the other wheel. This process, known as "taking the pattern through the rectangle", can then be repeated in the reverse direction, taking the pattern just found back again through the rectangle and so on, giving a sequence of approximations for the true difference patterns. This iterative process has the interesting property of frequently "converging" to two final patterns which are good approximations for the correct difference patterns for the K_1 and K_2 wheels. A further sequence of processes, which are closely related to the other wheel setting runs previously described, can then be used to obtain approximations for the patterns of the other K-wheels and to improve on the approximations already obtained.

At each stage of this work appropriate statistical procedures were used to estimate the strength of the evidence for the patterns derived. The whole process required about ten hours of operational time on Colossus, but the final results could be remarkably accurate. The final difference patterns for the five K-wheels were only accepted for use if the odds on any pattern element being wrong was of the order 1000 to 1 against, or using the logarithmic scale introduced by Alan Turing, "30 decibans up".

Split Knowledge Generation of RSA Parameters

Clifford Cocks

Communications-Electronics Security Group
PO Box 144, Cheltenham GL52 5UE
UK

Abstract. We show how it is possible for two parties to co-operate in generating the parameters for an RSA encryption system in such a way that neither individually has the ability to decrypt enciphered data. In order to decrypt data the two parties instead follow the co-operative procedure described.

1 Introduction

The parameters for the well known RSA system consist of a public modulus N which is a product of two primes, a public encipherment key e, and a secret decipherment key d. The factorisation of N is a secret parameter and the keys are related by the formula $de \equiv 1 \bmod \Phi(N)$, where $\Phi(N)$ is the order of the multiplicative group of integers modulo N. Then with knowledge of only the public parameters, any message x (represented as a positive integer less then N) can be enciphered as $y = x^e \bmod N$. The secret parameter d is needed to decipher the encrypted message y via the formula $x = y^d \bmod N$.

In this paper we show how it is possible for two participants to co-operate in generating the public parameters N and e, in such a way that individually neither knows the factorisation of N, and such that they each have a share d_1, d_2 respectively of the secret decrypt exponent d where $d = d_1 + d_2$. Hence neither participant has the ability to recover x from an enciphered message y, (where $y = x^e \bmod N$), but as we show they can enable recovery of x if they jointly agree and follow a specific decryption procedure. This procedure does not compromise knowledge of the secret parameters, and can be executed in such a way that a third party is involved, and that party alone is able to recover the value of x.

Boneh and Franklin in independent work [1] describe a method that is similar to ours. However, it differs principally in that they require the help of a third party to generate the RSA parameters. Our approach avoids the need for this - at the cost of increasing the amount of computation required.

2 Applications

One possible application of this method is in split escrow schemes, where a user will deposit an encryption key with two escrow agents, for possible retrieval later by a duly authorised entity [2]. The protocols described in this paper can be used to ensure that neither escrow agent acting alone has access to the encryption key, but that recovery is possible by co-operative action. Furthermore the escrow agents do not themselves need to gain access to the deposited key even when executing the recovery procedure.

Another application is to the Fiat Shamir signature scheme [3] which requires a trusted centre to issue secret identification data to new users on registration. This centre necessarily has all the information needed to allow it to masquerade as any registered user. We show how the the parameter generation method described in this paper allows for two centres to split the information so that neither of them individually can masquerade as a user or forge their signature.

3   A Simple Method

Before describing our method it is worth noting that there is a simple extension of the RSA system that can provide some of the desired functionality. In the case of a split escrow scheme suppose that the two agents separately generate moduli N_1 and N_2 and agree on a common public exponent e. The corresponding secret exponents will be d_1 and d_2, each known by only one agent. Then they make known the public modulus $N = N_1 N_2$. Now given an enciphered message y where $y = x^e \bmod N$, each agent is only able to recover $x \bmod N_1$ and $x \bmod N_2$ respectively, but if they share these then together they can easily recover x.

This approach has undesirable features which we avoid with the more complex protocol described in the paper. The first is that the length of the public modulus N will need to be twice as long as for a normal RSA system and this will make the system slower at the user level.

The second undesirable feature is the fact that each agent acting alone can recover one of $x \bmod N_1$ and $x \bmod N_2$. This fact places strong constraints on the way that x would need to be encoded to ensure that this information is of no value.

4   The New Method - Overview

The method proposed consists of three parts:

1. A procedure to enable two agents to generate a modulus N that is (to a high probability) the product of two primes P and Q, but neither agent can recover P or Q with a feasible amount of work.
2. A procedure for the two agents to generate their shares d_1 and d_2 of the secret recovery exponent d, given a public encryption exponent e.
3. A procedure to allow cooperative recovery of an encrypted message.

5   Generation of Modulus

In this section I will refer to the two cooperating agents as Alice and Bob. Alice will generate two numbers p_1 and q_1 (not necessarily prime) and Bob will generate p_2 and q_2. We first show how they can generate the number $N = PQ$ where $P = p_1 + p_2$ and $Q = q_1 + q_2$ in such a way that neither knows P or Q, and then show how they can test N for the condition that P and Q are probably prime. This procedure is repeated until they find an N for which the test is satisfied, and this N becomes the public modulus.

1. To begin with Alice chooses her own RSA modulus M and public exponent e_M, which she makes known to Bob. Alice's secret decrypt exponent will be d_M. The size of M must be at least as big as the largest size of the public modulus N they wish to generate. Then Alice sends to Bob the quantities $p_1{}^{e_M} \bmod M$ and $q_1{}^{e_M} \bmod M$.

2. Bob can now calculate the three numbers $(p_1 q_2)^{e_M} \bmod M$, $(p_2 q_1)^{e_M} \bmod M$ and $(p_2 q_2)^{e_M} \bmod M$. We will call these three quantities a_1, a_2 and a_3.

3. Bob also generates a set of numbers $b_{i,j}$ for $i = 1, 2, 3$ and $j=1, 2, ..., K$. The value of K will be discussed later in section 10 (Security Issues). The $b_{i,j}$ are chosen to be random modulo M, subject to the constraints: $\sum_j b_{1,j} = \sum_j b_{2,j} = \sum_j b_{3,j} = 1$

4. Bob then generates the $3K$ numbers $x_{i,j} = a_i b_{i,j}^{e_M} \bmod M$ and sends these to Alice **but in a new order**. The ordering must be such that it is computationally infeasible to recover the values of i and j from the set $x_{i,j}$ as sent. Thus a random order, or a sorted order are both acceptable.

5. Alice can now calculate $y_{i,j} = x_{i,j}^{d_M} \bmod M$. Thus $y_{i,j} = b_{i,j} a_i^{d_M} \bmod M$, and hence Alice can determine $N = (p_1 + p_2)(q_1 + q_2) = p_1 q_1 + \sum_{i,j} y_{i,j} \bmod M$, which she sends to Bob.

As it stands Alice could cheat by substituting any N of her choice at this point (although if she does, it is not clear that she will be able to complete the later steps of the method in a satisfactory fashion). Nevertheless, we can prevent such cheating by making the above procedure symmetrical. To do this Bob produces his own RSA modulus and executes the above exchange using the same values of p_1, p_2, q_1 and q_2, and for this exchange it will be Alice who produces the set of $3K$ random values. Then Alice and Bob will both know N, so they exchange a hash of N to confirm that they have recovered the same value.

6   Primality Testing

There is no particular reason to suppose that the P and Q produced in this way are prime, so it will be necessary for the above procedure to be carried out many times until an N is found that is likely to be the product of two primes and can be used as an RSA modulus. Thus we need a test for N being of the right form, and we propose to test for the condition that $x^{N+1} \equiv x^{P+Q} \bmod N$ for many x. (In practice almost all N not of the correct form will fail on the first x tested.) This condition does not guarantee that P and Q are prime, for example either of them could be Carmichael numbers. However, even if they are not prime we can use N as an RSA modulus, but being a product of more than two primes may make such N easier to factorise.

To test whether N is of the right form Alice and Bob will agree on a set of x values to use, and Alice will calculate $x^{N+1-p_1-q_1} \bmod N$ and Bob will calculate $x^{p_2+q_2} \bmod N$. They will exchange a hash (using a secure hash function) of these values and this will be sufficient to tell if they are equal.

It is possible to use the test of Boneh and Franklin [1] to increase confidence in the fact that P and Q are both prime, once an N has been found that passes the above test. To make use of this it is necessary that $P \equiv 3 \pmod 4$ and $Q \equiv 3 \pmod 4$. This can be achieved by agreeing the values of p_1, p_2, q_1 and q_2 modulo 4 in advance. The test, which is described below for the sake of completeness, will be executed k times, where k is set according to the level of confidence required. A number N of the correct form will always pass the test, whilst if either P or Q is composite the test will fail with probability at least $1/2$ at each of the k iterations.

Each test consists of two steps. At step 1, Alice and Bob will agree on a random integer x in the range 1 to $N-1$ such that the Jacobi symbol $(\frac{x}{N})$ equals $+1$. Alice then calculates a hash of each of the two values $\pm x^{\frac{(N+1+l-p_1-q_1)}{4}} \bmod N$, and Bob calculates a hash of $x^{\frac{(l+p_2+q_2)}{4}} \bmod N$. l will be 0,1,2 or 3 according to the agreed values of p_1, p_2, q_1 and q_2 modulo 4. The hashed values are compared, and there must be a match or the test fails.

At step 2, Alice and Bob agree on two random co-prime integers in the range 1 to $N-1$, u and v say. They work in the ring of polynomials $Z_N[X]$, and Alice computes the remainder when $(uX+v)^{N+1+p_1+q_1}$ is divided by X^2+1, and Bob computes the remainder when $(uX+v)^{p_2+q_2}$ is divided by X^2+1. Writing these polynomials as $u_1 X + v_1$ and $u_2 X + v_2$ respectively, Alice calculates a hash of $v_1/u_1 \bmod N$ and Bob calculates a hash of $-v_2/u_2 \bmod N$. These are compared and the test fails if the two values differ.

7 Generation of Exponent

Once an acceptable modulus has been found, the next step is to generate the public exponent e and the secret exponent d. This is done in such a way that d is held by the cooperating partners Alice and Bob in two parts d_1 and d_2, where $d_1 + d_2 = d$. The process of generating these parameters is as follows:

Firstly, the public exponent e is agreed. This should be chosen so that $P-1$ and $Q-1$ are likely to be coprime to e, but at the same time e should not be too large as Alice and Bob will have to share $(p_1+q_1) \pmod e$ and $(p_2+q_2) \pmod e$ with each other. A value such as $e = 2^{16}+1$ should be satisfactory.

Now Alice reveals $(p_1+q_1) \pmod e$ to Bob and Bob reveals $(p_2+q_2) \pmod e$ to Alice, so they can both calculate $f = P+Q-N-1 \pmod e$. If f is non-zero and co-prime to e then they can both calculate $g = f^{-1} \pmod e$. The secret decrypt exponent will be $d = \frac{(1+(N+1-P-Q)g)}{e}$, but

Alice will calculate : $d1 = \lfloor \frac{1+(\frac{(N+1)}{2}-p_1-q_1)g}{e} \rfloor$

Bob will calculate : $d2 = \lceil \frac{(\frac{(N+1)}{2}-p_2-q_2)g}{e} \rceil$

It can be easily verified that $d = d_1 + d_2$.

8 Data Recovery Procedure

We suppose that a third party, Carol say, is authorised to obtain the decrypt x from an enciphered message $y = x^e \bmod N$. Then she presents y to Alice and Bob and receives back $x_1 = y^{d_1} \bmod N$ and $x_2 = y^{d_2} \bmod N$ respectively. Thus Carol can determine $x = x_1 x_2 \bmod N$. Obviously, Alice and Bob could recover this information themselves by sharing x_1 and x_2 and then presenting the recovered value of x to Carol.

9 Use With Fiat Shamir Scheme

In the case of the Fiat Shamir scheme [3], where Alice and Bob take the place of the single trusted center, Carol will need to present values v_j, derived via her identity and a universal hash function, and obtain data that will allow her to calculate s_j, where $s_j^2 = v_j \bmod N$ whenever v_j has a square root.

To do this with our scheme, we require that $P \equiv 3 \pmod 4$ and $Q \equiv 3 \pmod 4$, and note that if $d = \frac{(P-1)(Q-1)+4}{8}$, then whenever v_j is a square modulo N then $s_j = v_j^d \bmod N$ is a square root. The exponent d will be held in two parts d_1 and d_2 by Alice and Bob respectively.

For Alice : $d1 = \lfloor \frac{(\frac{(N+5)}{2}) - p_1 - q_1)}{8} \rfloor$

For Bob : $d2 = \lceil \frac{(\frac{(N+5)}{2}) - p_2 - q_2)}{8} \rceil$

Then if presented with a value of v_j, Alice and Bob must check that the Jacobi symbol $\left(\frac{v_j}{N}\right)$ equals 1. If so Alice will calculate $w_{j,1} = v_j^{d1} \bmod N$ and Bob will calculate $w_{j,2} = v_j^{d2} \bmod N$. Carol will calculate $w_j = w_{j,1} w_{j,2} \bmod N$, which will either be the square root of $v_j \bmod N$, or will be the square root of $N - v_j \bmod N$. In the latter case Carol will reject this particular v_j.

10 Security Issues

A critical question is the size of K, the number of fragments into which Bob splits each of the following three quantitites, $(p_1 q_2)^{e_M} \bmod M$, $(p_2 q_1)^{e_M} \bmod M$ and $(p_2 q_2)^{e_M} \bmod M$. Alice receives (encrypted under her modulus) $p_1 q_2 b_{1,j}$, $p_2 q_1 b_{2,j}$ and $p_2 q_2 b_{3,j}$. She can recover the factorisation of N if she can identify which fragment is associated with each of the three quantities $p_1 q_2$, $p_2 q_1$ and $p_2 q_2$.

We propose that K be chosen so that the total number of possible arrangements $\frac{(3K)!}{(K!)^3}$, exceeds M^2. This ensures that for most guesses by Alice as to the value of $p_1 q_2$, $p_2 q_1$ and $p_2 q_2$ (subject to their sum being the value recovered) there will be a partition of the $3K$ fragments into 3 sets which produce these three values. In other words, Alice gains negligible additional information about the values of p_1q_2, p_2q_1 and p_2q_2 from the fact that they can be obtained from a partition of the $3K$ pieces. If M is 512 bits in size then K will need to be at least 218 to achieve this bound, and for 1024 bits K will need to be at least 433. In practice it is likely that smaller values of K will provided adequate security, but a detailed analysis is beyond the scope of this paper.

11   Computational Issues

Assuming that Alice's public encryption exponent e_M is small, the principal amount of work is performed by Alice, who calculates $3K$ decryptions for each trial N. As the probability that N is a product of two primes is about $\frac{1}{(logP)(logQ)}$, the total amount of work Alice will expect to have to do can be shown to be $0.75K(logN)^2$ decryptions, assuming that P and Q are of similar size. For numbers of size 512 bits, with K equal to 218, this is about 20.6 million decryptions. At 1024 bits, with K equal to 433, the expected number of decryptions is about 164 million. Clearly it is desirable to cut the work down if at all possible.

One way to do this is to increase the probability that P and Q will be prime. This can be achieved by taking a set of small primes: 3,5,7,...,R say and agreeing that both Alice and Bob will choose p_i mod S and q_i mod S to be less than $\lfloor S/2 \rfloor$ for each prime S in this range, and also agree that only one of them can choose numbers that are a multiple of S. For the prime 2, they must agree in advance who will select a multiple of 2 and who will choose an odd number. (If they are going to use the Boneh and Franklin primality test then they will control the values of p_i and q_i modulo 4 as well).

Each small prime places 1 bit of constraint for Alice and Bob on the choice of p_i and q_i, but significantly reduces the number of candidate moduli N that need to be tested. An alternative would be to use zero knowledge methods to ensure that p_1 mod $S \neq -p_2$ mod S and that q_1 mod $S \neq -q_2$ mod S, but this will not be necessary if the number of primes to be controlled is small. If the first 10 primes are controlled so that $R = 29$, the number of decryptions Alice needs to make drops by a factor of 40. Using the first 20 primes drops the expected number by a factor of about 60. This helps to make the work involved practical, at least for generating long term system wide RSA parameters. For example, if the first 20 primes are controlled, the calculations needed to generate a 512 bit modulus would take a little over one day to complete using MATHEMATICA on a SPARC10 workstation.

Acknowledgement

I thank my colleague Richard Smith for useful discussions on this work.

References

1. D. Boneh and M. Franklin *Efficient Generation of Shared RSA Keys.* In *Advances in Cryptology - Crypto 97* Lecture Notes in Computer Science vol 1294, pp 425-439.
2. S. Micali *Fair Cryptosystems.* MIT Technical report MIT/LCS/TR-579.h, November 1993.
3. A. Fiat and A. Shamir *How to prove yourself: Practical solutions to identification and signature problems.* In *Advances in Cryptology - Crypto 86* Lecture Notes in Computer Science vol 263, pp 186-194.

Analysis of Error Control in Digital Trunked Radio Systems

Pingzhi Fan

Institute of Mobile Communications
Southwest Jiaotong University
Chengdu, PR of China
Email: p.fan@ieee.org

1 Introduction

In order to provide a cost-effective facility for transmitting information from one end of the system at a rate and a level of reliability and quality that are acceptable to a user at the other end, suitable modulation and error control schemes are important. For a fixed signal-to-noise ratio, the only practical option available for changing data quality from problematic to acceptable is to use error-control coding. For a fixed bit error rate, the use of coding can also reduce the required signal-to-noise ratio.

TETRA, iDEN, APCO25 and FHMA are typical modern trunked professional mobile radio (PMR) systems. This paper will analyse, theorectically and by computer simulation, their error-control scheme and compare and comment on their error control techniques and relative performance.

2 Error Control Schemes

In modern digital trunked radio systems, each logical channel has its own error control scheme. The typical error control process is as follows:

1. the information bits are packed in medium access control blocks;
2. the medium access control blocks are encoded by a block code, providing block-encoded bits;
3. the block-encoded bits are encoded by a convolutional code, which provides the convolutionally-encoded bits;
4. the convolutionally-encoded bits are reordered and interleaved, into interleaved bits;
5. the interleaved bits are then scrambled and mapped into various multiplexed blocks.

Figure 1(a) gives a typical error-control structure and its interfaces. As a specific error control structure used in TETRA , Figure 1(b) gives examples of error-control for two types of logical channels, the broadcast synchronisation channel (BSCH) and traffic channel with net rate 7.2kbps(TCH/7.2).

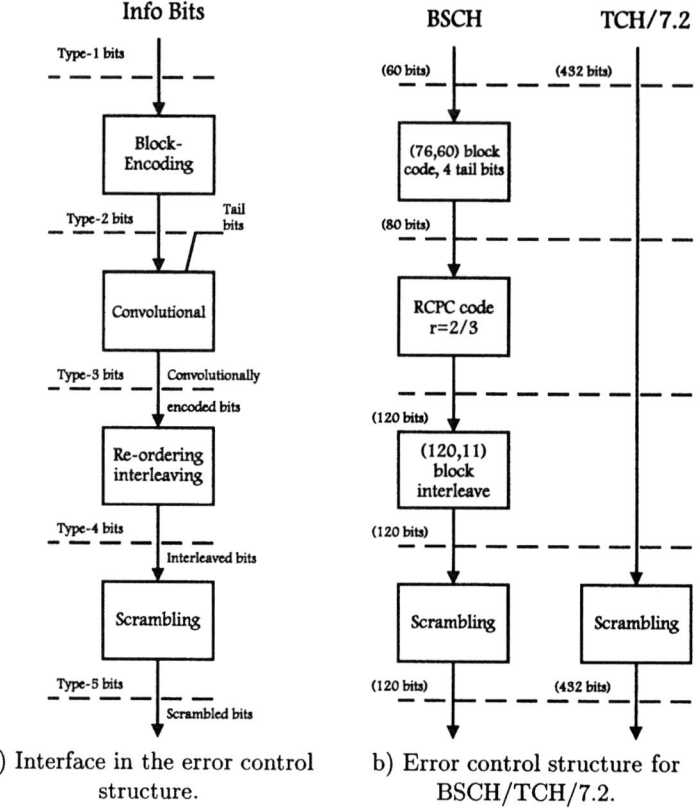

a) Interface in the error control
structure.

b) Error control structure for
BSCH/TCH/7.2.

Fig. 1. Error Control Schemes.

3   Error-Correcting Codes

There are three types of error-correcting codes: block codes, convolutional codes
and trellis codes. The binary block and convolutional codes are mainly used
for combating random errors, while m-ary codes can also be used for correcting
burst errors. To attain a more effective utilization of the available bandwidth
and power, coding and modulation can be treated as a single entity, i.e. trellis-
coded modulation(TCM). As the bursty errors are quite serious in mobile communication, interleaving and scrambling techniques are also frequently used to
randomize the error pattern caused by a bursty interference channel.

The following error-correcting codes are most commonly used in various digital trunked radio systems:

3.1   Convolutional codes

In addition to various convolutional codes with different r and k, the Rate-
Compatible Punctured Convoluational (RCPC) codes are also attractive in practice. The RCPC encoding is normally performed in two steps: encoding Type-2

bits $b_2(1), b_2(2), \ldots, b_2(K2)$ by a 16-state mother code of rate $1/4$,

$$V[4(k-i)+i] = \sum_{j=0}^{4} b_2(k-j)g_{i,j}$$

where $i = 1, 2, 3, 4$, $k = 1, 2, ..., K_2$ and g_{ij} is the coefficient of D^j in generator polynomial G_i.

$$G_1 = 1 + D + D_4$$
$$G_2 = 1 + D_2 + D_3 + D_4$$
$$G_3 = 1 + D + D_2 + D_4$$
$$G_4 = 1 + D + D_3 + D_4$$

and puncturing of the mother codes so as to obtain a 16-state RCPC code $b_3(1), b_3(2), \ldots, b_3(K_3)$ of K_2/K_3 rate:

$$b_3(j) = V[8((j-1)divt) + P(i - t((i-1)divt))], j = 1, 2, ..., K_3$$

The i and t are defined in Table 1.

rate	t	i	P(1)	P(2)	P(3)	P(4)	P(5)	P(6)
2/3	3	j	1	2	5			
1/3	6	j	1	2	3	5	6	7
292/432	3	j+(j-1) div 65	1	2	5			
148/432	6	j+(j-1) div 35	1	2	3	5	6	7

Table 1. The definition of i and t

3.2 Block codes

The common block codes used in digital PMR are BCH code, RS code, Golay code, Hamming code, shortened Reed-Muller(RM) code, $(K_1 + 16, K_1)$ block code, etc.

As an example, the shortened $(30, 14)$ RM code encodes 14 type-1 bits $b_1(1), b_1(2), \ldots, b_1(14)$ into 30 type-2 bits $b_2(1), b_2(2), ..., b_2(30)$.

$$[b_2(1), b_2(2), \ldots, b_2(30)] = [b_1(1), b_1(2), ..., b_1(14)]G$$

where G is the generator matrix.

The (K_1+16, K_1) code encodes type-1 bits $b_1(1), b_1(2), ..., b_1(K_1)$ into $(K_1 + 16)$ type-2 bits $b_2(1), b_2(2), \ldots, b_2(K_1 + 16)$:

$$b_2(k) = \begin{cases} b_1(k), & \text{when } k = 1, 2, \ldots, K_1 \\ f(K_1 + 16 - k), & \text{when } k = K_1 + 1, K_1 + 2, \ldots, K_1 + 16 \end{cases}$$

where $f(i)$ are the coefficients of x^i in polynomial $F(x)$ of degree 15:

$$F(x) = [(x^{16} \sum_{k=1}^{K_1} b_1(k) x^{K_1-k} + x^{K_1} \sum_{i=0}^{15} x^i) mod G(x)] + \sum_{i=0}^{15} x^i$$

$G(x)$ is the generator polynomial defined by $G(x) = x^{16} + x^{12} + x^5 + 1$.

3.3 Trellis coded modulation (TCM)

In an important class of trellis codes known as Ungerboeck codes, multilevel modulation is combined with convolutional coding. At the receiver, the Viterbi algorithm is used to perform a maximum-likelihood sequence detection and coding gains of 3 to 6 dB can be attained at bandwidth efficiencies equal to or larger than 2 bits per second per Hertz.

3.4 Interleaving

Block interleaving: a (K,a) block interleaving will re-order K_3 type-3 bits $b_3(1), b_3(2), \ldots, b_3(K_3)$ into K_4 type-4 bits $b_4(1), b_4(2), \ldots, b_4(K_4)$
$(K = K_3 = K_4)$

$$b_4(k) = b_3(i)$$

where $i = 1, 2, ..., K, k = 1 + [(a * i) mod K]$

Interleaving over N blocks: use two steps to interleave a sequence of M type-3 blocks $B_3(1), B_3(2), \ldots, B_3(M)$ of 432 bits each into a sequence of $(M + N - 1)$ type-4 blocks $B_4(1), B_4(2), ..., B_4(M + N - 1)$ of 432 bits each, where M is an integer and N has values 1,4 or 8. Firstly, a diagonal interleaver interleaves the M blocks $B_3(1), B_3(2), \ldots, B_3(M)$ into $(M+N-1)$ blocks $B'_3(1), B'_3(2), \ldots, B'_3(M)$

$$b'_3(m, k) = \begin{cases} b_3(m - j, j + 1 + i \times N), & \text{when } 1 \leq m - j \leq M \\ 0, & \text{otherwise} \end{cases}$$

where $j = (k - 1) div(432/N)$ and $i = (k - 1) mod(432/N)$. A block interleaver then interleaves each block $B'_3(m)$ into a type-4 block $B_4(m)$:

$$b_4(m, i) = b'_3(m, k)$$

where $k = 1, 2, \ldots, 432$ and $i = 1 + [(103 * k) mod 432]$

3.5 Scrambling

Scrambling transforms K_4 type-4 bits $b_4(1), b_4(2), \ldots, b_4(K_4)$ into K_5 type-5 bits $b_5(1), b_5(2), \ldots, b_5(K_5)$ with $K_4 = K_5$, as follows:

$$b_5(k) = b_4(k) + p(k)$$

for $k = 1, 2, \ldots, K_5$, $p(k)$ is the kth bit of the scambling sequence.

4 Performance Analysis and Comparison

Table 2 gives some basic characteristics of TETRA, iDEN, APCO25 and FHMA. The full paper will present the relative performance of various digital trunking radio systems. It will be shown that, although each system has its own unique features in error control and coding, TETRA provides the most powerful and flexible error protection capability.

Item	FHMA	IDEN	TETRA	APCO25
Multiaccess method	FHMA/ TDMA 3:1	TDMA 6:1	TDMA 4:1	FDMA 1:1
Channel spacing (kHz)	25	25(4x6.25)	25	12.5/ 6.25
Channel data rate (kbps)	36.9	64	36	9.6
Spectrum efficiency (bit/Hz)	1.48	2.56	1.44	0.77/ 1.54
Modulation Type	$\pi/4$ SQPSK	M16QAM(M=4)	$\pi/4$ DQPSK	C4FM/ CQPSK
Speech coding	AMBE	V-SELP(6:1)	ACELP	IMBE
Speech code rate (kbps)	2.4, 4.4, 5.55	4.2	4.567	4.4
Protected speech rate (kbps) and error control methods	2.28 $r=1/2, k=7$ conv. c.repeat import.bits CRC detection	3.177 multi-rate TCM 70-80 redund. interleaving	2.633 1/3 RCPC interleaving scrambling error detection	2.8 Golay Hamming
Protected data/signalling rate (kbps) and error control methods	9.6, 4.8, 2.4, 3.2 Conv. code with with r=1, 1/2, 1/4, 1/8, k=9 VDEC channel state analysis ARQ control repeated data	1.2, 2.4, 3.8, 3/4 TCM ARQ control	High protection: 9.6, 7.2, 4.8, 2.4 Med protection: 19.2, 14.4, 9.6, 4.8 No protection: 7.2,14.4, 21.6, 28.8 Conv. code with r=148/432, 292,/432, 1, 2/3 ARQ control	6.1, 9.6 BCH code TCM RS code
Error detection code	CRC	CRC	CRC	CRC
Interleaving type	Voice (12hops): 456bits, Short data block (12 hops): 456bits long data block (24 hops): 912bits, voice delay<100ms	time slot interleave data and voice interleave via mapping table voice delay<100ms	TCH interleave depth: N=1,4,8 BSCH interleave block (120,11) SCH interleave block: (216,101), (168,13), (432,103) voice delay=80ms	

Table 2. The definition of i and t

Reconstruction of Convolutional Encoders over $GF(q)$

Eric Filiol

INRIA, Domaine de Voluceau, Rocquencourt, BP 105,
78153 Le Chesnay Cedex, France

Abstract. Assuming we only have the coded sequence produced by a convolutional encoder at one's disposal, is it possible to recover all the parameters defining this encoder ? The implementation of such complete reconstruction for convolutional codes of any rate is presented in this paper, for both noiseless and noisy communications. Each time, several different solutions, *i.e.* alleged reconstructed encoders, are obtained, contrary to the awaited unicity. Their study yields a new characterization of convolutional codes along with some properties.

1 Introduction

Convolutional codes constitute an important class of very efficient error-correcting codes. They are widely implemented in civilian and military telecommunications systems, specially for deep-space and satellite comunications. Pioneer and Voyager missions in the 60s and 70s have almost uniquely used convolutional coding. At the present time, the burst of needs for modern telecommunications (cellular phones are the main example) makes it a very powerful and an increasingly useful mean of error-correction. As a significant example, three different convolutional codes (of rate $\frac{1}{2}, \frac{1}{3}$ and $\frac{1}{4}$) are implemented in the GSM norm. For a detailed list of applications see [2].

Proposed by Elias in 1955 as an alternative for the block codes, convolutional codes differ from them in the way codewords have not the same length. Decoding is consequently far different. Without entering into all details (see [2]), it is important to keep in mind that the Viterbi decoding (the most widely used) has an exponential complexity (in time and memory) in the constraint length of the code. Representing the size of the encoder, this parameter is therefore limited in most practical applications, to small values. This fact is important for the present work.

In an attack context, an important problem is to have access to the transmitted information (*the message*) without any knowledge on the encoder which produces the intercepted stream (*the coded sequence*). The only way is to reconstruct the encoder, that is to say to recover all its parameters. A simple decoding then gives access to the message. This paper describes the implementation of such a reconstruction for convolutional encoder of any rate. To be close to a real context, let us fix the following critical limitations:

- the available coded sequence can be short (about few kbits).
- the beginning or/and the end of the coded sequence can be missing.
- we cannot use in any way the message (or any information concerning it).

In this paper, Rice's algorithm [5] for $\frac{1}{2}$ rated convolutional encoder and noiseless communications is generalized to convolutional encoder of any rate, whether the communication is noiseless or not. In section 2, the basic concepts on convolutional coding and useful notations are given. Section 3 is devoted to the reconstruction in the noiseless case. Particularly the existence of several solutions, *i.e.* possible encoders producing the intercepted coded sequence, through their study, yields a new characterization of this kind of codes, based on the algebraic structure of modules and some properties have been derived. The notion of equivalent encoders has been precisely defined. Section 4 considers the same problem for the noisy case. Section 5 presents the implementation itself and some numerical results.

An extended version of this paper [7], including proofs, is available by contacting the author.

2 Basic Concepts and Notations

Basic description of convolutional coding along with notations used throughout this paper are given here to make the understanding of the next sections easier. For a complete and more accurate description see [1, 2, 3].

A convolutional encoder can be seen as an encoding system (based on a set of k shift-registers without feedback) such that, at each time instant, k information digits enter the encoder (one per register). Each information digit remains in the encoder for K time units and may affect each output during that time. The constant K is the constraint length or the *memory* of the encoder.

At each time instant, n information digits are output, each of them resulting from the xor of k digits produced by the action of n polynomials on each register. The encoder is thus said to be of rate $\frac{k}{n}$. The action of the kn polynomials and the shift are easily described by polynomial multiplications. So the polynomial representation will be used to represent the different streams.

A message will be composed of k interlaced input streams, each of them represented as a polynomial of degree $N + t$ noted $a_i(x)$, $i = 1, \ldots, k$. The kn polynomials are of degree N (hence $N = K - 1$) and will be noted $f_{i,j}(x)$. Then the encoder produces n output streams (of length t) represented as polynomials of degree t, $c_j(x)$, $j = 1, \ldots, n$ and we have:

$$\sum_{i=1}^{k} a_i(x) f_{i,j}(x) = u_{j,1}(x) + x^N c_j(x) + x^{N+t} u_{j,2}(x) \tag{1}$$

The polynomials $u_{j,1}(x)$ (resp. $u_{j,2}$) (the filling (resp. the emptying) of the registers) are of degree at most $N - 1$. Then the coded sequence is composed of the n interlaced output streams.

Thus the parameters (to be recovered) of a convolutional encoder are:

- k and n defining the rate and the number of polynomials.
- K the constraint length.
- the kn polynomials $f_{i,j}(x)$ of degree $N = K - 1$.

The convolutional encoder then describes a (n, k, N)-code. Generally, n and k are small integers with $k < n$. The most frequent case is $k = n - 1$. On the contrary, N must be made large enough to achieve low error probabilities. The symbols are usually elements of $GF(2)$ but generalization to $GF(q)$ where q is some prime power ($q = p^m$ for some positive integer m) can be easily done. We will only consider the case $q = 2$ but all the implementation and results, here presented can be extended to the general $GF(q)$ case.

3 The Noiseless Channel Case

A generalization of Rice's algorithm ([5]) for the rate $\frac{1}{2}$ is here presented.

3.1 The Rate $\frac{1}{n}$

Let us consider the basic case $n = 2$. Then the equations (1) become:

$$a(x)f_1(x) = u_{1,1}(x) + x^N c_1(x) + x^{N+t}u_{1,2}(x)$$
$$a(x)f_2(x) = u_{2,1}(x) + x^N c_2(x) + x^{N+t}u_{2,2}(x)$$

Since we don't use in any way $a(x)$ (the message), it must be eliminated. So we write:

$$a(x)f_1(x)f_2(x) = u_{1,1}(x)f_2(x) + x^N c_1(x)f_2(x) + x^{N+t}u_{1,2}(x)f_2(x)$$
$$a(x)f_2(x)f_1(x) = u_{2,1}(x)f_1(x) + x^N c_2(x)f_1(x) + x^{N+t}u_{2,2}(x).f_1(x)$$

or equivalently:

$$u_{1,1}(x)f_2(x) + u_{2,1}(x)f_1(x) + x^N(c_1(x)f_2(x) + c_2(x)f_1(x))$$
$$+ x^{N+t}(u_{1,2}(x)f_2(x) + u_{2,2}(x)f_1(x)) = 0$$

By identification, all the coefficients of power of x between N and $N + t$ in the polynomial $c_1(x)f_2(x) + c_2(x)f_1(x)$ are all equal to zero. Otherwise said, we have to solve the homogenous linear system:

$$C\underline{f} = \underline{0} \tag{2}$$

where $\underline{0}$ is the $(t - N)$-null vector and $\underline{f}$ is the $(2K)$-vector whose tranpose is $(f_{1,0}, f_{1,1}, f_{1,2}, \ldots, f_{1,N}, f_{2,0}, f_{2,1}, f_{2,2}, \ldots, f_{2,N})$. The unknowns $f_{1,i}$ and $f_{2,i}$ for $i = 0, \ldots, N$ represent the coefficients of the polynomials to be recovered. The $(t - N, 2K)$-matrix C is given by (recall $K = N + 1$):

$$C = \begin{pmatrix} c_{2,N} & c_{2,N-1} & \cdots & c_{2,0} & c_{1,N} & c_{1,N-1} & \cdots & c_{1,0} \\ c_{2,N+1} & c_{2,N} & \cdots & c_{2,1} & c_{1,N+1} & c_{1,N} & \cdots & c_{1,1} \\ c_{2,N+2} & c_{2,N+1} & \cdots & c_{2,2} & c_{1,N+2} & c_{1,N+1} & \cdots & c_{1,2} \\ \vdots & \vdots & \vdots & \vdots & \vdots & \vdots & \vdots & \vdots \\ c_{2,t-1} & c_{2,t-2} & \cdots & c_{2,t-N-1} & c_{1,t-1} & c_{1,t-2} & \cdots & c_{1,t-N-1} \end{pmatrix} \tag{3}$$

The $c_{1,i}$ and $c_{2,i}$ are known (coded sequence) so we can solve (2) and retrieve the two polynomials. For a non-trivial solution ($f_i(x) \neq 0$) the rank of C must be at most $2K - 1$ that is to say we must have:

$$t \geq 3N + 1 = 3(K - 1) + 1 = 3K - 2.$$

For any $n > 2$ the technique is the same. We only consider the $n - 1$ pairs of polynomials $(f_1(x), f_i(x))$ and the consistency of the solutions by coincidence on the polynomial $f_1(x)$. Finally the minimal number of bits must be:

$$L_{min} = n(3K - 2).$$

3.2 Algebraic Characterization and Properties

Many simulations were conducted and each time the awaited solution was succesfully retrieved but other solutions were also obtained. It seemed that different encoders could produce the same coded sequence. In fact, the study of these "different" solutions yiedeld the following algebraic characterization of convolutionnal encoders and some interesting properties have been derived. The proofs are here omitted but can be found in the full paper version.

Since an encoder can be seen as a n-tuple of polynomials, we define $\mathcal{C}_n = (\mathbb{F}_2[X])^n$ as the set of all convolutional encoders of rate $\frac{1}{n}$. In this section, we only consider the case $n = 2$ but all the following results can be extended to any n. Let us note $A = \mathbb{F}_2[X]$.

Proposition 1 $(\mathcal{C}_2, +, .)$ *is a A-module of dimension 2 where* $+$ *and* . *are so defined:*

$$\forall c = (f_1, f_2) \in \mathcal{C}_2, \quad \forall c' = (g_1, g_2) \in \mathcal{C}_2, \quad \forall h \in A$$

$$c + c' = (f_1 + g_1, f_2 + g_2),$$

and

$$hc = (hf_1, hf_2).$$

So the existence of several solutions comes from:

Proposition 2 *The set of the convolutional encoders producing the same coded sequence is a A-submodule S of dimension 1 in $\mathcal{C}_2$.*

In fact, for each of these elements of S, there exists a (polynomially noted) message such that each of these pairs (message and encoder) produces the same coded sequence. It can be shown that the relation between two encoders of the A-submodule S (a polynomial factor) links in an inverse way the two messages entering the encoders, that is to say they belong themselves to a A-submodule. The main result is given by the following theorem (we omit here the filling and emptying polynomials):

Theorem 1 *Let $(f_1(x), f_2(x))$ and $(g_1(x), g_2(x))$ be two encoders belonging to the same submodule S. Then there exists two messages $a(x)$ and $a'(x)$ such that if*

$$(a(x)f_1(x) = c_1(x)) \ (a'(x)g_1(x) = c_1'(x))$$
$$(a(x)f_2(x) = c_2(x)) \ (a'(x)g_2(x) = c_2'(x))$$

the output streams $c_i(x)$ and $c_i'(x)$ differ only on a finite number of bits $\mathfrak{b}$ such that

$$\mathfrak{b} < 2|deg(f_i(x)) - deg(g_i(x))|.$$

Then,

Proposition 3 *Two convolutional encoders are equivalent if and only if they verify theorem 1.*

The proof of the theorem is constructive: the $\mathfrak{b}$ bits are computable as soon as we know the polynomial factor between the two encoders. Then there is no information loss. If we want to transmit $a(x)$ we can send only $c_i'(x)$ that is to say, $c_i(x)$ truncated by $\mathfrak{b}$ bits. The decoding will give $a'(x)$ which is polynomially linked to $a(x)$. There exits an interesting possibility of information hiding since we can transmit a polynomially modified version of $a(x)$. The "information" is then the polynomial factor.

Proposition 4 *There exits a unique couple (resp. n-tuple) of polynomials of S, $(f_1(x), f_2(x))$ (resp. $(f_1(x), \ldots, f_n(x))$) such that $f_1(x)$ and $f_2(x)$ (resp. $(f_1(x), \ldots, f_n(x))$) are relatively prime. This couple (resp. n-tuple) is the generating element of S*

3.3 Generalization to Other Rate

Using the previous technique for the basic case of rate $\frac{1}{n}$, a recurring method is now presented for dealing with the general case of convolutional encoders of rate $\frac{k}{n}$. We can limit ourself only to the rate $\frac{n-1}{n}$ case, since it always possible to consider this case. Any rate $\frac{k}{n}$ encoder can be reconstructed by taking a "subencoder" of rate $\frac{k}{k+1}$ (since $k < n$) and recovering the parameters for the C_n^{k+1} possible subencoders. Then a simple coincidence checking on the common polynomials allows to test the consistency of the solutions and finally to completely reconstruct the whole encoder.

So let us consider the rate $\frac{n-1}{n}$ case. Such an encoder is defined by the n equations (1):

$$\sum_{i=1}^{k} a_i(x)f_{i,j}(x) = u_{j,1}(x) + x^N c_j(x) + x^{N+t} u_{j,2}(x) \qquad (4)$$

for $i = 1, \ldots, n-1$ and $j = 1, \ldots, n$. Since we cannot use in any way the different $a_i(x)$, we must eliminate them successively. As described in section 3.1, we (arbitrarily) choose first to eliminate $a_1(x)$. Then we obtain $(n-1)$ new equations involving only $(n-2)$ message polynomials $a_i(x)$ for $i \neq 1$. That is to say we are now facing the rate $\frac{n-2}{n-1}$ case. By repeating this, we finally deal with the rate $\frac{1}{2}$ case which is easily solved.

As an example, let us see the rate $\frac{2}{3}$ case. The encoder is defined by the three following equations:

$$a_1(x)f_{1,1}(x) + a_2(x)f_{1,2}(x) = u_{1,1}(x) + x^N c_1(x) + x^{N+t}u_{1,2}(x) \qquad (5)$$
$$a_1(x)f_{2,1}(x) + a_2(x)f_{2,2}(x) = u_{2,1}(x) + x^N c_2(x) + x^{N+t}u_{2,2}(x) \qquad (6)$$
$$a_1(x)f_{3,1}(x) + a_2(x)f_{3,2}(x) = u_{3,1}(x) + x^N c_3(x) + x^{N+t}u_{3,2}(x) \qquad (7)$$

We first choose to eliminate $a_1(x)$. By computing $f_{2,1}(x).(5) + f_{1,1}(x).(6)$ and $f_{3,1}(x).(5) + f_{1,1}(x).(7)$ we obtain the two equations:

$$a_2(x)\underbrace{(f_{1,2}(x)f_{2,1}(x) + f_{2,2}(x)f_{1,1}(x))}_{g_{1,2}(x)} =$$
$$f_{2,1}(x)u_{1,1}(x) + f_{1,1}(x)u_{2,1}(x) + x^N(c_1(x)f_{2,1}(x) + c_2(x)f_{1,1}(x))$$
$$+x^{N+t}(f_{2,1}(x)u_{1,2}(x) + f_{1,1}u_{2,2}(x)) \qquad (8)$$

$$a_2(x)\underbrace{(f_{1,2}(x)f_{3,1}(x) + f_{3,2}(x)f_{1,1}(x))}_{g_{1,3}(x)} =$$
$$f_{3,1}(x)u_{1,1}(x) + f_{1,1}(x)u_{3,1}(x) + x^N(c_1(x)f_{3,1}(x) + c_3(x)f_{1,1}(x))$$
$$+x^{N+t}(f_{3,1}(x)u_{1,2}(x) + f_{1,1}u_{3,2}(x)) \qquad (9)$$

Finally, computing $g_{1,3}(x).(8) + g_{1,2}(x).(9)$ allows us to eliminate $a_2(x)$ and:

$$\begin{aligned}
0 = \quad & g_{1,3}(x)f_{1,2}(x)u_{1,1}(x) + g_{1,3}(x)f_{1,1}(x)u_{2,1}(x) \\
& + g_{1,2}(x)f_{3,1}(x)u_{1,1}(x) + f_{1,1}(x)g_{1,2}(x)u_{3,1}(x) \\
& + x^N(c_1(x)f_{2,1}(x)g_{1,3}(x) + c_2(x)f_{1,1}(x)g_{1,3}(x)) \\
& + x^N(c_1(x)f_{3,1}(x)g_{1,2}(x) + c_3(x)f_{1,1}(x)g_{1,2}(x)) \\
& + x^{N+t}(f_{2,1}(x)g_{1,3}(x)u_{1,2}(x) + f_{1,1}(x)g_{1,3}(x)u_{2,2}(x)) \\
& + x^{N+t}(f_{3,1}(x)g_{1,2}(x)u_{1,2}(x) + f_{1,1}(x)g_{1,2}(x)u_{3,2}(x))
\end{aligned}$$

As previously, by identification the coefficients of

$$c_1(x)(f_{2,1}(x)g_{1,3}(x) + f_{3,1}(x)g_{1,2}(x)) + c_2(x)f_{1,1}(x)g_{1,3}(x) + c_3(x)f_{1,1}(x)g_{1,2}(x)$$

must be all equal to zero, that is to say the power of x between $4N$ and $N+t-1$. Then we have $9N+3$ unknowns to recover by solving a homogenous linear system.

Since we only recover the coefficients of a polynomial of degree at most $3N$, the different $f_{i,j}(x)$ must be separated. Simple polynomial operations (with Maple) give them easily. In the general case we summarize as follows:

Proposition 5 *The reconstruction of a convolutional encoder of rate $\frac{n-1}{n}$ can be done through the solving of a homogenous linear system involving $n(C_n N + 1)$ unknowns where $C_n = \frac{(n-2)(n+1)}{2} + 1$*

In terms of intercepted coded sequence length, we must have:

$$t \geq C_n N + 1 + n(C_n N + 1) - 2 \geq (n+1)(C_n N + 1) - 2$$

and

$$L_{min} = n(n+1)(C_n N + 1) - 2 \quad \text{bits}$$

For the practical values of n and N and with the fixed limitations, the reconstruction is quite possible.

4 The Noisy Channel Case

A noise is now added to the coded sequence. In a real context, the noise can be described as a gaussian process (*i.e.* Bernouilli trials) of probability p. Otherwise said, if b_t denotes a bit of noise:

$$P[b_t = 1] = p$$

and b_t is an independent, uniformly distributed random variable. In our case, this model is the most difficult to deal with, contrary to the decoding problem. As we experimented it, the burst-error model (see [4]) is for our problem far easier to consider even though it is difficult to treat for the decoding. So all our results must be seen as overestimations of the real difficulty since we use the most unfavourable model for the noise. The probability p of the noise is considered to be at most of 0.05 in practice. For more important values, the Viterbi decoding must use greater parameters and is *de facto* either too time consuming or impossible.

Our aim is to define the minimal length of noisy coded sequence we need to be sure that at least one run of noiseless bits of length L_{min} will be present, *i.e.* L_{min} successive noiseless bits. In fact we have to consider "unsuccess runs" of length L_{min} for the noise variable b_t since:

$$P[\hat{c}_t = c_t] = 1 - p = q$$

where $\hat{c}_t$ and c_t are respectively the received bit and the effectively emitted bit. An efficient and systematic, yet overestimating, approach is to use *recurrent events* (for details see [6, 7]). The definition is here just given:

Definition 1 *Let A a binary sequence containing as many runs of zeroes of length s as there are independent (non overlapping), ininterrupted blocks of exactly s zeroes. Each of these runs is then an occurence of a* recurrent event.

The use of recurrent events allows to greatly simplify calculations but it yields an (pessimistic) overestimation of the necessary sequence length. The main advantage is the possibility to use some results of the renewal theory:

Theorem 2 *[6] For large n, the number* $\mathbf{N}_n$ *of runs of length s produced in n trials is approximatively normally distributed, that is, for fixed* $\alpha < \beta$ *the probability that*

$$\frac{n}{\mu} + \alpha\sigma\sqrt{\frac{n}{\mu^3}} < \mathbf{N}_n < \frac{n}{\mu} + \beta\sigma\sqrt{\frac{n}{\mu^3}}$$

tends to $\mathcal{B}(\beta) - \mathcal{B}(\alpha)$ *where* $\mathcal{B}$ *is the normal distribution function and* μ, σ *the mean and variance of the recurrence time of runs of length s given by:*

$$\mu = \frac{1 - p^s}{qp^s} \qquad \sigma^2 = \frac{1}{(qp^s)^2} - \frac{2s+1}{qp^s} - \frac{p}{q^2}$$

Since we need $\mathbf{N}_n > 1$ and we fix the probability of error (*i.e.* there is no run of length L_{min}) to 0.001, we take $-\alpha = \beta = 3.29$. The table 1 gives some values for n and K (rate $\frac{1}{n}$) as examples. More detailed results (for any rate) can be found in [7]. The starting limitations are thus quite reasonable. It must kept

K	n (rec. events)	n'	$\mathbf{N}_n^-$	$\mathbf{N}_n^+$
5	1100	100	1	38
10	4100	1000	1	27
15	21000	7000	1	24
20	90000	44000	1	26

Table 1. Necessary minimal length sequence for rate $\frac{1}{n}$.

in mind that these are overestimations. Different simulations have proven this fact. A more accurate study allows to take $n' = \frac{s}{p^s}$ bits(see [7]). Although still overestimating (according to experiments), the values are far lower.

5   Implementation and Results

Recalling that the parameters (n, k, N) are generally small (due to the Viterbi decoding complexity), an exhaustive search allows to determine them. The recovering of the polynomials themselves is possible through the solving of a homogenous linear system, using gaussian elimination which is the most costly step.

Finally, our different programs yield complete reconstruction of the encoder with a complexity in $\mathcal{O}(K^4)$. The multiplicative constant varies according to the rate and whether the channel is noisy or not. Compared to the Viterbi complexity, we can see that "it is far easier to reconstruct than to decode" (in terms of complexity).

For the noisy channel case, two steps have been implemented. Each time one candidate (a set of kn polynomials) is proposed, through system solving and a statistical test of separation of two normal distributions (see [7]), a similar test

109

is conducted on an element of the submodule generated by the candidate. Thus an eventual false candidate will be rejected with high probability. For a complete description of the implementation see [7].

Many experiments have been conducted for all the cases, each time successfully. As examples computation times for rate $\frac{1}{n}$ on a D.E.C. Alpha workstation and using gcc are presented in table 2.

	max. n	max. K	Comput. Time
noiseless	7	20	$< 1sec.$
noisy	7	20	$< 1min.$

Table 2. Computation time for rate $\frac{1}{n}$

Acknowledgements

This work was supported by the Enseignement Militaire Supérieur Scientifique et Technique at Ecole Militaire, Paris and by the Centre de Recherches des Ecoles de Coëtquidan, Ecoles de Coëtquidan, Guer. I would like to thank the whole members of the projet Codes for their friendly encouragement and their hearty welcome and especially Pascale Charpin and Nicolas Sendrier.

References

1. D. G. Hoffman and al. *Coding Theory - The Essentials.* Dekker, 1991
2. S. Lin and D. J. Costello *Error Control Coding: Fundamentals and Applications* Prentice Hall, 1983
3. R. E. Blahut *Theory and Practice of Error Control Codes* Addison Wesley, 1983
4. J. H. van Lint *Coding Theory* Springer Verlag, 1971
5. B. Rice *Determining the parameters of a rate $\frac{1}{n}$ convolutional code over GF(q).* Third International Conference on Finite Fields and Applications. Glasgow, 1995.
6. W. Feller *An Introduction to Probability Theory*, Vol. 1 Wiley, 1966
7. E. Filiol *Reconstruction de codeurs convolutifs sur GF(q).* Mémoire de D.E.A., 1997

HCC: A Hash Function Using
Error Correcting Codes

Sami Harari

Universite de Toulon et du Var
MS/Laboratoire d'Informatique
B.P. 132, 83957 La Garde cedex
France Email : `harari@univ-tln.fr`

Abstract. After studying the constraints related to the use of error correcting codes as hash functions, this document introduces a hash function HCC (hash with codes and correlations) having a 'divide and conquer' approach for calculating the hash, while having an exponential complexity for the search of collisions. Possible generalizations are also sketched.

1 Introduction

Many computationally fast complete decoding algorithms are available for error correcting codes. By completing the decoding process, that is extracting of information bits of a decoded vector, the decoding process can be seen as a compressive function which could be used for hashing of data.

Good hash functions are compressive functions for which collisions are hard to find. This is not the case of the preceding scheme. However, the collisions can be characterized: they are all the error vectors of low weight. By using extra transformations on the data to be hashed it is possible to eliminate all the potentially weak collisions, why preserving the advantages of error correcting codes: the estimation of the complexity of finding collisions.

This work is a contribution for defining a hash function HCC (hash with codes and correlations) using the decoding function of error correcting codes having a 'divide and conquer' approach for calculating the hash, while having an exponential complexity for the search of collisions.

2 General Constraints

A possible way to define a hash function is to use a complete decoding algorithm of an error correcting code C. The data to be hashed is decoded and the hash of the data is the set of information bits. Operating in this way leads to a hash function for which some collisions are easy to find for two arguments.

- **The linear structure.** Given a string of data d_n and the corresponding hash $H(d_n)$ obtained in this manner, it is easy to obtain new sets of data with a known hash if the generator matrix G of the code is known.

If d_n and d'_n are bit strings which are codewords then

$$H(d_n + d'_n) = H(d_n) + H(d'_n)$$

Moreover, for any binary string s_k of length k, let $G.s_k$ be the corresponding codeword. Then the equation

$$H(d_n + G.s_k) = H(d_n) + s_k$$

yields other data strings for which the hash is known without any decoding computation.

- **The error correction property.** If the minimum distance δ of the code is greater than 3, the for any vector d'_n of weight less than $(\delta - 1)/2$ and any bit string d_n the following property holds:

$$H(d_n + d'_n) = H(d_n).$$

This shows that collisions are easy to find in this case. Moreover, for some codes all collisions are obtained by this type of consideration.

The use of certain non linear functions before applying the decoding algorithm solves most of the afore mentioned problems.

3   Description of the Hash Function HCC

The hash function uses a set C_i $0im - 1$ of binary linear error correcting codes, having all the same parameters (n, k) and a complete decoding algorithm.

The hash function associates to a set of n bits a set of k bits.

It also uses a set of non linear function f_i $0im - 1$ from n bits to n bits that have certain properties that will be studied later. Each function f_i is intimately related to the code C_i. The general flow of the hash function HCC is as follows.

- Let d_n be a set of n bits to be hashed. For each i in the set $0im - 1$ the set is transformed into a set of n bits by the non linear function f_i

$$d'_{i,n} = f_i(d_n) \quad 0im - 1$$

- Each resulting $d'_{i,n}$ is decoded with the complete decoding algorithm of the code C_i and the information bits are extracted to obtain $h_{i,k}(d_n)$ a set of binary string of length k.
- $H(d_n)$, the hash of d_n is the exclusive or of the m strings $h_{i,k}(d_n)$.

$$H(d_n) = \bigoplus_{i=0}^{m-1} h_{i,k}(d_n)$$

To be able to implement such a system the codes must be described, as well as the non linear functions.

4 Description of the Codes

Each code C_i $0 \le i \le m - 1$ is a (n, k) binary linear code which is an irregular concatenation of repetition codes.

• The generator matrix that is used for code C_i is a matrix G_i with each pair of row vectors $\mathbf{g}_{ij}$ and $\mathbf{g}_{il}$ $0 \le j, l \le k - 1$ $l \ne j$, $0 \le i \le m - 1$ having disjoint support:

$$supp(\mathbf{g}_{ij}) \cap supp(\mathbf{g}_{il}) = \emptyset \quad j \ne l$$

• Apart from the preceding constraint, the support of the row vectors are randomly distributed among the n possible positions.

• Denote by w_{ij} the weight of the j'th row of the generator matrix of code C_i. The integers w_{ij} must be odd integers.

The weights of the rows of the generator matrix have the property that

$$\sum_{j=0}^{j=k-1} w_{ij} = n, \quad i = 0, \ldots, m - 1.$$

In particular, each code has minimum weight equal to

$$\delta_i = min_j \{w_{ij}\}$$

• All these code have a complete decoding algorithm that is obtained by using the complete weight decoding algorithm, associated to each repetition code entering the concatenation of C_i, using its generator matrix:

To decode a vector $\mathbf{v}$ of length n with a code C of generator matrix G having row vectors $\mathbf{g}_i$, an iterative procedure is used:

for each i in the set $0, \ldots, k - 1$ do
- compute $\mathbf{v}_i$ the XOR of $\mathbf{v}$ and $\mathbf{g}_i$.
- compute the weight of $\mathbf{v}_i$.
- if this weight is greater than weight($\mathbf{g}_i$)
put the ith bit of the information set equal to 1
else
put the ith bit of the information set equal to 0.
- iterate on i.

• The weight of the rows are random variables submitted to the constraint that their weight be an odd integer greater than 3 and that the sum of the weights of the rows of a given matrix must be equal to n, the length of the code.

• A relation between the codes C_i has to be satisfied. It will be examined in section 6.

5 Description of the Non-Linear Function

The non linear part transforms a string of n bits to a string of n bits. It is the composition of two length preserving functions DA and WA which are applied in this order. These functions are specific for each code.

The function DA is a distance augmenting function applied on successive half bytes of string to be hashed which reduces their weight to 1 or 2.

The function WA is closely associated to the support of the vectors of the generator matrix of the code which is to be used for hashing.

The non linear function $f(d_n)$ on a bit string d_n is obtained by first applying the function DA, then applying the transformation WA which increases the weight. The resulting bit string is of weight close to $n/2$.

$$f(d_n) = WA(\tilde{d_n}) = WA(DA(d_n))$$

5.1 The Distance Augmenting (DA) Function

An error correcting code algorithm of a linear error correcting code decodes low vectors into the zero code word. A bit string of low weight can be detected locally, by examining the successive half bytes which compose it.

The function DA is a distance augmenting function that assigns half bytes to half bytes submitted to the constraint :

1. All half bytes b_1 and b_2 of hamming distance $d(b_1, b_2)$ less than 2 are transformed into half bytes such that $d(DA(b_1), DA(b_2)) > 1$
2. For any half byte b weight of $DA(b)$ is non zero.
3. The weight of $DA(b)$ is less than 3.

This function induces collisions since not all the half bytes are represented. However it has the property that it transforms all strings d_n of weight smaller than $n/4$ in strings of weight close to $n/2$.

Finding such functions by computer search is quite easy. Though there are 15345 such functions this number is not high enough for cryptographic purposes.

Example The set of values
    $3, 4, 4, 1, 4, 1, 1, 2, 4, 1, 1, 6, 1, 8, 12, 1$
represent the table of such a function.

5.2 A Weight Augmenting (WA) Function

The non linear function WA is used to enhance the preceding one. It is specific to each code C that is used in the algorithm. A template for the function WA is first described:

The 'constant weight' transformation on n bits is a mapping ϕ from the integers less than 2^n to the set of n-tuples of bits E_n.

$$\phi : \{0, \ldots, 2^n\} \longrightarrow E_n$$

It maps the first n integers to n-tuples of weight 1, the next $\binom{n.(n-1)}{2}$ ones to n-tuples of weight 2, etc ...

This mapping has an inverse ψ which can be interpreted as a mapping from n bits to n bits, if the corresponding integer is developed into an n bit sequence.

$$\psi : E_n \longrightarrow E_n$$

This mapping ψ has the property that it increases the weight of the output if the input is restricted in weight.

It is highly non linear on the bits. It is the mapping ψ that is used for the function WA

Example In the case where $n = 8$, if the input of the function is restricted to bytes of weight 1 or 2, (average weight 1.79) the average weight of the output is 4.19.

5.3 The Weight Augmenting Function

For each index i in the set $0 \ldots k - 1$ Let $S_i = supp(\mathbf{g}_i)$ be the support of the i-th row vector of the generator matrix of an allowable code C.

The WA function is the concatenation of copies of the preceding mapping ψ:

$$WA = (\psi_0, \ldots, \psi_{k-1}).$$

For each index i the set of variables of ψ_i is the set corresponding to the support of S_i. Since the supports of the row vectors cover the set of indices and are disjoint, the mapping is well defined.

6 Cryptoanalysis of the Algorithm

One of the advantages of using error correcting codes is that their structure allow a fairly precise cryptanalysis of the problem. To obtain the general complexity of finding collisions structurally reduced cases will be examined, gradually increasing the structure until the complexity of the complete system is obtained.

Let

$$H(d_n) = \bigoplus_{i=0}^{m-1} h_{i,k}(d_n)$$

be the hash of a bit string d_n. A collision for the function H is a bit string d'_n , having for hash $H(d_n)$.

This is a non linear equation E in n binary variables which can be decomposed into a set of k simultaneous non linear equations in n variables. Cryptanalysing the system is estimating the complexity of finding the solution.

The non linear functions $WA(DA)$ eliminate the possibility of constructing such a vector d'_n in polynomial time by using a constraint on the weight of d'_n thereby reducing the size of the set of vectors to be explored.

Let us suppose that only one code C is used for hashing ($m = 1$)

6.1 Case of a Simplified Non-Linear Function

If the non linear function was reduced to DA then the single equation equation E with n variables could be transformed into $n/4$ equations with 4 variables, allowing a divide and conquer approach to the problem, with a complexity $O(n)$.

6.2 CASE of a COMPLETE NON LINEAR FUNCTION

If the non linear function is equal to $WA(DA)$, denote by r the quantity n/k. It can be interpreted as the mean weight of a code vector of the generator matrix C. It can also be understood as the mean number of variables which are linked together in equation E.

Therefore solving equation E for code C would be equivalent to solving k independent equations, each one having r binary variables. The complexity of finding collisions in this case is equal to $O(2^r)$.

If solving equation E is to be of greatest complexity, that is that the only method left is to try all bit strings of length n and apply the non linear function and check the weight of the result, one must require that all binary n variables must be linked together.

6.3 The General Case

The requirement to link all variables together leads to a relation between n the length of the string to be hashed, k the length of the hash and m the number of codes that has to be satisfied, as well as a link between the codes.

- Each variable being linked to r other variables in a single code, there must be at least k codes to obtain maximal complexity, inducing the condition mk.
- The support of the row vectors of the generator matrices of the codes C_i must chosen in such a way that for each couple of indices (u, v) in the set $\{0, \ldots, n-1\}$ one of these two conditions hold:
  1. There exists a code C_i and a row vector g_{ij} such that (u, v) belongs to the support of g_{ij}.
  2. There exists a set index w_i $0 in - 3$ such that the preceding property holds for $(u, w_0), (w_0, w_1), \ldots, (v, w_{n-3})$.

Another way of representing the requirements is to consider the graph Γ having for vertices the set of n variables and having an edge between two vertices if and only if there exists an equation in one of the codes C_i linking these two variables.

The hash function has maximal complexity if the graph Γ has a path of length n between any two vertices.

If all the preceding requirements are satisfied the complexity of the task of finding collisions is exponential in n, the length of the bit strings.

7 Realistic Parameters

In the preceding section a link on the parameters of the system was developed to obtain highest possible complexity for finding collisions.

These conditions can be relaxed to obtain a faster hash function, while preserving a guaranteed complexity against finding collisions.

Let $\kappa = 2^{100}$ be the complexity to be maintained.

Let (n, k) be the parameters of the codes, $r = n/k$ and m be the cardinality of the set of codes. If m is chosen in such a way that $r.m > 100$ and such that each set of $r.m$ variables are linked in the equations, then the preceding analysis show that the complexity of finding collisions is greater than κ.

Another way of representing the requirements is to consider the graph Γ having for vertices the set of n variables and having an edge between two vertices if and only if there exists an equation in one of the codes C_i linking these two variables.

The system has complexity κ if any two vertices are on a path of length greater or equal to $r.m$.

If $n = 1000$ and $k = 100$ then a set of $m = 10$ appropriately chosen allowable codes is sufficient to obtain a system that has complexity κ.

If $n = 500$ and $k = 100$ then a set of $m = 20$ appropriately chosen allowable codes is sufficient to obtain a system that has complexity κ.

With these last parameters, computer simulations show that obtaining a hash takes the same amount of time as computing a hash with MD5.

8 The Set of Codes

There is obviously a tradeoff between the number of codes to be used m and the dimension k of the codes. The complexity of decoding is a polynomial function in k, with degree greater then 1. This shows that to obtain a fast system the hash length k must be chosen as small as possible.

The number m can be seen as a index of the gain in complexity: the higher the integer m, the faster the computation of the hash.

9 Conclusion

The preceding developments show how to design efficient hash functions using a set of decodable error correcting codes, which have fast complete decoding algorithms by using a 'divide and conquer' approach to the computation of the hash. The set of codes used for HCC are low rate.

Higher rate codes can be used, if they have the appropriate combinatorial properties on the generator matrices and fast decoding algorithms, to obtain more compact hash functions.

References

[1] M.R. Garey, D.S. Johnson *Computers and intractability. A guide to the theory of NP completeness.* Freeman and Co, 1979.

[2] D.R. Stinson *Cryptography theory and practice.*, CRC Press 1995.

[3] D. Chaum, E. Van Heist and B. Pfitzman *Cryptographically strong undeniable signatures, unconditionnaly secure for the signer.* LNCS 576 (1992) CRYPTO '91.

Public-Key Cryptosystems Based on Elliptic Curves

- An Evolutionary Approach -

Erwin Hess

SIEMENS AG, Corporate Technology, Information and Communications,
Email:erwin.hess@mchp.siemens.de, D-81730 Munich, Germany

The famous RSA scheme still is a de-facto-standard in all branches of public-key applications, but it is rapidly loosing its attractiveness. This is mainly due to the enormous key length that is necessary to make the RSA system secure. With current knowledge, public key systems based on the *elliptic curve discrete logarithm problem* are an alternative superior to the RSA system as they can offer a very high level of cryptographic strength with rather short parameters. E. g., an elliptic curve based system defined over a finite field of cardinality $\sim 2^{160}$ can have roughly the same level of security as the RSA system with a modulus of length 1024 bits. Elliptic curve based public-key cryptosystems where the underlying field is a *finite prime field* are of special interest. These systems can easily be implemented by using only ordinary modular arithmetic. Hence, it is possible to use arithmetic devices originally developed for the RSA system for these modern public-key systems, too. Furthermore, in contrast to other elliptic curve based public-key systems, the use of this type of curves is widely free of patents.

It is the purpose of this contribution to describe the basic constituents of a public-key system using elliptic curves defined over a finite prime field. Starting with an overview of the best currently known algorithms to tackle the elliptic curve discrete logarithm problem, we give security and efficiency criteria for an elliptic curve to be used in a public-key application. Then, we present some elliptic curve based cryptographic schemes. This comprises the new German digital signature standard based on elliptic curves defined over finite prime fields. We further show how elliptic curves appropriate for these public-key schemes are easily found using the theory of complex multiplication. Despite the fact that the method is of high mathematical level, the practical implementation is rather simple. We describe the various steps involved in this process and give examples of curves constructed using this method.

Finally, we report on an implementation of the German digital signature standard on the Siemens smart card SLE44CR80S. This smart card was originally designed to support RSA. Nevertheless, it is possible to run an elliptic curve based public-key system on it. Moreover, both the RSA and the elliptic curve based scheme can "co-exist" on one single card. We give the times necessary to generate or verify elliptic curve based digital signatures for various field sizes. The possibility of supporting security applications using RSA or elliptic curve technology with one single security token makes it possible to pass easily and nearly cost free from RSA to the modern elliptic curve based technique.

Novel Application of Turbo Decoding for Radio Channels

B.Honary *,B.Thomas *, P.Coulton * and M.Darnell**

*   Lancaster Communications Research Centre, Lancaster University, UK.
**   HW Communications Ltd, UK.

Abstract

Since their introduction by Berrou et al [1] in 1993, Turbo codes have generated considerable interest within the coding community as they offer near-Shannon limit performance. As yet, there is little application of these codes in practical systems and the vast majority of results thus far have been from simulations. The paper will discuss Turbo codes with particular reference to their application in radio channels and offers practical solutions for future implementation.

1. Introduction

Turbo decoding is performed using a combination of four basic principles: recursive systematic convolutional or block codes, combined with extrinsic soft output decoding of component codes, interleaving and iterative stage decoding. This paper develops the basic elements required for the development of any Turbo decoding system and gives examples of the iterative decoding procedure and the operation of the Maximum Aposteriori Probability (MAP) algorithm. It then analyses the requirements for the application of Turbo decoding to a radio system and offers solutions developed by Lancaster Communications Research Centre (LCRC) to the problems of channel evaluation, synchronisation and equalisation. Further applications of Turbo decoding is discussed for transmission of speech and images and discusses the advantages offered by such a system.

2. Basic elements required for Iterative Decoding

The iterative decoding stage requires a decoding technique which will not only accept soft channel inputs but will also furnish soft decoded outputs, i.e. provide reliability information associated with each decoded bit. The two algorithms most commonly employed are the MAP [2] and the Soft Output Viterbi Algorithm (SOVA) [3]; the MAP is generally preferred, even though it is more complex, since it offers improved performance over SOVA.

A Maximum Likelihood (ML) decoder furnishes at its output the most probable output sequence, given the received sequence, together with its associated Aposteriori probability (APP). All other codeword APPs are lost, except perhaps for that of the second most likely path. A MAP decoder, on the other hand, seeks to find the ML information symbol d_t, given the received vector $\mathbf{Y}$, for each of the K symbols. The information bit APPs are conveniently expressed in terms of their Log-Likelihood Ratio (LLR) as:

$$L(d_t | \mathbf{Y}) = \log\left(\frac{P(d_t = 0 | \mathbf{Y})}{P(d_t = +1 | \mathbf{Y})}\right)$$

(1)

Working with the LLR is useful because the ratio allows the cancelling out of an awkward probability during the MAP algorithm.

It has been shown [9] that

$$L(u_1 \oplus u_2) = sign(L(u_1)).sign(L(u_2)).min(|L(u_1)|,|L(u_2)|)$$

(2)

It has also been shown in [9] that the *Aposteriori* LLR that code symbol x was transmitted given that we have received y is,

$$L(x | y) = L_c y + L(x)$$

(3)

where L_c is called the channel reliability and is equal to 4.*Eb/No*(where Eb/No is the bit energy to noise ratio). When no *Apriori* knowledge of x is assumed, the *Aposteriori* probability $P(x|y)$ can be evaluated (via a rearrangement of equation (1)), by

$$P(x | y) = \exp(1/2 \sum_{v=1}^{V} L_c y_v x_v)$$

(4)

for a V bit branch symbol. Note $P(x|y)$ is not the true probability since a constant (normalising) factor should be included. This factor however is common to all paths in the code trellis and hence can be ignored.

Let $Y=Y_1^N=Y_1,Y_2,Y_3,...,Y_N$ and $Y_a^b=Y_a,Y_{a+1},Y_{a+2},...,Y_b$ and $s_t=\{m|m'$ is a possible state at depth t$\}$

In order to calculate the branch APPs, we define the quantity $\sigma_t(m',m)$ where,

$$\sigma_t(m',m) = \Pr(s_{t-1} = m', s_t = m, \mathbf{Y}_1^N) = \Pr(s_{t-1} = m', s_t = m, Y_1^{t-1}, Y_t, Y_{t+1}^N)$$
$$= \Pr(s_{t-1} = m', Y_1^{t-1}).\Pr(s_t = m, Y_t | s_{t-1} = m').\Pr(Y_{t+1}^N | s_t = m)$$

(5)

i.e. the joint probability of making the transition from state m' at time $t-1$ to state m at time t and receiving the sequence Y_1^N. It is now convenient to define the probabilistic functions $\alpha_t(m),\beta_t(m)$ and $\gamma_t(m'm)$ where:

$$\alpha_t(m) = \Pr(s_t = m, Y_1^t)$$
$$\beta_t(m) = \Pr(Y_{t+1}^N | s_t = m)$$
$$\gamma_t(m',m) = \Pr(s_t = m, y_t | s_{t-1} = m')$$

(6)

Hence,

$$\sigma_t(m',m) = \alpha_{t-1}(m').\gamma_t(m',m).\beta_t(m) \tag{7}$$

It can be shown [6] that $\alpha_t(m)$ and $\beta_t(m)$ and $\gamma_t(m',m)$ can be calculated by the recursion;

$$\alpha_t(m) = \sum_{m'}\alpha_{t-1}(m').\gamma_t(m',m)$$

$$\beta_t(m) = \sum_{m'}\beta_{t+1}(m').\gamma_t(m,m')$$

$$\gamma_t(m',m) = \Pr\!\left(S_t = m \mid S_{t-1} = m'\right)R\!\left(y_t \mid x_t\right) \tag{8}$$

Assuming an uncorrelated data source all $Pr(s_t=m|s_{t-1}=m')$ are equally likely. For M length branch sub codeword,

$$R(\mathbf{y}_t,\mathbf{x}_t) = \prod_{m=0}^{M-1}R(y_{t,m} \mid x_{t,m}) \tag{9}$$

where all possible R(a|b) are known Apriori at the receiver. Let $S_t^{(i)}$ represent a set of ordered pairs representing the transitions (m,m') at time t induced by $d_t=i$. The LLR can then be evaluated as follows:

$$L(d_t \mid Y) = \log\!\left(\frac{\sum_{S_t^{(1)}}\sigma_t(m',m)}{\sum_{S_t^{(0)}}\sigma_t(m',m)}\right) \tag{10}$$

2.1 Example of MAP Decoding for a (5,2,3) block code

One particular encoding configuration for a (5,2,3) block code is depicted in figure 1, utilising a filter whose structure is similar to that of a constraint length 2 convolutional code.

Fig. 1. Encoding filter

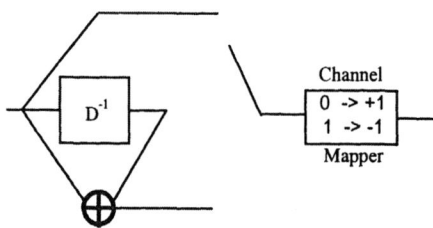

The filter is initially held in the zero state and is flushed to the zero state by the inclusion of a zero (which is not transmitted) after the 2 info bits. The information sequence drives the filter through a state transition sequence whose trellis is shown in figure 2.

Fig. 2. Decoding Trellis

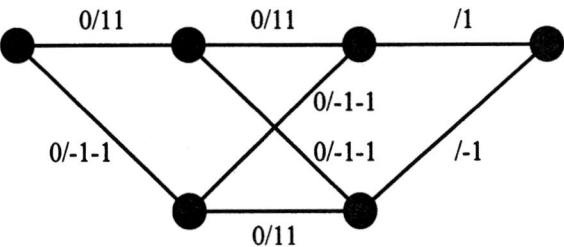

Consider the information sequence **d** = {0,0} is encoded to the channel sequence **x** = {1,1,1,1,1} which is then corrupted by additive memoryless noise resulting in a received sequence Lc.**y** = {-0.2, -0.1, 1.0, 0.3, 0.9}. We shall now utilise the trellis of figure 2 and apply equations (4,7,8) to calculate the APPs of the 2 data bits.

2.1.1 Calculation of α and β values

Initialisation: $\alpha_0(0) = \beta_0(0) = 1$

Fig. 3. Calculation of α and β values for end sections of trellis

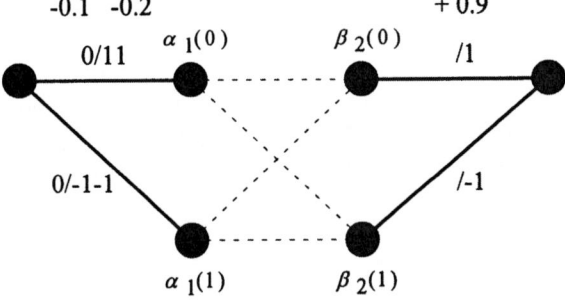

$$\alpha_1(0) = \alpha_0(0)\gamma_1(0,0) = 1*\exp(0.5*(-0.1-0.2)) = 0.86$$
$$\alpha_1(1) = \alpha_0(0)\gamma_1(0,1) = 1*\exp(0.5*(0.1+0.2)) = 1.16$$
$$\beta_2(0) = \beta_3(0)\gamma_3(0,0) = 1*\exp(0.5*(0.9)) = 1.57$$
$$\beta_2(1) = \beta_3(0)\gamma_3(1,0) = 1*\exp(0.5*(-0.9)) = 0.64$$

Fig. 4. Calculation of α and β values for middle section of trellis

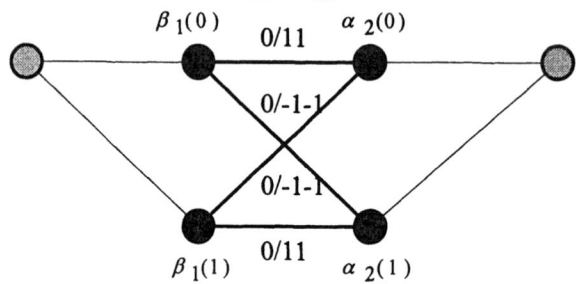

$$\beta_1(0) = \beta_2(0)\gamma_2(0,0) + \beta_2(1)\gamma_2(0,1) = 3.334 \qquad \beta_1(1) = 2.046$$
$$\alpha_2(0) = \alpha_1(0)\gamma_2(0,0) + \alpha_1(1)\gamma_2(0,1) = 2.253 \qquad \alpha_2(1) = 2.67$$

2.1.2 Calculation of APP soft output

Utilising equation (10) we have:

$$L(\hat{d_1}) = \ln\left(\frac{\alpha_1(1)*\beta_1(1)}{\alpha_1(0)*\beta_1(0)}\right) = -0.18$$

$$L(\hat{d_2}) = \ln\left(\frac{\alpha_1(1)*\gamma_2(1,0)*\beta_2(0)+\alpha_1(0)*\gamma_2(0,1)*\beta_2(1)}{\alpha_1(0)*\gamma_2(0,0)*\beta_2(0)+\alpha_1(1)*\gamma_2(1,1)*\beta_2(1)}\right)$$

$$L(\hat{d_2}) = -0.92$$

Note that we have less confidence in d_1 than d_2, which is to be expected since the channel errors occurred in the visinity of d_1.

3. Practical Turbo Decoding

Practical implementation of Turbo decoders with particular reference to radio channels requires factors to be considered. This paper describes briefly a number of design elements for Turbo decoders developed at LCRC that make practical implementation possible. These are:

- Channel Estimation;
- Synchronisation;
- Equalisation.

3.1 Channel Estimation

The availability of an accurate estimate of the noise variance is a requirement of both MAP and SOVA, recently a method has been introduced [4] which uses soft maximum likelihood trellis decoding (SMLTD) to provide channel information for use in a Turbo decoder for very little degradation in performance. The basic configuration is a shown in figure 5 and represents a typical Turbo decoding system employing horizontal and vertical decoding where the component codes could be either systematic convolutional or block codes.

Fig. 5.Combined Turbo Decoder And Channel Estimator

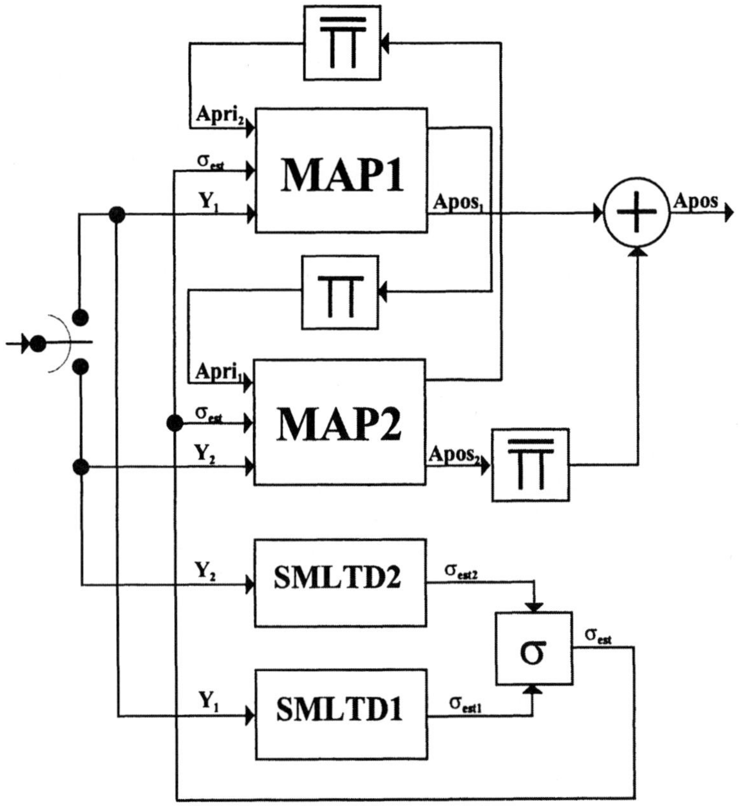

The scheme provides noise estimation with only a small loss of performance i.e. less than 0.25 dB. It is expected that this could be further reduced by incorporating the channel estimation into the feedback loop, which would improve the estimation with each iteration and subsequently the system performance.

The channel estimator uses three different metrics, obtained from the trellis decoding procedure, to obtain an estimate of channel conditions.

They can be defined as follows. Let d_i represent the minimum Euclidean distance between the received vector and the ith transmitted vector such that.

$$d_i^2 = \sum_{j=1}^{j=n} \left(x_j - a_j^i \right)^2$$

(11)

where $x = (x_1, x_2, x_3, ..., x_n)$ is the received vector of scalars and a_j^i is the jth element of the ith code word. A metric(a) can be defined by :

$$\Delta_a d_i^2 = (\min d_i^2 - \overline{\min d_i^2}) , i = 1,2,3,...,N_s$$

(12)

where Ns is the number of states in the trellis, $\min d_i^2$ is the minimum distance for the current received vector in the trellis and $\overline{\min d_i^2}$ is next minimum distance. Consider the case where dm_i is the maximum Euclidean distance between the received vector and the ith transmitted vectors such that a metric (b) can be defined as:

$$\Delta_b di^2 = (\max dm_i^2 - \min d_i^2) , i = 1,2,3,...,N_s$$

(13)

where Ns is the number of states in the trellis, $\max d_i^2$ is the maximum distance for the current received vector in the trellis and $\min d_i^2$ is minimum distance. Using the previous definition for minimum (12) and maximum (13) Euclidean distances, a new metric (c) can be defined as follows:

$$\Delta_c = (\Delta b - \Delta a) , i = ,2,3,...,N_s$$

(14)

The performance of the metrics is shown in figure 6 and as can be seen they provide sufficient information over a wide range to suit any application and any component code. The metrics are read from a pre-stored table to relate current metric value to channel noise.

Fig. 6. Decoder metrics Vs Noise

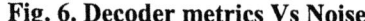

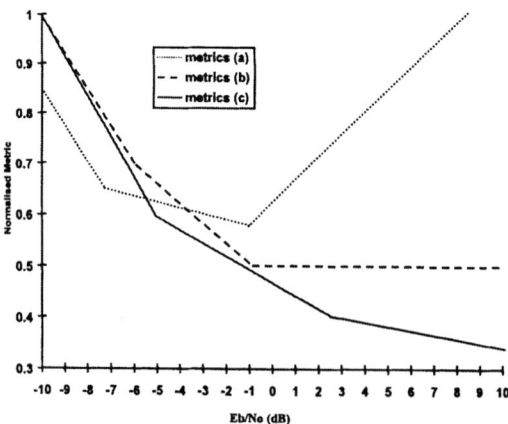

3.2 Synchronisation

The metrics previously defined may also be used to provide synchronisation [5] using Trellis Extracted Synchronisation Techniques [6]. These techniques have been applied to both block a convolutional codes to provide a very robust synchronisation scheme using a purely digital method which makes it particularly suitable for implementation on a Digital Signal Processing (DSP) device. The SMLTD element shown in figure 5 would be modified as shown in figure 7.

Fig. 7. Channel Estimation And Synchronisation Element

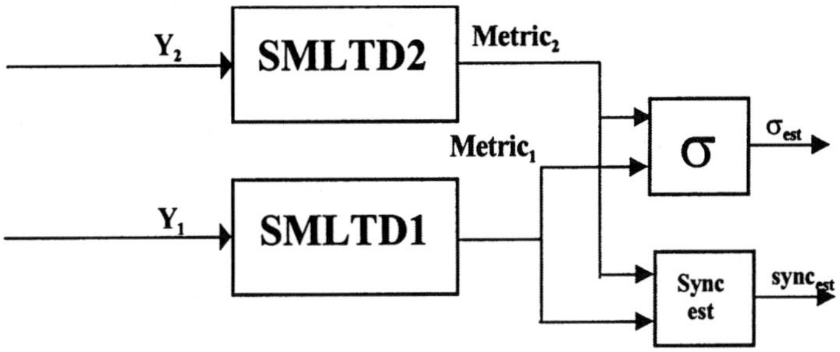

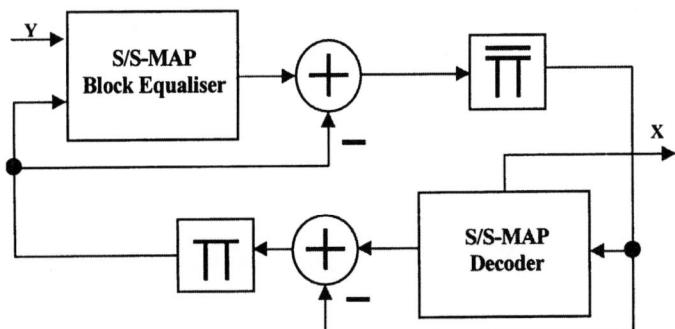

3.3 Equalisation

Turbo decoding principle, along with suitable orthogonal transformation techniques [7] can yield simultaneous equalisation and coding gain, over time and frequency selective channels. The orthogonal transform effectively converts a Rayleigh fading channel into a set of parallel AWGN channels, enabling the utilisation of channel codes designed for the AWGN channel.

The effect of the channel however, on the transformed code vector, is to destroy the orthogonality and hence equalisation methods are required. The Turbo-Equalizer (TE) represents an elegant solution [8]; both equalising the effects of the channel and providing coding gain. The block diagram for the equaliser is shown in figure8.

Fig. 8. Block Turbo Equaliser Structure

4. Applications of Turbo Decoding

Generally the present methods for the transmission of speech and images use as much compression as possible and the resultant data is transmitted or in the case of a noisy channel redundancy is added for error correction and detection. The high levels of compression mean that very long, powerful codes are required to protect the compressed data, as it is now highly susceptible to errors. The use of Turbo decoding and the incorporation of Apriori information now means that less severe levels of compression may be employed without a loss of performance. Consider the cases for speech and image data.

4.1 Combined speech and Turbo coding

The transmission of digital speech over voice-band radio links becomes a particularly problematic exercise when the channel is very noisy. Such links can support a channel bit-rate of 2.4kb/s, necessitating the use of LPC (Linear Predictive Coding) type vocoders, whose intelligibility starts to degrade rapidly at SNR's below 6dB. A technique for permitting intelligible communication at lower SNR levels is to compress the speech further, allowing the introduction of forward error correction codes. For example, the speech can be compressed to 1200b/s in order to introduce a 1/2 rate convolutional code [10]. The speech could be further compressed to 600b/s and combined with a concatenated convolutional code and Golay/RS code [11]. Though these techniques reduce the raw error-rate presented to the speech coder, they have several disadvantages, i.e.

Being computationally complex. The lion's share of a vocoder's data bits is used to model the spectral shaping of the vocal tract. The theory of linear predictive coding [12] is used to derive an optimum all-pole, digital filter to model the vocal tract, whose vector of taps must be quantised and coded. In order to reduce the number of bits required to code this vector, rather than individually quantise each tap (scalar quantisation), the complete vector is quantised as a single entity (vector quantisation, VQ). This greatly increases the complexity since a large codebook of representative vectors must be stored in memory and this codebook must be searched each frame to locate the 'best-match' vector.

Increasing the speech codecs vulnerability to channel errors. The greater a signal is compressed, the greater is its vulnerability to channel errors. This is particularly the case when utilising VQ since a channel error in this parameter will result in the receiver selecting a wrong codebook vector. The receiver will therefore synthesise speech whose spectrum is very different to that at the transmitter. This will essentially completely destroy the intelligibility of that section of the signal.

The greater a signal is compressed, the greater will be the introduced distortion. We propose a solution, which involves the concatenation of a low complexity 1200b/s, combined with a 1/2 rate Turbo code (possessing constraint length 3, recursive systematic convolutional component codes).

The speech coder is in essence very similar to the LPC10 algorithm operating at 2.4kb/s but including improved spectral parameter coding utilising the Line Spectral Frequency (LSF) representation [11]. A bit rate of 1200b/s is achieved by periodically deleting every odd frame's parameter set. At the receiver, the missing frames' parameters are approximately recovered through interpolation. Since the spectral parameters are scalar quantised, the error vulnerability of VQ is removed.

One practical problem associated with Turbo codes is the delay introduced by the interleaving process.

Fig. 9. Probability of Error Vs Interleaver Length at Eb/No=2dB

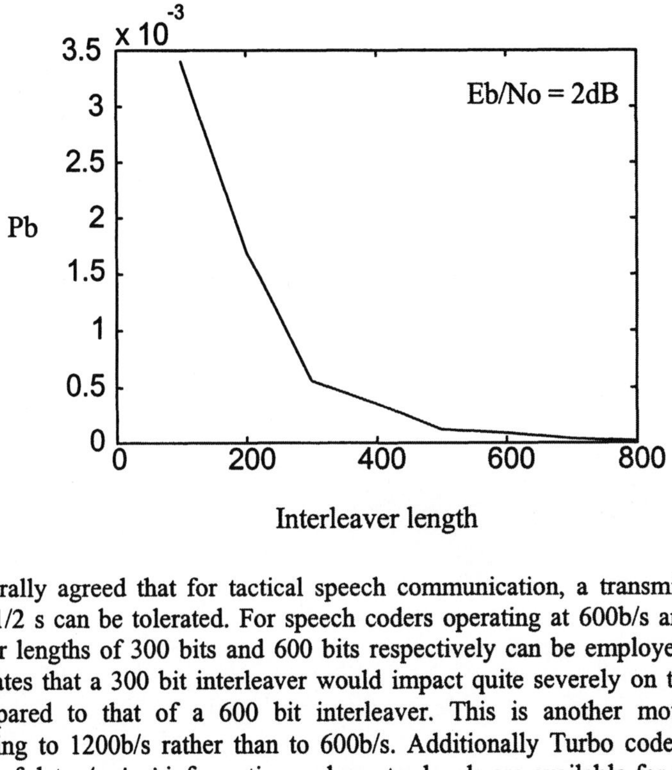

It is generally agreed that for tactical speech communication, a transmitter coding delay of 1/2 s can be tolerated. For speech coders operating at 600b/s and 1200b/s, interleaver lengths of 300 bits and 600 bits respectively can be employed. Figure 9 demonstrates that a 300 bit interleaver would impact quite severely on the bit error rate compared to that of a 600 bit interleaver. This is another motivation for compressing to 1200b/s rather than to 600b/s. Additionally Turbo codes allow the inclusion of data *Apriori* information and greater levels are available for the 1200b/s coder relative to the that operating at 600b/s.

4.2 Transmission of Images using Turbo decoding

Figure10 shows how the performance of the Turbo decoder improves the probabilities of data with each subsequent iteration. As can be seen from the result the main improvement is obtained after the second iteration and the image quality has been improved considerably by the fourth iteration. The images where transmitted using a k=3 recursive systematic convolutional code with an iterleaver size of 1000 and at Eb/No=2dB.

Fig. 10. Received Image Data after 1,2,4 and 8 Iterations

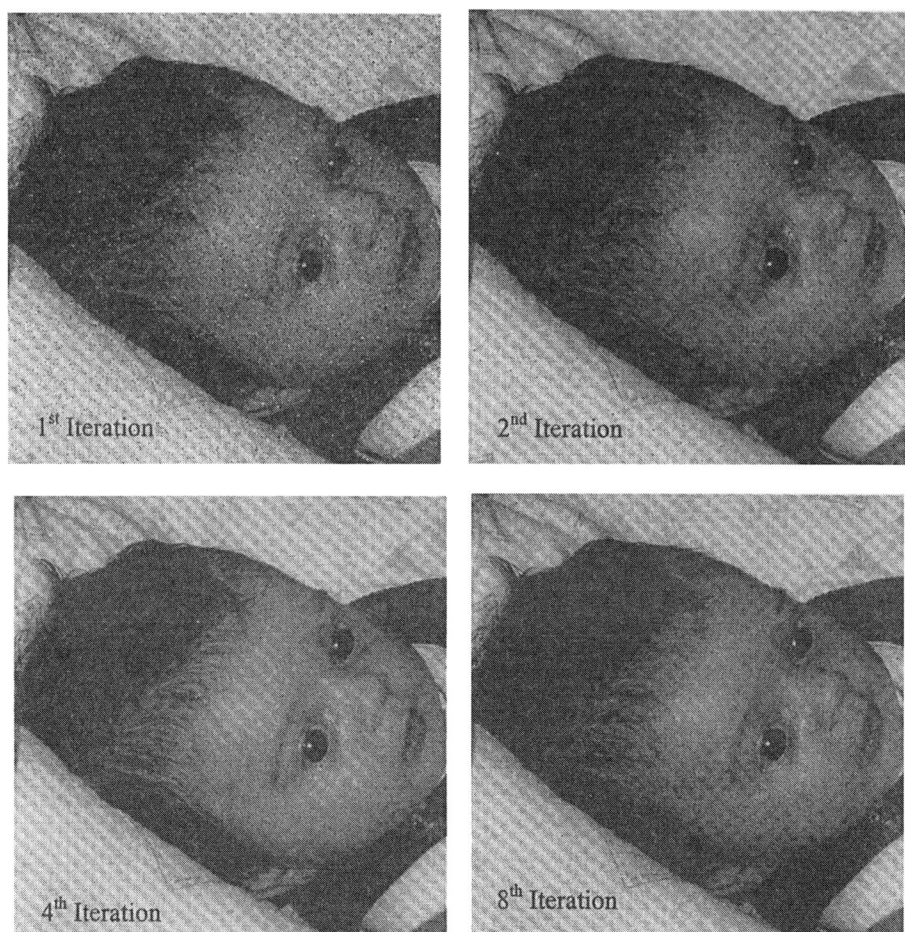

5. Conclusions

In this paper we have shown the basic elements required for iterative decoding using the MAP algorithm and have provided an example of MAP decoding of a Trellis. Further we have introduced methods for obtaining channel estimation, synchronisation and equalisation in order that a practical system utilising Turbo decoding can be implemented. Finally we have shown how Turbo decoding can be used to enhance the performance of systems required for the transmission of speech and images.

Turbo decoding will become an increasingly powerful tool for systems designers employing error correcting codes for the transmission of data. In this paper we have brought together the theory of this decoding process and provided solutions that will allow practical implementation.

6. References

1. Berrou,C. Glavieux,A. and Thitimajshima,P. " *Near Shannon Limit Error Correcting Coding And Decoding : TURBO-CODES (1)* ". In Proc, I.E.E.E Int. Conf. Commun., Geneva, Switzerland, pp 1740-1745, May 1993.

2. Bahl,B.R, et al., *"Optimal Decoding Of Linear Codes For Minimising Symbol Error Rate"*, I.E.E.E Trans.Inf.Theory, Vol. 20, No 2, March 1973, pp 284-287.

3. Hagenauer,J. and Hoer,P. *"A Viterbi Algorithm with soft-decision outputs and its applications"*, Proc I.E.E.E Globecom Conf, Dallas, Texas, November 1989, pp 1680-1686.

4. Coulton,P. Honary,B. Darnell,M. and Wicker,S. B *"Application Of Turbo Codes To HF Data Transmission "* Proceedings of 7th Int. Conf. on HF Radio Systems And Techniques. Nottingham UK 7-10 July 1997, pp 95-99.

5. Coulton,P. Hannaford,C. and Honary,B. " *Coding For Both Protection And Synchronisation* ". I.E.E Proc-I-Communications. December 1995. Vol-142. No 6. pp 352-356.

6. Honary,B. Darnell,M. and Markarian,G. " *Trellis Extracted Synchronisation Techniques (TEST)* ". UK Patent Application No 9414275-7. 14 July 1994

7. Reinhardt M., Linder J., *"Transformation of a Rayleigh fading channel into a set of parallel AWGN channels and its advantage for coded transmission"*, Electronic Letters, Vol. 31 (1995).

8. Reinhardt M., Englehart J., Linder J., Markarian G., Honary B., " *Transformation methods and iterative equalisation and decoding for symbol spread transmission"*, Procs of 4th Int Conf on Comms Theory and Applications, Ambleside, Lake District, UK 13-18 July 1997, pp 228-229.

9. Hagenauer J., Offer E., and Papke L., ``*Iterative Decoding of Binary Block and Convolutional Codes,"* IEEE Trans. on IT, vol. 42, no.2, pp. 429-445, March 1996.

10. Thomas B. and Honary B., *"Adaptive speech communication system for time-varying, narrowband channels"*, accepted for The Society for Computer Simulation International Journal (SIMULATION) for publication November 1997.

11. Kang G., and Jewett W., *"Error-Resistant Narrowband Voice Encoder"*, NRL Report 9018, Dec 25, 1986.

12. Rabiner,L.R, and Schafer,R.W., *"Digital Processing of Speech Signals"*, Prentice-Hall, 1978.

Finding Small Roots of Univariate Modular Equations Revisited

Nicholas Howgrave-Graham

University of Bath, `nahg@maths.bath.ac.uk`

Abstract. An alternative technique for finding small roots of univariate modular equations is described. This approach is then compared with that taken in (Coppersmith, 1996), which links the concept of the dual lattice (see (Cassels, 1971)) to the LLL algorithm (see (Lenstra *et al.*, 1982)). Timing results comparing both algorithms are given, and practical considerations are discussed. This work has direct applications to several low exponent attacks on the RSA cryptographic scheme (see (Coppersmith, 1996)).

1 Introduction

Let $p(x)$ be a univariate modular polynomial of degree k;

$$p(x) = x^k + a_{k-1}x^{k-1} + \ldots + a_1 x + a_0 \pmod{N}. \tag{1}$$

It is assumed that $p(x)$ is monic and irreducible, and that N is not prime, but hard to factorise.

In this paper we describe a new method for finding all the small integer roots, $|x_0| < N^{1/k}$, of equation 1, and show the relationship between the approach taken here, and that taken in (Coppersmith, 1996). It will be proved, via a general result on dual lattices that these two algorithms are in fact equivalent, though the present approach may be preferred for computational efficiency.

It has been shown in (Coppersmith, 1996) how finding small solutions to equation 1 can lead to various attacks on the RSA cryptographic scheme when using a small encrypting exponent.

Since both approaches employ lattice basis reduction, the remainder of this section deals with the notation and technical results that will be required.

Sections 2 and 3 give expositions of the algorithms in question, together with proofs of their validity; examples of both algorithms are shown in section 4.

Section 5 proves a technical result about dual lattices with respect to the LLL algorithm, whilst section 6 shows that it is indeed this theory that links the two methods. Section 7 then discusses practical issues relating to the algorithms and gives relevant timing results.

1.1 Notation

For the sake of consistency, all the results stated in this paper will be with respect to the *rows* of the relevant matrices. We shall denote the i'th row of a matrix M by m_i, and the i'th element of a vector v by v_i.

The sum and Euclidean norm are denoted by $||v||_1 = \sum |v_i|$ and $||v||_2 = \sum v_i^2$ respectively. If no subscript is present then the norm should be taken to be Euclidean. The dot product of two vectors will be denoted $v \cdot w = \sum v_i w_i$.

The set of all $(n) \times (n)$ matrices of determinant ± 1, and with integer coefficients will be denoted by $GL_n(Z)$. The symbol M^r (resp. M^c) is used to denote a matrix M that has had its rows (resp. columns) reversed.

1.2 Lattice reduction

For a thorough grounding on lattices see (Cassels, 1971), however for our purposes the following will suffice.

For a given basis $B = \{b_1, \ldots, b_n\}$ of R^n a lattice L is defined to be the set of points

$$L = \left\{ y \in R^n \;\middle|\; y = \sum_{i=1}^{n} a_i b_i, \; a_i \in Z \right\}.$$

Clearly many matrices will generate the same set of lattice points, in fact if we represent a basis B by a matrix with rows $\{b_1, \ldots, b_n\}$, then it is exactly the bases $B' = HB$ where $H \in GL_n(Z)$ that generate L. All these matrices have the same (absolute) determinant $d(L) = |\det B|$, which is referred to as the determinant of the lattice.

The landmark paper of (Lenstra *et al.*, 1982) gives a definition of an *LLL-reduced* basis of L, and more importantly an effective way of computing one. The method is closely connected to the Gram-Schmidt orthogonalisation procedure which, given a basis B, forms an orthogonal basis $B^* = \{b_1^*, \ldots, b_n^*\}$, where

$$b_i^* = b_i - \sum_{j=1}^{i-1} \mu_{i,j} b_j^*,$$

and $\mu_{i,j} = (b_i \cdot b_j^*)/||b_j^*||^2$. Note that span $\{b_1^*, \ldots, b_i^*\}$ = span $\{b_1, \ldots, b_i\}$ for all $1 \leq i \leq n$, but B^* is not typically a basis for L.

With this notation an LLL-reduced basis of L is defined to be one where, if the Gram-Schmidt orthogonalisation procedure were applied to it, then the following conditions would hold.

$$|\mu_{i,j}| \leq 1/2 \quad \forall\; 1 \leq j < i \leq n, \tag{2}$$

$$||b_i^* + \mu_{i,i-1} b_{i-1}^*||^2 \geq (3/4) \left||b_{i-1}^*\right||^2. \tag{3}$$

Together these conditions imply that $||b_i^*||^2 \geq (1/2)||b_{i-1}^*||^2$ which enables one to prove the following essential results.

$$||b_1|| \leq 2^{(n-1)/4} d(L)^{1/n} \tag{4}$$

$$||b_1|| \leq 2^{(n-1)/2} ||x|| \quad \forall\, x \in L \tag{5}$$

$$||b_n^*|| \geq 2^{-(n-1)/4} d(L)^{1/n} \tag{6}$$

In fact these results do not rely on condition 2 being quite so strict, only that

$$|\mu_{i,i-1}| \leq 1/2 \quad \forall \, 1 \leq i \leq n. \tag{7}$$

For this reason we will refer to a basis that satisfies conditions 3 and 7 as being *effectively* LLL-reduced. To turn an effectively LLL reduced basis into an LLL reduced basis is very simple, and akin to one application of the Gram-Schmidt orthogonalisation procedure.

When the so called *LLL algorithm* for lattice reduction is applied to an integer basis $b_i \in Z^n$, $1 \leq i \leq n$, it has complexity $O(n^6 \log^3 R)$ where n is the dimension of the basis, and $R = \max_{1 \leq i \leq n}\{||b_i||^2\}$. This complexity however, is typically quite pessimistic, and faster times are often achieved in practice. If the entries of the basis are rational, then one can clear denominators before applying the LLL algorithm.

2　The new method

In this section we give an exposition of a new method for finding the small roots of a (monic) univariate modular equation $p(x) = 0 \pmod{N}$.

Observe that for any polynomial $r(x)$, and natural number X,, we have the following upper bound on the absolute size of $r(x)$ in the region $|x| \leq X$.

$$|r(x)| \leq |x^k| + |a_{k-1}x^{k-1}| + \ldots + |a_1 x| + |a_0|$$
$$\leq |X^k| + |a_{k-1}X^{k-1}| + \ldots + |a_1 X| + |a_0| \quad \text{for all } |x| \leq X.$$

For some integer $h \geq 2$, and natural number X we define a lower triangular $(hk) \times (hk)$ matrix $M = (m_{i,j})$. The entry $m_{i,j}$ is given by $e_{i,j}X^{j-1}$, where $e_{i,j}$ is the coefficient of x^{j-1} in the expression

$$q_{u,v}(x) = N^{(h-1-v)}x^u(p(x))^v, \tag{8}$$

with $v = \lfloor (i-1)/k \rfloor$, and $u = (i-1) - kv$. Notice that $q_{u,v}(x_0) = 0 \pmod{N^{h-1}}$ for all $u, v \geq 0$. All other entries of the matrix are zero, so it has determinant $X^{hk(hk-1)/2}N^{hk(h-1)/2}$.

Let B be an LLL-reduced basis of the rows of M, and denote the first (small) row vector of B by b_1. Equation 4 implies that

$$||b_1||_2 \leq 2^{(hk-1)/4}X^{(hk-1)/2}N^{(h-1)/2}. \tag{9}$$

Letting $b_1 = cM$ for some $c \in Z^n$ also gives

$$||b_1||_2 \geq \frac{1}{\sqrt{hk}}||b_1||_1$$

$$= \frac{1}{\sqrt{hk}}\left(\left| \sum_{i=1}^{hk} c_i m_{i,1} \right| + \left| \sum_{i=1}^{hk} c_i m_{i,2} \right| + \ldots + \left| \sum_{i=1}^{hk} c_{hk} m_{i,hk} \right| \right)$$

$$= \frac{1}{\sqrt{hk}}\left(\left| \sum_{i=1}^{hk} c_i e_{i,1} \right| + \left| \left(\sum_{i=1}^{hk} c_i e_{i,2} \right) X \right| + \ldots + \left| \left(\sum_{i=1}^{hk} c_{hk} e_{i,hk} \right) X^{hk-1} \right| \right)$$

$$\geq \frac{1}{\sqrt{hk}} \, |r(x)| \quad \text{for all } |x| \leq X, \tag{10}$$

where

$$r(x) = \sum_{i=1}^{hk} c_i e_{i,1} + \left(\sum_{i=1}^{hk} c_i e_{i,2}\right) x + \ldots + \left(\sum_{i=1}^{hk} c_{hk} e_{i,hk}\right) x^{hk-1}$$

$$= c_1 \sum_{j=1}^{hk} e_{1,j} x^{j-1} + c_2 \sum_{j=1}^{hk} e_{2,j} x^{j-1} + \ldots + c_{hk} \sum_{j=1}^{hk} e_{hk,j} x^{j-1}. \quad (11)$$

So $\|b_1\|$ is "almost" an upper bound for the polynomial $r(x)$ in the entire range $|x| \leq X$. Notice also that $r(x_0) = 0 \pmod{N^{h-1}}$ since each sum in equation 11 is zero modulo N^{h-1}.

Combining equations 9 and 10 implies that, from making the matrix M with a natural number X, one can form a polynomial $r(x)$ that satisfies $r(x_0) = 0$ $\pmod{N^{h-1}}$ and

$$|r(x)| \leq \left(2^{(hk-1)/4}\sqrt{hk}\right) X^{(hk-1)/2} N^{(h-1)/2} \quad \text{for all } |x| \leq X.$$

Thus choosing

$$X = \left\lceil \left(2^{-1/2}(hk)^{-1/(hk-1)}\right) N^{(h-1)/(hk-1)} \right\rceil - 1 \quad (12)$$

shows that one can form a polynomial $r(x)$ such that $r(x_0) = 0 \pmod{N^{h-1}}$ and $|r(x)| < N^{h-1}$ for all $|x| \leq X$. This implies that $r(x_0) = 0$ over the integers as well, for any x_0 such that $|x_0| \leq X$, and $p(x_0) = 0 \pmod{N}$. Solving this univariate equation over the integers can be done in polynomial time (for instance by Hensel lifting the linear factors, or by finding small factors of the trailing coefficient), and then one can test each solution to see if it satisfies $p(x_0) = 0 \pmod{N}$. Notice that the bound $X \rightarrow 2^{-1/2}N^{1/k}$ as $h \rightarrow \infty$.

The polynomial $r(x)$ can be formed from equation 11 or the coefficients may be obtained by dividing the entries of the vector b_1 by appropriate powers of X.

3 A review of Coppersmith's method

Below we outline the approach given in (Coppersmith, 1996) for finding small roots of univariate modular equations. One firstly chooses a natural number X, and forms the upper triangular $(2hk - k) \times (2hk - k)$ matrix

$$M = \left(\begin{array}{c|c} D & A \\ \hline O_{hk} & D' \end{array}\right),$$

where $D = (d_{i,j})$ is an $(hk \times hk)$ diagonal matrix with entries $d_{i,i} = X^{1-i}$, $A = (a_{i,j})$ is an $(hk \times (h-1)k)$ matrix, where the entry $a_{i,j}$ is the coefficient of x^i in the expression $x^u((p(x))^v$, with $v = \lfloor (k+j-1)/k \rfloor$, and $u = (j-1) - k(v-1)$, and $D' = (d'_{i,j})$ is an $((h-1)k \times (h-1)k)$ diagonal matrix with entries $d'_{i,i} = N^v$ where $v = \lfloor (k+i-1)/k \rfloor$. This matrix has determinant $N^{hk(h-1)/2} X^{-hk(hk-1)/2}$.

Since there is a triangular sub-matrix of A with 1's on the diagonal it is possible to transform the matrix M (using integral elementary row operations implied by a matrix $H_1 \in GL_n(Z)$), to

$$\tilde{M} = H_1 M = \left(\begin{array}{c|c} \widehat{M} & 0_{(hk \times (h-1)k)} \\ \hline A' & I_{(h-1)k} \end{array} \right).$$

This means that the absolute value of the determinant of both $\tilde{M}$ and $\hat{M}$ are the same as M. We then reduce $\hat{M}$ using lattice basis reduction to give a matrix $B = H_2 \hat{M}$. If B^* (with row vectors b_i^*) denotes this basis after Gram-Schmidt orthogonalisation, then equation 6 implies

$$\|b_{hk}^*\| \geq 2^{-(hk-1)/4} N^{(h-1)/2} X^{-(hk-1)/2}. \tag{13}$$

Assume that there exists an x_0 such that $p(x_0) = 0 \pmod{N}$ and let $y_0 = p(x_0)/N \in Z$. Define the following vector of length $(2hk - k)$,

$$c_0 = \left(1, x_0, \ldots, x_0^{hk-1}, -y_0, -y_0 x_0, \ldots, -y_0 x_0^{k-1}, -y_0^2, -y_0^2 x_0, \ldots, -y_0^{h-1} x_0^{k-1} \right).$$

Further, when given a vector v of length $(2hk - k)$ that has 0's for the last $(h - 1)k$ entries, then denote by $[v]_{\text{sh}}$ this vector "shortened" to just the first (hk) elements.

Assuming that $|x_0| \leq X$, the above implies

$$\begin{aligned}
\sqrt{hk} &\geq \left\| (1, x_0/X, \ldots (x_0/X)^{hk-1}, 0, \ldots, 0) \right\| \\
&= \|c_0 M\| \\
&= \|c_0 H_1^{-1} \tilde{M}\| \\
&= \|[c_0 H_1^{-1}]_{\text{sh}} \hat{M}\| \quad \text{since } (c_0 H_1^{-1})_i = 0 \text{ for } i > hk \\
&= \|[c_0 H_1^{-1}]_{\text{sh}} H_2^{-1} B\| \\
&= \|c' B\| \quad \text{where } c' = [c_0 H_1^{-1}]_{\text{sh}} H_2^{-1} \\
&= \|c'' B^*\| \quad \text{for some } c'' \in R^{hk} \\
&\geq \|c_{hk}'' b_{hk}^*\| \\
&= \|c_{hk}' b_{hk}^*\| \quad \text{since } c_{hk}'' = c_{hk}' \in Z \\
&= |c_{hk}'| \|b_{hk}^*\| \\
&\geq |c_{hk}'| 2^{-(hk-1)/4} N^{(h-1)/2} X^{-(hk-1)/2}, \tag{14}
\end{aligned}$$

which means, since $c_{hk}' \in Z$, that $c_{hk}' = 0$ for any

$$X < \left(2^{-1/2} (hk)^{-1/(hk-1)} \right) N^{(h-1)/(hk-1)}. \tag{15}$$

If instead of c_0 we consider the variable vector

$$\begin{aligned}
c(x) = \big(&1, x, \ldots, x^{hk-1}, -p(x)/N, -xp(x)/N, \ldots, -x^{k-1}p(x)/N, \\
&-(p(x)/N)^2, -x(p(x)/N)^2, \ldots, -x^{k-1}(p(x)/N)^{h-1} \big), \tag{16}
\end{aligned}$$

which satisfies $c(x_0) = c_0$, then $c'_{hk}(x)$ is a univariate polynomial given by

$$c'_{hk}(x) = [c(x)H_1^{-1}]_{\text{sh}} \cdot ((H_2^{-1})^t)_{hk}. \tag{17}$$

This has integer coefficients after multiplying through by N^{h-1}, and with X chosen as large as possible (from equation 15), this polynomial must satisfy $c'_{hk}(x_0) = 0$ for any $|x_0| < X$.

The polynomial $c'_{hk}(x)$ is not identically zero since it is the sum of integer multiples of polynomials of differing degrees, and not all these multiples can be zero otherwise H_2 would be singular.

4   Examples

We examine the approach used by both methods to solve the equation $p(x) = x^2 + 14x + 19 = 0 \pmod{35}$ with $h = 3$ (thus we are guaranteed of finding any solutions of absolute size at most $X = 2$). Actually this polynomial has a solution $x_0 = 3$, but as we will see the methods still find it even though $x_0 > X$. It is often the case that the theoretical X given in the previous two sections is a little pessimistic.

4.1   Coppersmith's method

Coppersmith's method would firstly form the $(10) \times (10)$ matrix below.

$$M = \begin{pmatrix} 1 & & & & & & 19 & 0 & 361 & 0 \\ & 2^{-1} & & & & & 14 & 19 & 532 & 361 \\ & & 2^{-2} & & & & 1 & 14 & 234 & 532 \\ & & & 2^{-3} & & & & 1 & 28 & 234 \\ & & & & 2^{-4} & & & & 1 & 28 \\ & & & & & 2^{-5} & & & & 1 \\ & & & & & & 35 & & & \\ & 0 & & & & & & 35 & & \\ & & & & & & & & 1225 & \\ & & & & & & & & & 1225 \end{pmatrix}$$

This is transformed (using elementary row operations) to $\tilde{M} = H_1 M$ given below.

$$\begin{pmatrix} 1 & 0 & -19 \times 2^{-2} & 266 \times 2^{-3} & -3363 \times 2^{-4} & 42028 \times 2^{-5} & & & & \\ 2^{-1} & -14 \times 2^{-2} & 177 \times 2^{-3} & -2212 \times 2^{-4} & 27605 \times 2^{-5} & & & & \\ & -35 \times 2^{-2} & 490 \times 2^{-3} & -5530 \times 2^{-4} & 58800 \times 2^{-5} & 0 & & & \\ & & -35 \times 2^{-3} & 980 \times 2^{-4} & -19250 \times 2^{-5} & & & & \\ 0 & & & -1225 \times 2^{-4} & 34300 \times 2^{-5} & & & & \\ & & & & -1225 \times 2^{-5} & & & & \\ & 2^{-2} & -14 \times 2^{-3} & 158 \times 2^{-4} & -1680 \times 2^{-5} & 1 & & & \\ & & 2^{-3} & -28 \times 2^{-4} & 550 \times 2^{-5} & & 1 & & \\ 0 & & & 2^{-4} & -28 \times 2^{-5} & & & 1 & \\ & & & & 2^{-5} & & & & 1 \end{pmatrix}$$

We then examine (after clearing denominators and swapping the rows and columns for efficiency), the top left $(6) \times (6)$ sub-matrix below.

$$\hat{M} = \begin{pmatrix} -1225 & & & & & 0 \\ 34300 & -1225 \times 2 & & & & \\ -19250 & 980 \times 2 & -35 \times 2^2 & & & \\ 58800 & -5530 \times 2 & 490 \times 2^2 & -35 \times 2^3 & & \\ 27605 & -2212 \times 2 & 177 \times 2^2 & -14 \times 2^3 & 2^4 & \\ 42028 & -3363 \times 2 & 266 \times 2^2 & -19 \times 2^3 & 0 & 2^5 \end{pmatrix}$$

This is LLL reduced to $B_2 = H_2 \hat{M}$ where (the inverses of) the relevant matrices are given below.

$$H_1^{-1} \qquad\qquad\qquad\qquad H_2^{-1}$$

$$\begin{pmatrix} 1 & & 19 & 0 & 361 & 0 \\ & 1 & & 14 & 19 & 532 & 361 \\ & & & 1 & 14 & 234 & 532 \\ & & & & 1 & 28 & 234 \\ & & & & & 1 & 28 \\ & & & & & & 1 \\ 1 & & 35 & & & & \\ & 1 & & 35 & & & \\ & & 1 & & 1225 & & \\ & & & 1 & & 1225 & \end{pmatrix} \qquad \begin{pmatrix} -5 & 4 & -2 & 1 & -1 & -2 \\ 138 & -109 & 56 & -18 & 31 & 57 \\ -77 & 60 & -32 & 8 & -18 & -32 \\ 231 & -171 & 104 & -7 & 59 & 98 \\ 109 & -82 & 48 & -6 & 27 & 46 \\ 166 & -125 & 73 & -9 & 41 & 70 \end{pmatrix}$$

The vector $[c(x)H_1^{-1}]_{\text{sh}}$ has (as is typical) the following form

$$\left(1, x, \frac{-p(x)}{35}, \frac{-xp(x)}{35}, \frac{-p^2(x)}{1225}, \frac{-xp^2(x)}{1225} \right),$$

and so taking the dot product of $[c(x)H_1^{-1}]_{\text{sh}}$ with the last column of H_2^{-1} (and then multiplying by 1225) gives the polynomial

$$r(x) = 2x^5 - x^4 - 8x^3 - 24x^2 + 8x + 3,$$

which evaluates to zero over the integers at the root of $(p(x) \pmod{35})$, $x_0 = 3$.

4.2 The new method

The approach given in section 2 would immediately form the $(6) \times (6)$ matrix below.

$$M = \begin{pmatrix} 1225 & & & & & 0 \\ 0 & 1225 \times 2 & & & & \\ 665 & 490 \times 2 & 35 \times 2^2 & & & \\ 0 & 665 \times 2 & 490 \times 2^2 & 35 \times 2^3 & & \\ 361 & 532 \times 2 & 234 \times 2^2 & 28 \times 2^3 & 2^4 & \\ 0 & 361 \times 2 & 532 \times 2^2 & 234 \times 2^3 & 28 \times 2^4 & 2^5 \end{pmatrix}$$

This is then LLL reduced to

$$B = \begin{pmatrix} 3 & 8 \times 2 & -24 \times 2^2 & -8 \times 2^3 & -1 \times 2^4 & 2 \times 2^5 \\ 49 & 50 \times 2 & 0 & 20 \times 2^3 & 0 & 2 \times 2^5 \\ 115 & -83 \times 2 & 4 \times 2^2 & 13 \times 2^3 & 6 \times 2^4 & 2 \times 2^5 \\ 61 & 16 \times 2 & 37 \times 2^2 & -16 \times 2^3 & 3 \times 2^4 & 4 \times 2^5 \\ 21 & -37 \times 2 & -14 \times 2^2 & 2 \times 2^3 & 14 \times 2^4 & -4 \times 2^5 \\ -201 & 4 \times 2 & 33 \times 2^2 & -4 \times 2^3 & -3 \times 2^4 & 1 \times 2^5 \end{pmatrix},$$

where $B = HM$, and

$$H = \begin{pmatrix} 70 & 46 & -98 & 32 & -57 & 2 \\ 73 & 48 & -104 & 32 & -56 & 2 \\ 55 & 36 & -74 & 27 & -50 & 2 \\ 125 & 82 & -171 & 60 & -109 & 4 \\ -175 & -115 & 254 & -74 & 126 & -4 \\ 41 & 27 & -59 & 18 & -31 & 1 \end{pmatrix}.$$

The polynomial relationship required can be obtained by dividing the entries of b_1 by $1, 2, \ldots 2^5$; this gives the polynomial $r(x) = 2x^5 - x^4 - 8x^3 - 24x^2 + 8x + 3$. Alternatively one may form the sum

$$r(x) = \alpha_1 N^2 + \alpha_2 N^2 x + \alpha_3 N p(x) + \alpha_4 N x p(x) + \alpha_5 p^2(x) + \alpha_6 x p^2(x),$$

where the α_i are the elements of h_1.

This new method may be thought of as "flattening" the polynomial $p(x)$ around the origin, making it continuous in this region even modulo N^{h-1}.

5 The dual lattice and LLL

The dual (or polar) lattice, as given in (Cassels, 1971), is defined as the following.

Definition 5.1 *If $\{b_1, \ldots, b_n\}$ is a basis for a lattice L, then there do exist orthogonal vectors $\{d_1, \ldots, d_n\}$ such that*

$$d_j \cdot b_i = \begin{cases} 1 \ if \ i = j \\ 0 \ otherwise. \end{cases} \tag{18}$$

The lattice which is spanned by $\{d_1, \ldots, d_n\}$ is called the dual *lattice of L.*

In terms of matrices, if the rows of B form a basis for a lattice L, then the rows of $(B^{-1})^t$ form a basis (the *dual* basis) for the dual lattice of L. In (Cassels, 1971) the notation L^* and B^* are used for the dual lattice and basis respectively, however to avoid confusion with the Gram-Schmidt procedure we shall adopt the notation L^{-t} and B^{-t} for these concepts. Notice $B^{-t} = (B^{-1})^t = (B^t)^{-1}$. We now give a theorem linking the dual lattice and the LLL algorithm.

139

Theorem 5.2 *Let the rows of an $(n) \times (n)$ matrix A form a basis for a lattice L, and let B be an effectively LLL reduced basis for this lattice. Further let A^{-t} denote the dual basis, and A^r denote the matrix A with it's rows reversed. Then the rows of the matrix $D = (B^{-t})^r$ form an effectively LLL reduced basis for the dual lattice L^{-t} generated by the rows of A^{-t}.*

Moreover, if $\{b_1,\ldots,b_n\}$ and $\{d_1,\ldots,d_n\}$ denote the rows of B and D respectively, then the following relationships hold for all $1 \le i \le n$;

$$b_i^* = \frac{d_{n+1-i}^*}{||d_{n+1-i}^*||^2}, \tag{19}$$

and

$$\frac{b_i \cdot b_{i-1}^*}{||b_{i-1}^*||^2} = \frac{d_{n+2-i} \cdot d_{n+1-i}^*}{||d_{n+1-i}^*||^2}. \tag{20}$$

Proof. It is relatively easy to see that D is a basis for L^{-t}, since $(H^{-t})^r \in GL_n(Z)$. To show that it is effectively LLL reduced consider the definition of the dual lattice;

$$b_i \cdot d_j = \begin{cases} 1 \text{ if } i+j = n+1, \\ 0 \text{ otherwise.} \end{cases}$$

By induction on j we have $b_i \cdot d_j^* = 0$ for all $j \le n-i$, and $b_i \cdot d_{n+1-i} = 1$. Further, since $b_1 = b_1^* = \sum \alpha_{1,i} d_i^*$ with $\alpha_{1,i} = (b_1 \cdot d_i^*)/||d_i^*||^2$ this gives $b_1^* = d_n^*/||d_n^*||^2$.

Now assume $b_i^* = d_{n+1-i}^*/||d_{n+1-i}^*||^2$ and induct on i. Thus we write $b_{i+1}^* = \sum \alpha_{i+1,j} d_j^*$ where

$$||d_j^*||^2 \alpha_{i+1,j} = b_{i+1}^* \cdot d_j^*$$

$$= \left(b_{i+1} - \sum_{k=1}^i \mu_{i+1,k} b_k^* \right) \cdot d_j^*$$

$$= b_{i+1} \cdot d_j^* - \sum_{k=1}^i \mu_{i+1,k} b_k^* \cdot d_j^*.$$

If $j < n-i$ then both terms on the right hand side are 0, so $\alpha_{i+1,j} = 0$. If $j = n-i$ then $b_{i+1} \cdot d_j^* = 1$ and the terms in the sum are 0, so $\alpha_{i+1,n-i} = 1/||d_{n-i}||^2$. Finally if $j > n-i$ then $d_j^* = ||d_j^*|| b_{n+1-j}^*$ by the inductive hypothesis (since $(n+1-j) \le i$) which implies $\alpha_{i+1,j} = 0$. Thus only $\alpha_{i+1,n-i}$ is non-zero, and so equation 19 is true.

With this result we have

$$\frac{d_{n+2-i} \cdot d_{n+1-i}^*}{||d_{n+1-i}^*||^2} = d_{n+2-i} \cdot b_i^* = d_{n+2-i}^* \cdot b_i^* = d_{n+2-i}^* \cdot b_i = \frac{b_i \cdot b_{i-1}^*}{||b_{i-1}^*||^2},$$

which shows equation 20 is valid, and hence equation 7 holds for the basis D of L', assuming B is itself effectively LLL reduced.

Finally to show equation 3 also holds for the basis D of L' when B is effectively LLL reduced, observe that this condition is equivalent to

$$||b_i^*||^2 \geq \left(\frac{3}{4} - \left(\frac{b_i \cdot b_{i-1}^*}{||b_{i-1}^*||^2}\right)^2\right) ||b_{i-1}^*||^2,$$

which implies

$$\frac{1}{||d_{n+1-i}^*||^2} \geq \left(\frac{3}{4} - \left(\frac{d_{n+2-i} \cdot d_{n+1-i}^*}{||d_{n+1-i}^*||^2}\right)^2\right) \frac{1}{||d_{n+2-i}^*||^2},$$

$$||d_{n+2-i}^*||^2 \geq \left(\frac{3}{4} - \left(\frac{d_{n+2-i} \cdot d_{n+1-i}^*}{||d_{n+1-i}^*||^2}\right)^2\right) ||d_{n+1-i}^*||^2$$

as required.

This theorem implies that a vector satisfying condition 6 can alternatively be found by LLL-reducing the dual basis.

Corollary 5.3 *Let the rows of a matrix C form a basis for a lattice L'. A vector d_n^* such that $||d_n^*|| \geq 2^{-(n-1)/4}|\det C|^{1/n}$ for some basis D of L' can be found by LLL reducing the matrix C^{-t}.*

Proof. Let $A = C^{-t}$, and LLL-reduce this to form the matrix B. From theorem 5.2 we know that $D = (B^{-t})^r$ is an effectively LLL reduced basis for C (where $d_n^* = b_1/||b_1||^2$), so condition 4 implies $||d_n^*||^2 \geq 2^{-(n-1)/4}|\det C|^{1/n}$.

If, as in the method in section 3, it is not explicitly the vector d_n^* that is required but a coefficient γ such that $||vC|| \geq |\gamma| 2^{-(n-1)/4}d(L')^{1/n}$, then the following corollary is more useful.

Corollary 5.4 *Given a basis C of a lattice L' and a vector $v \in Z^n$, one can find a constant $\gamma \in Z$ such that*

$$||vC|| \geq |\gamma| 2^{-(n-1)/4}d(L')^{1/n}, \tag{21}$$

by LLL reducing the matrix C^{-t}.

Proof. As the theory in section 3 shows, the normal way to find such a γ is to form an LLL reduced basis D from the initial basis C, and then $\gamma = (v(H')^{-1})_n$ will satisfy equation 21, where $D = H'C$.

Instead if we LLL reduce $A = C^{-t}$ to form a basis B, where $B = HA$, and H has rows $\{h_1, \ldots h_n\}$, then

$$||vC|| = ||vA^{-t}|| = ||vH^tB^{-t}|| = ||v(H^t)^c(B^{-t})^r||,$$

where $(H^t)^c$ is H^t with its *columns* reversed, and we know $D = (B^{-t})^r$ is an effectively LLL-reduced basis for L'. Thus

$$||vC|| \geq ||(v(H^t)^c)_n d_n^*||$$
$$\geq |\gamma| 2^{-(n-1)/4}d(L')^{1/n},$$

where $\gamma = (v(H^t)^c)_n = v \cdot h_1$.

This theory suggests that if the LLL algorithm is being used for a purpose concerning a large vector d_n^*, it may be better to reduce the dual lattice searching for a small vector b_1. The advantage of this is that the LLL algorithm (since it works its way up through the vectors) can have an "early exit" when it has found a small enough b_1, rather than reducing the whole basis to find a large d_n^*.

6 The connection between the methods

We must actually show that it is the theory in section 5 that links the lattices produced by the two methods given in sections 2 and 3.

Define the $(hk) \times (hk)$ matrix,

$$E = \left(\begin{smallmatrix} I_k \\ 0_{(h-1)k} \end{smallmatrix} \middle| A \right),$$

where A is defined as in section 3. By the process also given in section 3, the matrix $\hat{M}$ is actually $PE^{-1}Q$, where $P = (p_{i,j})$ is diagonal and has entries $p_{i,i} = -N^v$ ($v = \lfloor (j-1)/k \rfloor$), and $Q = \text{diag}\{1, X^{-1}, \ldots, X^{-(hk-1)}\}$.

This implies that $\hat{M}^{-t} = P^{-t}E^tQ^{-t}$, with $P^{-t} = (p'_{i,j})$ diagonal and such that $p'_{i,i} = -N^{-v}$, and $Q^{-t} = \text{diag}\{1, X, \ldots, X^{hk-1}\}$. After clearing denominators we verify that $\hat{M}^{-t} = M'$, where M' is the matrix formed by the method given in section 2.

7 Practical implementations and results

There are (at least) two relatively small improvements that can be made to the algorithm given in section 2. Firstly it can be shown that the column corresponding to the linear terms (i.e. the left hand one) can be removed because of it's small contribution to $||b_1||$. Secondly the following lemma can be utilised to increase the permissible bound X slightly, by including rows corresponding to different polynomials.

Lemma 1. *Given a polynomial $p(x)$ modulo N of degree k, and provided that N and $k!$ are coprime, then one can produce a polynomial $q(x)$ modulo $(k!)N$ also of degree k, which shares the same roots as $p(x)$.*

Proof. Simply change each coefficient of $p(x)$ by multiples of N (using the Euclidean algorithm) until the resulting polynomial is congruent to $\prod_{i=1}^k (x-i)$ (mod $k!$), and then apply the Chinese remainder theorem.

Implementations of both algorithms have been written in C using Gnu MP as a multi-precision integer package (source code available on request). The timing results are from runs on a SGI Indy with one 100MHz IP22 processor (further results are also available on request). The main part of the program was an efficient implementation of the integral LLL algorithm, details of which may be found in (Cohen, 1991).

It should be stated that the algorithms only find solutions of univariate modular equations up to $O(N^{1/k})$, and that the time to find these solutions is of complexity $t = O(h^9 k^6 \log^3 N)$. Therefore as the degree of the polynomial k increases, less possible solutions are checked in greater time, i.e. the method becomes increasingly bad compared to a brute force search.

When $h = 3$ we find solutions up to $O(N^{2/(3k-1)})$. In this case average times (in seconds) for polynomials of degree k and various N are shown below on the left. On the right we give average times for cubic polynomials, but varying h and N.

$k\backslash \log_{10}(N)$	50	100	120	200
2	0.68	3.3	5.5	19
3	8.4	52	83	320
4	47	290	470	1900
5	170	1100	1900	–

$h\backslash \log_{10}(N)$	50	80	150	200
2	0.29	0.80	3.4	7.3
3	8.4	28	150	320
4	89	320	1700	–
5	560	1900	–	–

Comparing the above tables shows (as expected) that the algorithm is far more sensitive to an increase in h than one in k. Furthermore since $X \to O(N^{1/k})$ as $h \to \infty$ (i.e. $t \to \infty$), there must be a compromise as to which h to use to maximise the number of X checked per unit time. For cubic polynomials modulo $N = 10^{50}$ this turns out to be at $h = 4$.

The effect of using the dual approach as opposed to that taken in (Coppersmith, 1996) gave a saving of about 5%, when the root x_0 was approximately as large as the bound X; it is thought that this saving can be attributed, in part, to the fact that the theoretical X is actually a little pessimistic. In the cases that x_0 was significantly smaller than X, the dual approach became increasingly preferable.

8   Acknowledgements

I would like to thank Professor James Davenport for his supervision and his help in the production of this report. Thanks also go to Dr. Richard Pinch for pointing me to references on the dual lattice.

References

Cassels, J. W. S. 1971. *An introduction to the geometry of numbers.* Springer.

Cohen, H. 1991. *A Course in Computational Algebraic Number Theory.* Springer-Verlag.

Coppersmith, D. 1996. Finding a small root of a univariate modular equation. *In: Proceedings of Eurocrypt 96.*

Lenstra, A. K., Lenstra, H. W., & Lovasz, L. 1982. Factoring polynomials with integer coefficients. *Mathematische Annalen*, **261**, 513–534.

Robust Reed Solomon Coded MPSK Modulation

E. Husni and P. Sweeney

Centre for Communication Systems Research
University of Surrey
Guildford GU2 5XH, United Kingdom
Phone: +44 (0) 1483 259800   FAX: +44 (0) 1483 259504
email: p.sweeney@ee.surrey.ac.uk

Abstract

In this paper, construction of partitioned Reed Solomon coded modulation (RSCM) which is robust for the additive white Gaussian noise channel and a Rayleigh fading channel is investigated. By matching configuration of component codes with the channel characteristics, it is shown that this system is robust for the Gaussian and a Rayleigh fading channel.

This approach is compared with non-partitioned RSCM, a Reed Solomon code combined with an MPSK signal set using Gray mapping; and block coded MPSK modulation (BCM) using binary codes, Reed Muller codes, (RMCM). All codes use hard decision decoding algorithm. Simulation results for these schemes show that RSCM based on set partitioning performs better than not based on set partitioning and RMCM across a wide range of conditions.

The novel idea here is that in the receiver, we use a rotated 2^{m+1}-PSK detector if the transmitter uses a 2^m-PSK modulator.

1 Introduction

Much work has been done on design of efficient coded modulation schemes for improving the performance of digital transmission systems since the publication of Ungerboeck's paper [1] for trellis coded modulation (TCM) and Imai and Hirakawa paper [2] for BCM. Recently, the increasing interest for digital mobile radio or indoor wireless systems has led to the consideration of coded modulation design for combating fading channels.

In a number of previous papers [7]-[9], codes were designed for the Rayleigh fading channel so as to maximize their diversity by not using coded modulation techniques for the Gaussian channel.

Here, an alternative approach for combating the Rayleigh fading channel is proposed. The coded modulation system is based on partitioned BCM using Reed Solomon codes which is optimum BCM for the Gaussian channel. By using different configuration of component codes which is matched with the channel characteristic, it is shown that this approach yields a better coding gain over a Gaussian and Rayleigh channel as compared to the previous approach.

There are several reasons for using Reed Solomon codes, such as:

- These codes are maximum distance separable codes, and hence, they make highly efficient use of the redundancy
- Reed Solomon codes are burst error correcting codes, which are suitable for non-Gaussian channels.
- Reed Solomon codes provide a wide range of code rates that can be chosen such that the coded scheme has bandwidth efficiency compatible with the reference uncoded system.

For each code, we use (n,k) Reed Solomon codes over $GF(2^q)$ having code symbol length $n = 2^q - 1$, minimum Hamming distance $(n - k + 1)$ and error correcting capability $(n - k)/2$.

In partitioned RSCM, each of the m bits defining an MPSK symbol, where $M = 2^m$ is coded and decoded by different Reed Solomon codecs. The set partitioning principle is applied to define subsets with distances Δ_i, (i = 1 to m) that are nondecreasing with i. Each of the m bits defines a subset and is decoded in multistage decoding schemes. This method was first proposed by Cusack [3], who used Reed Muller codes and a QAM signal set. Later Sayegh [4] generalised Cusack's work to other signal constellations by using known binary codes. Here, we use Reed Solomon codes combined with an MPSK signal set.

In section 2, we define a baseline coded modulation approach to which the new coded modulation system is compared. The baseline system uses a Reed Solomon code which is combined with an MPSK signal set using Gray code mapping, this approach is called non-partitioned RSCM.

Specific designs of partitioned RSCM are given in section 3. Performance analysis over the Gaussian channel and the Rayleigh fading channel are dealt with in section 4. Finally, conclusions are given in section 5.

2 Non-Partitioned Reed Solomon Coded Modulation

Here, we address the issue of designing these schemes based on maximizing the time diversity (the effective length) of the code. We consider two methods generalising this approach.

2.1 Method 1

The first method is that a Reed Solomon code, defined over $GF(2^{ym})$, are mapped to the signal points of a 2^m-PSK signal set such that each symbol of the code consists of the concatenation of y channel symbols. In this combination the code rate is chosen such that the rate of the coded scheme is the same as the uncoded one (usually 2^{m-1}-PSK). This method is based on Jamali and Le-Ngoc's work in [7]. The proposition in [7] indicates that the effective order of time diversity in such a mapping is at least d, the minimum Hamming distance of the Reed Solomon code.

As an example we consider a Reed Solomon (63, 42) code, defined over $GF(2^6)$, combined with 8-PSK, and hence, each code symbol consists of two concatenated 8-

PSK symbols. The rate of the code is 2/3 which translates into 2 bits/s/Hz throughput, equivalent to that of uncoded QPSK.

2.2 Method 2

The second method is that a Reed Solomon code, defined over $GF(2^z)$, having code rate $R_c = {^m/_{(m+1)}}$ combined with a 2^{m+1}-PSK signal set. In this combination the code rate is chosen such that the rate of the coded scheme is the same as the uncoded one (2^m-PSK). In this case, the MPSK signal set used for modulation does not correspond to finite field over which the code is defined.

For example, a (30, 15) Reed Solomon code, defined over $GF(2^5)$, is combined with QPSK signalling. The overall coded QPSK throughput is comparable to that of uncoded BPSK, i.e., 1 bit/s/Hz.

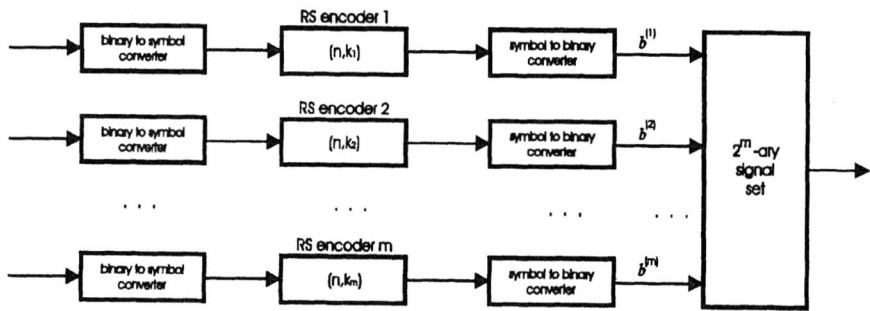

Fig. 1 A Reed Solomon coded modulation based on set partitioning encoder

3 Partitioned Reed Solomon Coded Modulation

The block coded modulation encoder consists of m Reed Solomon codes (called Reed Solomon component codes); this is illustrated in Fig. 1. Binary digits are assigned to each point in the signal space according to Ungerboeck's set partitioning scheme [1]. We now form an array of v·n columns and m rows, where v denotes bit length of one symbol and n coded symbol length.

$$
\begin{array}{cccc}
b_1^{(1)} & b_2^{(1)} & \cdots & b_{v \cdot n}^{(1)} \\
b_1^{(2)} & b_2^{(2)} & \cdots & b_{v \cdot n}^{(2)} \\
\cdots & \cdots & \cdots & \cdots \\
b_1^{(m)} & b_2^{(m)} & \cdots & b_{v \cdot n}^{(m)}
\end{array}
$$

where $b_j^{(i)} \in \{0,1\}$ for $1 \leq j \leq v \cdot n$. Each column of the array will correspond to one point in the signal space, with the bit in the first row corresponding to the leftmost digit and the bit in the last row corresponding to the rightmost digit in the representation of the signal space points. The array will be transmitted one column at a time, each column being represented by the corresponding signal space point.

The array contains $m \cdot n \cdot v$ bits of which $v \cdot n \cdot m \cdot R_c$ bits are information bits. Denoting the number of information bits in ith row by $v \cdot k_i$, we can write

$$v \cdot k_1 + v \cdot k_2 + \cdots + v \cdot k_m = v \cdot m \cdot n \cdot R_c$$

For a given rate R_c, the values of k_i's are chosen subject to the above conditions in such a way as to maximise the minimum Euclidean distance between the codewords of the code.

3.1 Multistage Decoding

A multistage decoding approach has been used for partitioned RSCM. Multistage decoding of multilevel trellis modulation codes has been recently studied and analysed in a number of papers [5, 6]. The main case of interest here is using a block encoder and block decoding algorithm for each component code of a multilevel modulation code.

The novel idea here is that in the receiver, a rotated 2^{m+1}-PSK detector will be used if the transmitter uses a 2^m-PSK modulator. This is illustrated in Figure 2(a), a detector of a QPSK modulator, and Figure 2(b), a detector of an 8-PSK modulator. This ensures that the received level does not fall on a decision boundary when decoding any of the bits in the symbol. It can be seen that there are two signal points of the rotated 2^{m+1}-PSK signal set in each signal point's region of the 2^m-PSK signal set.

The block diagram for multistage decoding of partitioned RSCM using 8-PSK is illustrated in Fig. 3. Let L be the labeling for a two dimensional signal set S with 2^m signal points. Let $C \equiv C^{(1)}C^{(2)} \cdots C^{(m)}$ be an m-level code of length n over L with m component codes where $C^{(i)}$ is a code over the sublabeling $L^{(i)}$. In multistage decoding of C, component codes are decoded sequentially one at a time, stage by stage. The decoded information at each stage is passed to the next stage.

Suppose a codeword in C is transmitted and $z = (z_1, z_2, \cdots, z_{v \cdot n})$ is the received sequence at the demodulator, where z_j is the rotated 2^{m+1}-PSK signal point. Given a hard decision output at the ith stage of the demodulator, the decoder performs a decoding for $C^{(i)}$. If the decoding is successful, the decoder puts out a decoded codeword $v_D^{(i)}$ which is in $C^{(i)}$. Otherwise, $v_D^{(i)}$ is a null array.

The demodulator makes a hard decision as follows. For the given received point z_j and decoded sublabels $v_D^{(g)} \in L^{(g)}$ with $1 \leq g < i$ and $i < g \leq m$, find the label

$v = v_{Dj}^{(1)} \cdots v_{Dj}^{(i-1)} v^{(i)} v_{Dj}^{(i+1)} \cdots v_{Dj}^{(m)}$ in L with $v_{Dj}^{(1)} \cdots v_{Dj}^{(i-1)}$ as a prefix and $v_{Dj}^{(i+1)} \cdots v_{Dj}^{(m)}$ as a suffix such that the norm $\left\| z_j - s(v) \right\|^2$ is minimised, where $s(v)$ denote the signal point in 2^m-PSK represented by v. If there is no decoded sublabels v is a null string.

Either one or two iterations may be used for decoding. In the first iteration, level 1 is first decoded, then level 2 is estimated based on the result of level 1. Finally level 3 is determined based on levels 1 and 2. For the second (optional) iteration, the decoding process start from level 2 which is again decoded using the result of levels 1 and 3 from the first iteration. The corresponding is done for levels 1 and finally for level 3.

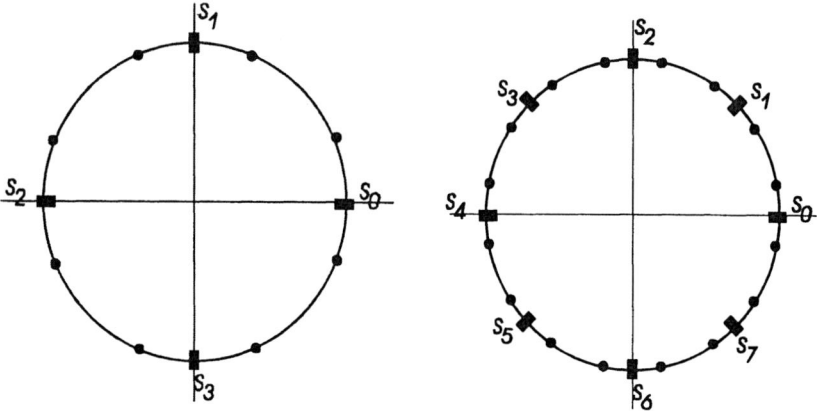

Fig. 2(a). A rotated 8-PSK signal set for a QPSK modulator

Fig. 2(b). A rotated 16-PSK signal set for an 8-PSK modulator

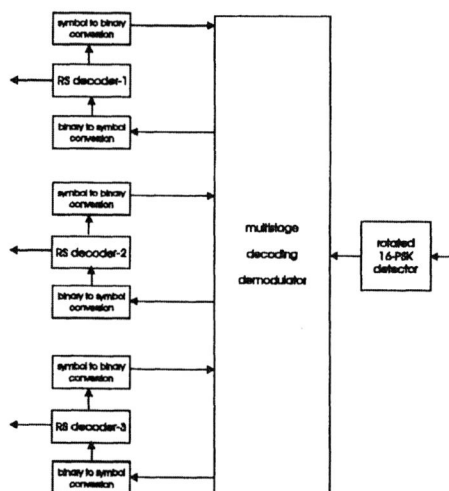

Fig. 3. Multistage decoding of RS coded 8-PSK modulation based on set partitioning.

3.2 Distance Considerations

The minimum squared Euclidean distance of binary block coded 8-PSK scheme is obtained as

$$D_{min}^2 = \min\left(\Delta_0^2 d_1, \Delta_1^2 d_2, \Delta_2^2 d_3\right) \qquad (1)$$

Thus in binary block coded scheme design, the minimum Hamming distance of each component code can be determined by set $\Delta_0^2 d_1 \cong \Delta_1^2 d_2 \cong \Delta_2^2 d_3$.

The optimum values for minimum Hamming distance of each component code of partitioned RSCM can not be determined. Firstly Reed Solomon codes are non-binary codes so minimum Hamming distance is not a binary measure and Equation (1) is no longer valid for this code. Secondly in multistage decoding the first stage decoding gives an output codeword to the second stage decoding, and so on. Thus the second stage decoding depends on and takes advantages from the first stage decoding, and so on. Thirdly Equation 1 is valid for a Gaussian channel only. Consequently, equation 1 can not be used to determine exactly the minimum Hamming distance of each component code.

4 Error Performance

In this section, the error performance of partitioned RSCM over the Gaussian channel and the Rayleigh fading channel are compared with non-partitioned RSCM and RMCM using computer simulations.

BCM using Reed Muller codes require the same approaches. At the first approach, Reed Muller code having code rate $R_c = {}^m\!/_{(m+1)}$ combined with a 2^{m+1}-PSK signal set. In this combination the code rate is chosen such that the rate of the coded scheme is the same as the uncoded one (2^m-PSK). At the second approach, Reed Muller codes are used for component codes. It uses similar multistage decoding procedure for decoding the received codewords.

4.1 Over the Gaussian Channel

Reed Solomon codes provide a wide range of code rates, thus there are many configurations of component codes for RSCM based on set partitioning. Table 1 gives a list of good codes for each specified coded symbol length for coded 8-PSK modulation and Table 2 for coded QPSK modulation.

level	k	
	n = 63	n = 127
1	7	11
2	59	119
3	61	125
throughput (bits/s/Hz)	2.01	2.007
infor-mation-bit length	(7+59+ 61)·6 = 762	(11+119 +125)·7 = 1785

level	k	
	n = 31	n = 63
1	7	11
2	25	53
throughput (bits/s/Hz)	1.03	1.01
information-bit length	(7+25)·5 = 160	(11+53)·6 = 384

Table 1. Configuration of good codes for RS coded 8-PSK modulation for coded symbol length of 63 and 127.

Table 2. Configuration of good codes for RS coded QPSK modulation for coded symbol length of 31 and 63.

As previously mentioned for the binary block coded scheme, the minimum Hamming distance of each component code can be determined using eq. 1. This however does not hold for multilevel codes; for example, for the best multilevel code of length 63 from Table 1

$$\Delta_0^2 \cdot d_1 = 0.586 \cdot 57 = 33.4$$
$$\Delta_1^2 \cdot d_2 = 2 \cdot 5 = 10$$
$$\Delta_2^2 \cdot d_3 = 4 \cdot 3 = 12$$

From the results we can see that eq. 1 is not valid for the best multilevel code of each code rate and code symbol length.

Figs. 4-5 show block error probability for Reed Solomon coded QPSK modulation and coded 8-PSK modulation. In these figures, the error performances are compared with those of some uncoded reference modulation systems for transmitting the same (or almost the same) number of information bits.

We can see that for coded QPSK modulation, partitioned RSCM has improvement on non-partitioned RSCM by an amount approximates equivalently to doubling code length. For coded 8-PSK modulation at block error probability of 10^{-4}, set partitioning produces 1 dB more coding gain than doubling the code length.

Fig. 6 shows bit error rates of various coded modulation schemes for coded QPSK modulation. Reed Muller codes used here are however more complex to decode than the Reed Solomon codes used. At bit error probability of 10^{-4}, partitioned RSCM has 0.75 dB more coding gain than non-partitioned RSCM with the same code length. It appears that for coded QPSK modulation, Reed Muller coded modulation based on set partitioning is worse than Reed Muller coded modulation not based on set partitioning. For example, at bit error rate 10^{-4}, Reed Muller coded modulation based on set partitioning, length 256, is 0.5 dB worse than Reed Muller coded modulation not based on set partitioning, length 128.

Fig. 7 shows bit error rates of various coded modulation schemes of coded 8-PSK modulation. For partitioned coded modulation, it turns out that all codes have about the same performances. At a bit error rate of 10^{-4}-10^{-5}, partitioned coded modulation has at least 1 dB coding gain over non-partitioned coded modulation for the same code length.

In this channel, it appears that one iteration and two iterations multistage decoding have about the same performances. In [6] it was claimed that in the AWGN channel, two iterations will give better performance than one iteration if there is an interleaver between the coded bits of all stages or we use modified generalized-minimum-distance-decoding. They used a three dimensional block interleaver for three levels coded modulation with memory cells as many as coded bit length in every dimension. Every memory cell contains one coded bit from every level. The coded bits of the first level are always written in the cells corresponding to the direction of the first dimension. The coded bits of the second level are written in according to the direction of the second dimension. The corresponding is done for the third level.

They used binary convolutional codes for the first and second levels and a single parity check for the third level. If we use the Viterbi algorithm for decoding, the estimated information bit sequence is liable to contain error bursts. This sequence is re-encoded and fed into the decoder of the next level. Thus, the re-encoded sequence also contains error bursts. If this decoder also uses the Viterbi algorithm then it is very sensitive to these error bursts, because the algorithm is designed to deal with independent errors in the input stream.

In order to avoid this error propagation effect, they introduced interleaving between the coded bit streams of each level as explained above. The interleaving spreads the re-encoded bit streams of any two decoders for the third decoder. So it will be hardly influenced by error bursts from any of the other two decoders.

Reed Solomon codes have powerful error detection capability which are different from binary convolutional codes. In the simulations if the decoder detected the errors but it could not correct the errors, the decoder passed the input sequence to the decoder output. Therefore, the possibility of an error propagation effect because of undetected errors is very small.

4.2 Over the Rayleigh Fading Channel

In this subsection, we analyse the error performance of partitioned RSCM compared with non-partitioned RSCM and RMCM over the nonselective slow Rayleigh fading channel. Here 'slow' means that the fading bandwidth is small compared to the signal bandwidth so that the receiver will be able to track the phase variations.

For all schemes of Reed Solomon coded modulation, the code symbols are interleaved before modulation in order to destroy the memory of the fading channel.

Figs. 8 and 9 show bit error rates of various coded modulation schemes for coded QPSK and 8-PSK modulation. We can see that in this channel, RSCM schemes have a large coding gain to Reed Muller coded modulation schemes. We also see that the error performances of partitioned RSCM using 8-PSK are better than non-partitioned

RSCM at high BER, and finally they become the same at low BER; and the error performances of partitioned RSCM using QPSK are better than those of non-partitioned RSCM. Therefore, the error performances partitioned RSCM are never worse than those of non-partitioned RSCM for the same code length.

This scheme is different from the Gaussian channel in that good configurations have the same component codes. Therefore, it seems that all levels have the same distance. In other words, the fading phase is a uniformly distributed random process. By matching configuration of component codes to the channel characteristic, it is shown that partitioned RSCM can be robust codes for the Rayleigh fading channel.

The error performances of RSCM not based on set partitioning using the first approach whose code symbol consists of one channel symbol can be seen in [7]. These schemes are not simulated because they have limited configurations.

Figs. 10-11 compare error performances between one iteration and two iterations multistage decoding. We can see that one and two iterations have different good codes. The good codes of one and two iterations differ by 0.5 dB at a bit error rate of 10^{-4}.

In Fig. 10, a good code of one iteration decoding has component codes RS(31,11) and RS(31,21), and in Fig. 11 the good code has component codes RS(63,19), RS(63,51) and RS(63,57).

Therefore we can conclude that in this channel, two iteration multistage decoding must be used. This because the good codes have the same component codes, thus the first level component code also needs decoded codewords of other levels.

5 Conclusion

In this paper, construction of partitioned Reed Solomon coded modulation (RSCM) which is robust for the additive white Gaussian noise channel and a Rayleigh fading channel is proposed. By matching configuration of component codes with the channel characteristics, it is shown that this system is robust for the Gaussian and a Rayleigh fading channel.

Its error performances were compared with those of non-partitioned RSCM and coded MPSK modulation using binary codes, Reed Muller codes. It appears that partitioned RSCM performs better than non-partitioned RSCM and RMCM over the Gaussian and a Rayleigh fading channel.

References

[1] G. Ungerboeck, *Channel coding with multilevel/phase signals*, IEEE Trans. Inf. Theory, IT-28, pp. 55-67, Jan. 1982.

[2] H. Imai and S. Hirakawa, *A new multilevel coding method using error correcting codes*, IEEE Trans. Inf. Theory, IT-23, pp. 371-376, May 1977.

[3] E.L. Cusack, *Error control codes for QAM signalling*, Electron. Lett., 20, pp. 62-63, Jan. 1984.

[4] S.I. Sayegh, *A class of optimum block codes in signal space*, IEEE Trans. Commun., COM-30, pp. 1043-1045, Oct. 1986.

[5] T. Takata, S. Ujita, T. Kasamiand S. Lin, *Multistage decoding of multilevel block M-PSK modulation codes and its performance analysis*, IEEE Trans. Inf. Theory, 39, pp. 1204-1218, July 1993.

[6] T. Woerz and Fazel, *Comparison of different decoding strategies for block coded modulation*, in Proceedings of 1992 URSI Int. Symp. Signals, Sys. and Electronics, Paris, pp. 117-120, Sept., 1992.

[7] S.H. Jamali, *Coded modulation techniques for fading channels*, Klower Academic Publishers, 1994.

[8] E. Biglieri, G. Caire, G. Taricco and J. Ventura-Traveset, *Coding and modulation for the fading channel: the quest for robustness*, in Proceedings of the Fifth ESA Intl. Workshop on DSP techniques Applied to Space Communication, Barcelona, pp. 8.1-89, Oct., 1996.

[9] E. Zehavi, *8-PSK trellis codes for a Rayleigh channel*, IEEE Trans. on Comm., 40, pp. 873-884, May 1992.

[10] E. Husni and P. Sweeney, *Block coded MPSK modulation using Reed Solomon codes*, in Proceedings of the Fifth ESA Intl. Workshop on DSP techniques Applied to Space Communication, Barcelona, pp. 1-8, Oct., 1996.

[11] P. Sweeney, *Error control coding an introduction*, Prentice Hall, 1991.

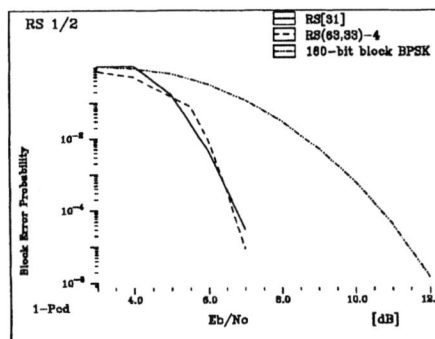
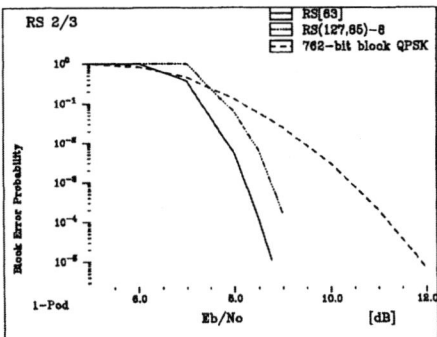

Fig. 4. Error performances of coded QPSK modulation; partitioned RSCM **RS[63]** of length 63 listed in Table 2 and non-partitioned RS(127, 65) with information-bit length of 65·7 = 455.

Fig. 5. Error performances of coded 8-PSK; partitioned RSCM **RS[127]** of length 127 listed in Table 1 and non-partitioned RS(255,171) with information-bit length of 171·8 = 1368.

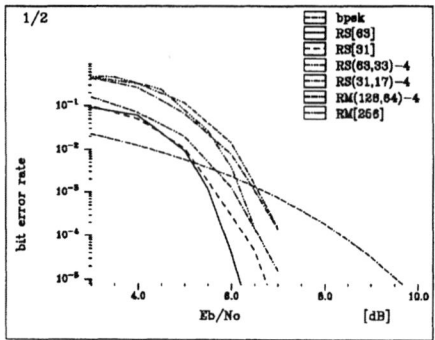

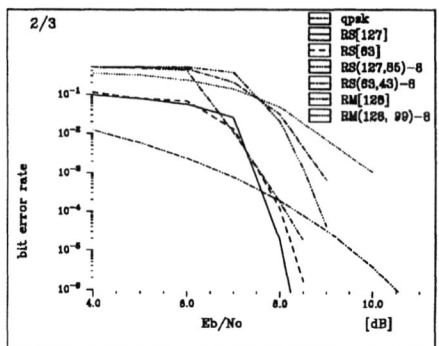

Fig. 6. Error performances of coded QPSK modulation; partitioned RSCM: **RS[63]** of length 63 listed in Table 2, **RS[31]** of length 31 listed in Table 2, non-partitioned: RS(63,33), RS(31,17), and partitioned RMCM **RM[256]** of length 256, non-partitioned 3rd-order RM(128,64).

Fig. 7. Error performances of coded 8-PSK modulation; partitioned RSCM: **RS[127]** of length 127 listed in Table 1, **RS[63]** of length 63 listed in Table 1, non-partitioned: RS(127,85), RS(63,43), and partitioned RMCM **RM[128]** of length 128, non-partitioned 4th-order RM(128,99).

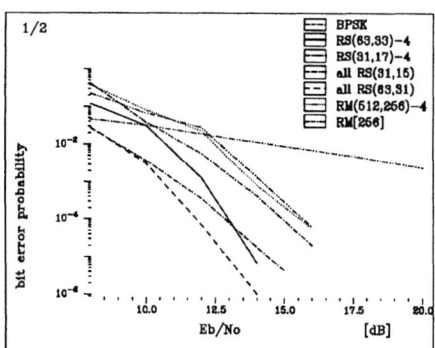

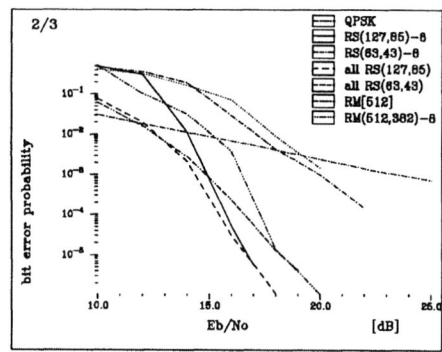

Fig. 8. Error performances of coded QPSK modulation over a Rayleigh fading channel; partitioned RSCM: **all RS(63,31)** of length 63 whose component codes are all RS(63,31), **all RS(31,15)** of length 31 whose component codes are all RS(31,15), non-partitioned: RS(63,33), RS(31,17), and partitioned RMCM **RM[256]** of length 256, non-partitioned 4th-order RM(512, 256).

Fig. 9. Error performances of coded 8-PSK modulation over a Rayleigh fading channel; partitioned RSCM: **all RS(127,85)** of length 127 whose component codes are all RS(127,85), **all RS(63,43)** of length 63 whose component codes are all RS(63,43), non-partitioned: RS(127,85), RS(63,43), and partitioned RMCM **RM[512]** of length 512, non-partitioned 5th-order RM(512, 382).

154

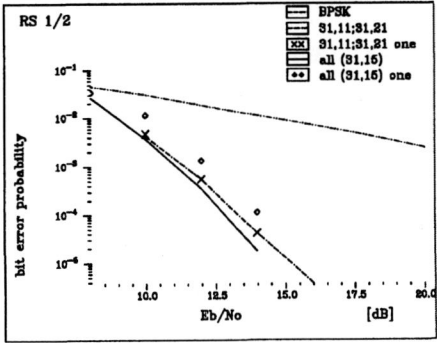

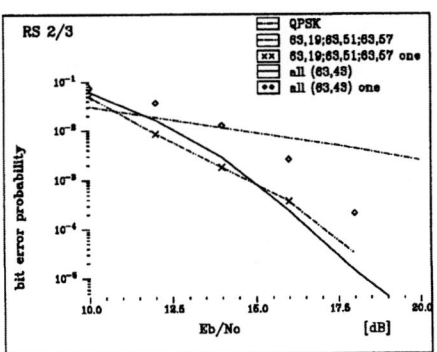

Fig. 10. Error performances over a Rayleigh fading channel of RS coded QPSK modulation based on set partitioning using two iterations multistage decoding: **all (31,15)** of length 31 whose component codes are all RS(31,15), **31,11;31,21** of length 31 whose component codes are RS(31,11) and RS(31,21), and one iteration multistage decoding: **all (31,15) one, 31,11;31,21 one.**

Fig. 11. Error performances over a Rayleigh fading channel of RS coded 8-PSK modulation based on set partitioning using two iterations multistage decoding: **all (63,43)** of length 63 whose component codes are all RS(63,43), **63,19;63,51;63,57** of length 63 whose component codes are RS(63,19), RS(63,51), and RS(63,57), and one iteration multistage decoding: **all (63,43) one, 63,19;63,51;63,57 one.**

RSA-type Signatures in the Presence of Transient Faults

Marc Joye[1], Jean-Jacques Quisquater[2], Feng Bao[3] and
Robert H. Deng[3]

[1] UCL Crypto Group, Dept of Mathematics, University of Louvain
Chemin du Cyclotron 2, B-1348 Louvain-la-Neuve, Belgium
Email: joye@agel.ucl.ac.be
[2] UCL Crypto Group, Microelectronics Labs, University of Louvain
Place du Levant 3, B-1348 Louvain-la-Neuve, Belgium
Email: jjq@dice.ucl.ac.be
[3] Institute of Systems Science, National University of Singapore
Kent Ridge, Singapore 119597
Email: {baofeng, deng}@iss.nus.sg

Abstract. In this paper, we show that the presence of transient faults
can leak some secret information. We prove that only one faulty RSA-
signature is needed to recover one bit of the secret key. Thereafter, we
extend this result to Lucas-based and elliptic curve systems.

Keywords. RSA, Lucas sequences, elliptic curves, transient faults.

1 Introduction

At the last Workshop on Security Protocols, Bao, Deng, Han, Jeng, Narasimhalu
and Ngair from the Institute of Systems Science (Singapore) exhibited new attacks against several cryptosystems [2]. These attacks exploit the presence of
transient faults. By exposing a device to external constraints, one can induce
some faults with a non-negligible probability [1].

In this paper, we show that these attacks are of very general nature and
remain valid for cryptosystems based on other algebraic structures. We will
illustrate this topic on the Lucas-based and elliptic curve cryptosystems. Moreover, we will focus on the signatures generation, reducing the number of required
signatures for a successful attack to one.

The paper is organized as follows. In Section 2, we recall the RSA-type sig-
nature schemes. Next, we present the basic attack in Section 3, and extend it to
Lucas sequences and elliptic curves. In Section 4, we give some further results.
Finally, we conclude in Section 5.

2 RSA-type signatures

Let $n = pq$ be the product of two large primes. Let G be the (multiplicatively
written) group given by the direct product of groups

$$G = G_p \times G_q. \tag{1}$$

The public verification key v is chosen relatively prime to $\#G$, and the secret signature key s is chosen according to

$$vs \equiv 1 \pmod{\#G}. \tag{2}$$

To sign a message $M \in G$, Bob computes the corresponding signature

$$S = M^s \tag{3}$$

with his secret key s. Then, he transmits the pair (M, S) to Alice. In order to verify the signature, Alice checks if $S^v = M$, where v is the public key of Bob.

Remark. For security and efficiency purposes, many variations of this signature scheme were proposed ; however, since our attack does not depend on these modifications, we decided to work with this elementary model.

The original RSA cryptosystem [9] works in the group $G = \mathbb{Z}_n^* (= \mathbb{F}_p^* \times \mathbb{F}_q^*)$, the group of units of integers modulo n. It was later extended to Lucas sequences by Smith and Lennon [11] to produce LUC. In this case, we have $G = \{V_j(P, 1) = r^j + r^{-j} \bmod n\}$ where r is a root of $x^2 - Px + 1 = 0$. Elliptic curves were also proposed to implement analogues of RSA. The first scheme, called KMOV, was presented by Koyama, Maurer, Okamoto and Vanstone [7]. It is based on elliptic curves of the form $E_n(0, b) (= E_p(0, b) \times E_q(0, b))$ or $E_n(a, 0) (= E_p(a, 0) \times E_q(a, 0))$ in order to have a message-independent group order. Another scheme was proposed by Demytko [5]. It has the particularity to only use the first coordinate of the points of elliptic curves. It relies on the fact that an integer x must be the x-coordinate of a fixed elliptic curve $E_p(a, b)$ or of its twist $\overline{E_p(a, b)}$.

We refer to the original papers for a complete description of these systems.

3 Transient faults

We assume that an error occurs during the computation of the signature S. More precisely, if $s = \sum_{i=0}^{t-1} s_i 2^i$ denotes the binary expansion of the secret key s, we suppose that the jth bit of s, namely s_j, flips to its complementary value. Therefore, Bob will obtain the faulty signature

$$\hat{S} = M^{\hat{s}} \tag{4}$$

instead of S, where

$$\hat{s} = \begin{cases} s + 2^j & \text{if } s_j = 0, \\ s - 2^j & \text{if } s_j = 1. \end{cases} \tag{5}$$

3.1 Attacking the RSA

Let $S = M^s \bmod n$ and $\hat{S} = M^{\hat{s}} \bmod n$ be the correct and the faulty signatures corresponding to message M, respectively. Mimicking the attack in [2], we find the flipped bit of s by computing

$$\frac{\hat{S}}{S} \equiv M^{\hat{s}-s} \equiv \begin{cases} M^{2^j} & (\bmod\ n) \quad \text{if } s_j = 0, \\ \frac{1}{M^{2^j}} & (\bmod\ n) \quad \text{if } s_j = 1. \end{cases} \tag{6}$$

However, this formulation is not optimal in signature context. Putting Eq. (6) to the v, we obtain

$$\frac{\hat{S}^v}{M} \equiv \begin{cases} (M^v)^{2^j} & (\bmod\ n) \quad \text{if } s_j = 0, \\ \frac{1}{(M^v)^{2^j}} & (\bmod\ n) \quad \text{if } s_j = 1. \end{cases} \tag{7}$$

So only the faulty signature $\hat{S}$ is required to recover the flipped bit. The attack can thus be summarized as follows. The attacker randomly chooses a message M. He computes $F := M^v \bmod n$ and $\alpha_j := F^{2^j} \bmod n$. Then, inducing a physical effort, he asks the device to sign message M. If one bit of s has flipped, he easily recover it by comparing $\hat{S}^v/M$ to α_j and $1/\alpha_j$.

3.2 Attacking LUC

The attack on LUC works similarly. The attacker chooses a random message M. He computes $F := V_v(M, 1) \bmod n$, $G := (M^2 - 4)U_v(M, 1) \bmod n$, $\alpha_j := V_{2^j}(F, 1) \bmod n$ and $\beta_j := U_{2^j}(F, 1) \bmod n$. Now, inducing a external effort to the signing device, he gets the faulty signature of message M,

$$\hat{S} = V_{\hat{s}}(M, 1) \bmod n.$$

Finally, he finds the flipped bit of s as

$$2V_v(\hat{S}, 1) \equiv \begin{cases} \alpha_j M + \beta_j G & (\bmod\ n) \quad \text{if } s_j = 0, \\ \alpha_j M - \beta_j G & (\bmod\ n) \quad \text{if } s_j = 1. \end{cases} \tag{8}$$

Proof. From the properties of Lucas sequences [3, Chapter 12], it follows that

$$\begin{aligned} 2V_v(\hat{S}, 1) &\equiv 2V_v\big(V_{\hat{s}}(M, 1), 1\big) \equiv V_{v(\hat{s}-s+s)}(M, 1) \equiv 2V_{v(\hat{s}-s)+1}(M, 1) \\ &\equiv V_{v(\hat{s}-s)}(M, 1)V_1(M, 1) + \Delta U_{v(\hat{s}-s)}(M, 1)U_1(M, 1) \\ &\equiv V_{\hat{s}-s}\big(V_v(M, 1), 1\big)M + (M^2 - 4)U_v(M, 1)U_{\hat{s}-s}\big(V_v(M, 1), 1\big) \\ &\equiv V_{\hat{s}-s}(F, 1)M + U_{\hat{s}-s}(F, 1)G \quad (\bmod\ n). \end{aligned}$$

Furthermore, since $V_{-k}(P, 1) = V_k(P, 1)$ and $U_{-k}(P, 1) = -U_k(P, 1)$, we have the required result. $\qquad\square$

An efficient implementation of Lucas sequences may be found in [6]. Notice also that the computation of α_j and β_j is not expensive. They can recursively be evaluated by

$$\begin{cases} \alpha_0 = F \\ \alpha_j = \alpha_{j-1}^2 - 2 \bmod n \end{cases} \quad \text{and} \quad \begin{cases} \beta_0 = 1 \\ \beta_j = \alpha_{j-1}\beta_{j-1} \bmod n \end{cases}. \tag{9}$$

3.3 Attacking KMOV

In the KMOV system, the messages are represented as points of elliptic curves. Let $\mathbf{M} = (M_1, M_2)$ be the message being signed, and let $\mathbf{S} = [s]\mathbf{M}$ and $\hat{\mathbf{S}} = [\hat{s}]\mathbf{M}$ respectively be the correct and the faulty signatures. We still assume that bit j of s flipped during the computation of the signature, *i.e.* $\hat{s} = s \pm 2^j$.

To recover this flipped bit, the attacker has just to compare $[v]\hat{\mathbf{S}} - \mathbf{M}$ to $[\hat{s} - s]([v]\mathbf{M})$. More precisely,

$$[v]\hat{\mathbf{S}} - \mathbf{M} = \begin{cases} [2^j](F_1, F_2) & \text{if } s_j = 0 \\ [2^j](F_1, -F_2) & \text{if } s_j = 1 \end{cases}, \tag{10}$$

where $(F_1, F_2) := [v]\mathbf{M}$.

Proof. Straightforwardly, we have

$$[v]\hat{\mathbf{S}} = [v\hat{s}]\mathbf{M} = [v(\hat{s} - s + s)]\mathbf{M} = [v(\hat{s} - s)]\mathbf{M} + \mathbf{M}$$
$$= [\hat{s} - s]([v]\mathbf{M}) + \mathbf{M}.$$

Moreover, if $\mathbf{P} = (P_1, P_2)$ is a point on an elliptic curve, then its inverse is given by $-\mathbf{P} = (P_1, -P_2)$. $\qquad\square$

3.4 Attacking Demytko

The Demytko's system only uses the x-coordinate of points of elliptic curves. Let M the message to be signed. This message is represented as a point $\mathbf{M} = (M, \cdot)$, or equivalently $M = x(\mathbf{M})$. As before, we suppose that a bit of s flipped during the computation of the signature. So, the resulting signature is $\hat{S}$, which is represented by $\hat{\mathbf{S}} = (\hat{S}, \cdot)$.

It is useful to introduce some notation. Let $\mathbf{P} = (x, y)$ be a point of an elliptic curve and let $\mathbf{Q} = [k]\mathbf{P}$. Using division polynomials [8, Chapter II], we can compute $x(\mathbf{Q}) = x([k]\mathbf{P})$ and $y(\mathbf{Q})/y(\mathbf{P}) = y([k]\mathbf{P})/y(\mathbf{P})$ from the only knowledge of $x = x(\mathbf{P})$ [8, Theorem 2.1]. These latter functions will respectively be denoted $\mathcal{X}_k(x)$ and $\mathcal{Y}_k(x)$.

To recover the flipped bit of s, the attacker computes $F := x([v]\mathbf{M}) = \mathcal{X}_v(M)$, $G := y([v]\mathbf{M})/y(\mathbf{M}) = \mathcal{Y}_v(M)$, $H := M^3 + aM + b \bmod n$, $\alpha_j := \mathcal{X}_{2^j}(F)$ and $\beta_j := \mathcal{Y}_{2^j}(F)$. Then,

$$\mathcal{X}_v(\hat{S}) + M \equiv \begin{cases} \dfrac{(\beta_j G - 1)^2 H}{(\alpha_j - M)^2} - \alpha_j \pmod{n} & \text{if } s_j = 0, \\ \dfrac{(\beta_j G + 1)^2 H}{(\alpha_j - M)^2} - \alpha_j \pmod{n} & \text{if } s_j = 1. \end{cases} \tag{11}$$

Proof. The chord-and-tangent addition formulae on elliptic curves [10, Algorithm 2.3] yields

$$\mathcal{X}_v(\hat{S}) \equiv x([v]\hat{\mathbf{S}}) \equiv x([\hat{s} - s]([v]\mathbf{M}) + \mathbf{M})$$
$$\equiv \left[\frac{y([v(\hat{s} - s)]\mathbf{M}) - y(\mathbf{M})}{x([v(\hat{s} - s)]\mathbf{M}) - x(\mathbf{M})}\right]^2 - x([v(\hat{s} - s)]\mathbf{M}) - x(\mathbf{M})$$

$$\equiv \frac{M^3 + aM + b}{(x\left([\hat{s}-s]([v]\mathbf{M})\right) - M)^2}\left[\frac{y\left([\hat{s}-s]([v](\mathbf{M}))\right)}{y(\mathbf{M})} - 1\right]^2$$
$$- x\left([\hat{s}-s]([v]\mathbf{M})\right) - M$$

$$\equiv \frac{H\left[\frac{y([\hat{s}-s]([v](\mathbf{M}))}{y([v]\mathbf{M})}\frac{y([v]\mathbf{M})}{y(\mathbf{M})} - 1\right]^2}{(\mathcal{X}_{\hat{s}-s}(F) - M)^2} - \mathcal{X}_{\hat{s}-s}(F) - M$$

$$\equiv \frac{H\left(\mathcal{Y}_{\hat{s}-s}(F)G - 1\right)^2}{(\mathcal{X}_{\hat{s}-s}(F) - M)^2} - \mathcal{X}_{\hat{s}-s}(F) - M \pmod{n}.$$

Hence, since $\mathcal{X}_{-k}(x) = \mathcal{X}_k(x)$ and $\mathcal{Y}_{-k}(x) = -\mathcal{Y}_k(x)$, the proof is complete. □

4 Further results

4.1 Generalizing the number of faulty bits

Assume that two bits of s are faulty when performing the signature of a message M with the RSA. Then, the signature is $\hat{S} = M^{\hat{s}} \bmod n$, where

$$\hat{s} = s \pm 2^j \pm 2^k. \tag{12}$$

As before, the attacker computes $F := M^v \bmod n$ and $\alpha_j := F^{2^j} \bmod n$, and compares $\hat{S}^v/M$ to $(\alpha_j)^{\pm 1}(\alpha_k)^{\pm 1}$ until a match is found. In this case, he finds two bits of the secret key s.

This method naturally extends to multiple faulty bits and to Lucas-based and elliptic curve systems.

4.2 Davida's attack

Our attack also applies for encryption process. This can been considered as a special case of the Davida's attack [4]. We shall illustrate this attack on RSA, but it remains valid for Lucas-based and elliptic curve systems.

Let e and d be a pair of public encryption key and secret decryption key related by

$$ed \equiv 1 \pmod{\operatorname{lcm}(p-1, q-1)}. \tag{13}$$

The attacker chooses a random message M, computes the ciphertext $C = M^e \bmod n$ with the public key e of Bob. Then, inducing a external constraint, he asks Bob to decipher C. If Bob does so, he obtains $\hat{M} = C^d \bmod n$. Since $\hat{M}$ is meaningless, Bob will discard it. Suppose that the attacker can get access to this discard and that only one bit of d is faulty, i.e. $\hat{d} = d \pm 2^j$. So,

$$\frac{\hat{M}^e}{C} \equiv \begin{cases} (C^e)^{2^j} & \pmod{n} \quad \text{if } d_j = 0, \\ \frac{1}{(C^e)^{2^j}} & \pmod{n} \quad \text{if } d_j = 1. \end{cases} \tag{14}$$

5 Conclusion

We have shown that the presence of transient faults gives the bit-values of the secret signature key s in RSA-type signature schemes. As in [2], this attack may of course be generalized to discrete log based cryptosystems.

References

1. ANDERSON, R., and KUHN, M. Tamper resistance – a cautionary note. In *Proceedings of the Second USENIX Workshop on Electronic Commerce* (1996), USENIX Association, pp. 1–11.
2. BAO, F., DENG, R. H., HAN, Y., JENG, A., NARASIMHALU, A. D., and NGAIR, T. Breaking public key cryptosystems on tamper resistant devices in the presence of faults. In *Pre-proceedings of the 1997 Security Protocols Workshop* (1997).
3. BRESSOUD, D. M. *Factorization and primality testing*. Undergraduate Texts in Mathematics. Springer-Verlag, 1989.
4. DAVIDA, G. Chosen signature cryptanalysis of the RSA (MIT) public key cryptosystem. Tech. Report TR-CS-82-2, Dept. of Electrical Engineering and Computer Science, University of Wisconsin, Milwaukee, USA, Oct. 1982.
5. DEMYTKO, N. A new elliptic curve based analogue of RSA. In *Advance in Cryptology – Eurocrypt '93* (1994), T. Helleseth, Ed., vol. 765 of *Lectures Notes in Computer Science*, Springer-Verlag, pp. 40–49.
6. JOYE, M., and QUISQUATER, J.-J. Efficient computation of full Lucas sequences. *Electronics Letters 32*, 6 (Mar. 1996), 537–538.
7. KOYAMA, K., MAURER, U. M., OKAMOTO, T., and VANSTONE, S. A. New public-key schemes based on elliptic curves over the ring $\mathbb{Z}_n$. In *Advance in Cryptology – Crypto '91* (1992), J. Feigenbaum, Ed., vol. 576 of *Lectures Notes in Computer Science*, Springer-Verlag, pp. 252–266.
8. LANG, S. *Elliptic curves: Diophantine analysis*, vol. 231 of *Grundlehren der mathematischen Wissenschaften*. Springer-Verlag, 1978.
9. RIVEST, R. L., SHAMIR, A., and ADLEMAN, L. A method for obtaining digital signatures and public-key cryptosystems. *Communications of the ACM 21*, 2 (Feb. 1978), 120–126.
10. SILVERMAN, J. H. *The arithmetic of elliptic curves*, vol. 106 of *Graduate Texts in Mathematics*. Springer-Verlag, 1986.
11. SMITH, P. J., and LENNON, M. J. J. LUC: A new public key system. In *Ninth IFIP Symposium on Computer Security* (1993), E. G. Douglas, Ed., Elsevier Science Publishers, pp. 103–117.

A Digital Signature Scheme Based on Random Error-Correcting Codes

G Kabatianskii[1], E. Krouk[2], B. Smeets[3]

[1] Institute of Problems of Information Transmission,
Russian Academy of Sciences, Moscow, Russia
[2] St Petersburg State Academy of Aerospace Instrumentation,
St Petersburg, Russia
[3] Department of Infomation Technology,
Lund University, Sweden
kaba@ippi.ac.msk.su, ben@it.lth.se

Abstract. Over the past years there have been few attempts to construct digital signature schemes based on the intractability of the decoding of linear error-correcting codes. Unfortunately all these attempts failed. In this paper we suggest a new approach based on a seemingly unknown before fact that the set of correctable syndroms being nonlinear nevertheless contains a rather large linear subspace.

Key words: digital signatures, public-key cryptography, linear error-correcting codes, intractability, correctable syndromes.

1 Introduction

There has been few attempts [1], [2], [3], [4], [5] to construct digital signature schemes based on the intractability of the decoding of linear error-correcting codes. However all these attempts failed. Yet it is interesting to explore further the use of this intractability feature of the decoding problem. Using a seemingly unknown fact about the set of syndromes we suggest a new approach and provide some constructions. Before we present the details of our construction we first recall some relevant facts about two public-key cryptosystems based on linear error-correcting codes.

Presently, there are two known ways of constructing public-key cryptosystems based on linear error-correcting codes; the McEliece cryptosystem and the Niederreiter cryptosystem. In both cases a secret key is a q-ary (n,k)-code V, which is randomly chosen from a family of linear error-correcting codes for which we know a rather simple, i.e., of polynomial (in n) complexity, algorithm for the decoding t or less errors. The public key is a generator matrix $G_{pub} = A'GP$ for the McEliece cryptosystem [6] and a parity-check matrix $H_{pub} = AHP$ for the Niederreiter cryptosystem [7], where A' and A are arbitrary nonsingular $k \times k$ and $r \times r$-matrices, respectively, P is an arbitrary $n \times n$ permutation matrix, G and H are a generator and a parity-check matrices, respectively, of the code V. Matrices P, G and A' or H and A, respectively, form the secret key. Both cryptosystems (we call them McENi, for shortness) are equivalent [8] and based on the unproven conjecture that decoding of arbitrary linear code up to half of its minimal Hamming distance is an NP-complete problem (it becomes NP-complete for the "total" minimal distance decoding, i.e., for maximum likelihood decoding [9]). Nevertheless for some classes of good codes the systems can be broken by revealing G (or H) from the public key (see [10])

Let us recall some more facts about Niederreiter's construction [7]. Denote by $\mathbb{F}_q$ a finite field of q elements and denote by $\mathbb{F}_q^n$ the set of all q-ary vectors of length n. Let the set M of possible messages be the set $\mathbb{F}_q^{n,t}$ of all q-ary vectors of length n, whose Hamming weight equals t (there is a well-known algorithm [11] for an enumeration of this set). Encryption produces a cyphertext (a syndrom) $s = H_{pub}m^T$ that will be sent for the message $m \in \mathbb{F}_q^{n,t}$, where $H_{pub} = AHP$. A legal user decrypts s in the following way : $m = \varphi_{V,t}(A^{-1}s)P$, where $\varphi_{V,t}$ is a decoding algorithm of the code V capable to correct t or less errors. As we mentioned above, the system is based on the assumption that the problem of finding an error vector of weight t for the given value s of its syndrom has for large n and t a very large complexity, and is based on the possible hardness of revealing H from H_{pub} (this is not always true as was shown in [10] for generalized Reed-Solomon codes).

There is a natural way of constructing a digital signature scheme in a similar way as Niederreiter's cryptosystem. Let now the set $\mathcal{M}$ of messages be the set of correctable syndroms, i.e., $M = S_t(H_{pub}) = \{H_{pub}e^T : e \in \mathbb{F}_q^{n,t}\}$. Then the sender signs a message s by e, where a signature e is a solution of the following equation:

$$H_{pub}e^T = s : e \in \mathbb{F}_q^{n,t}. \tag{1}$$

He evaluates the signature as $e = \varphi_{V,t}(A^{-1}s)P$ as he knows the algorithm $\varphi_{V,t}$, but everybody who wants to forge the signature should solve the equation (1). A receiver can check the validity of a received word $[s;e]$ by checking the equation (1). The description of this scheme is not yet complete because usually a set M of messages represents either by the segment of natural number $N_M = \{1,\ldots,M\}$ or by q-ary (mainly, binary) vectors of length k. Therefore users should have an efficient algorithm for the enumeration of the set of messages, i.e., the set $S_t(H_{pub})$ of correctable syndroms. Consider as the first candidate the following enumeration $\beta(i) = H_{pub}\Psi(i)$, where Ψ is the known enumeration $\Psi : N_M \to \mathbb{F}_q^{n,t}$. It is clear that such a scheme will be immediately broken because an opponent can create a false message just by setting $e := \Psi(m)$ and $s = \beta(m)$. An "opposite" candidate is a random choice of such a map among all $M!$ maps : $N_M \to S_t(H_{pub})$. To forge such a scheme seems to be as hard as breaking the general McENi cryptosystem. However, the scheme is not efficient because it requires a huge public key (an enumeration map becomes a common part of the public keys and for storing it one needs roughly $M \log M$ bits).

In this paper we use a probably unknown fact that for every linear code the set of its correctable syndroms contains a linear subspace of relatively large dimension L. The syndrome decoding problem was also used in the work of Stern [14] to obtain an efficient identification scheme. We restrict a set of messages only to correctable syndroms and generate them efficiently due to their linear structure. Unfortunately, due to the same linear structure the proposed scheme can be broken after approximately (roughly) L of its uses. Nevertheless, as this scheme is more than only a one-time scheme, it can be useful, for instance, for improving the parameters of the systems proposed in [12]. The main difference from previous attempts of constructing digital signature schemes based on error-correcting codes as well as from McENi systems, is that we use *arbitrary linear codes*, i.e., *we do not use some classes of codes (like Goppa codes) with known decoding algorithm. Instead of doing this we construct a set of syndroms which can be simply decoded for arbitrary linear codes if to know trapdoors.*

2 The Basic Scheme

Let V be a q-ary $(n, n-r)$-code with minimal Hamming distance $d(V) > 2t$ and let C be an equidistant (n', k')-code with minimal Hamming distance $d(C) = t$, where $n \geq n' = \frac{q^{k'}-1}{q-1}$ and $d(C) = t = q^{k'-1}$. Let $r \times n$-matrix $H_V = [h_1, \ldots, h_n]$ be a parity-check matrix of the code V and let $k' \times n'$-matrix G_C be a generator matrix of the code C. Define an $r \times k'$-matrix $F = H(J)G_C^T$, where J is a subset of the set $\{1, \ldots, n\}$ of cardinality n', and $H(J)$ is a submatrix of the matrix H_V consisting of the columns $h_j, j \in J$. Recall that $S_t(H) = \{He^T : e \in \mathbb{F}_q^{n,t}\}$ is the set of syndroms corresponding to errors of weight t. It is easy to prove the following

Proposition 1. $Fx^T \in S_t(H_V)$ *for any* $x \in \mathbb{F}_q^{k'} \setminus \{0\}$.

Now one can define a signature scheme in the following way. There are two public matrices: H and F. The set J and the matrix G are secret (private) keys. The sender signs a message $m \in \mathbb{F}_q^{k'} \setminus \{0\}$ by a signature e, where the vector e equals mG on positions of the set J and equals 0 outside of J. It is easy to see that e is a (unique) solution of the following signature equation:

$$H_V e^T = Fm^T : wt(e) = t, \tag{2}$$

Such a signature scheme is not resistant against homomorphism attack, since after observation two signed messages $[m_1; e_1]$ and $[m_2; e_2]$ an opponent can create a false but a valid word $[m_1 + m_2; e_1 + e_2]$. To avoid homomorphism attack we do the same as usually is done in such a case (for instance, for RSA signatures). Namely, consider any "good" (i.e., "enough nonlinear", simple to evaluate, hard to invert, "no collisions" etc.) *public* hash function $f : M = \mathbb{F}_q^K \to \mathbb{F}_q^{k'} \setminus \{0\}$, and modify the definition of the signature equation

$$H_V e^T = Ff(m)^T : wt(e) = t, \tag{3}$$

For any given message m the signer evaluates the signature $e = f(m)G$. Now the opponent's attempt to solve the equation (3) should fail because of the hardness of decoding of an *arbitrary* linear code. To find trapdoors, i.e., the set J and the matrix G, also seems to be as difficult as to decode the code V because every column of the public matrix F is a linear combination of exactly t columns of the matrix H (taken from positions of J) and to find this linear combination is the same as to decode V for some particular value of the syndrom. Unfortunately, the straight application of this signature scheme demands a too large code length.

Example 1. Consider binary codes. We desire that working factor for the opponent should not be less than 2^{50}. It implies that the number of values of possible signatures is at least 2^{50}. Hence $k' \geq 50$ and $n \geq n' = 2^{k'} - 1 \approx 10^{15}$, but this is too large for any practical applications.

In order to improve the parameters of the scheme we replace the equidistant code by any (n', k')-code C, whose nonzero codewords c have weight $t_1 \leq wt(c) \leq t_2$. This leads to the following modification of the signature equation

$$H_V e^T = F f(m)^T : t_1 \leq wt(e) \leq t_2 \qquad (4)$$

Let us remark that the condition $d(V) > 2t_2$ which guarantees the uniqueness of a signature, i.e., a solution of the equation (4), is not so important for the proposed scheme as for the McENi scheme, because one cyphertext has to correspond to only one message, but one message can have few signatures. We need only that it should be difficult to solve the equation (4), i.e., to decode the code V if the number of errors lies in the interval $[t_1, t_2]$. It leads us to the following modification of the initial scheme.

Consider the code dual to binary BCH code of length $n' = 2^l - 1$ and designed distance $2s + 1$ as an (n', k')-code C. It is known that $k' = sl$ (mainly) and $| wt(c) - \frac{n'+1}{2} | \leq (s-1)\sqrt{n'+1}$ for any nonzero codeword c (see [13]).

Consider a *random* systematic binary (n, k)-code as a code V. We assume that the first k positions of the code are information positions, hence the code can be defined by its parity-check $r \times n$-matrix $H_V = [E_r|D]$, where E_r is the unit $r \times r$-matrix and D is a random $r \times k$-matrix. Hence for describing of the corresponding part (i.e. the matrix H_V) of the public key one needs only $r \times k$ bits instead of $r \times n$ bits for a truly random codes. Usual counting arguments show that nevertheless the lower bound on the minimal code distance for this smaller ensemble is the same as for ensemble of all binary codes.

Proposition 2. *The probability that a random $r \times n$-matrix $H_V = [E_r|D]$ generates binary code V with minimal distance $d_V \geq d$ is at least $1 - 2^{-r+nh(\frac{d-1}{n})}$, where $h(x) = -x \log_2 x - (1-x) \log_2(1-x)$.*

Now the description of the system is as follows. The signer chooses *randomly* : a binary $r \times k$-matrix D; a nonsingular $k' \times k'$matrix A; an n'-subset $J \subset \{1, \ldots, n\}$. He forms a binary $r \times n$-matrix $H_V = [E_r|D]$. He takes a parity-check matrix of the binary BCH code with designed distance $2s + 1$ as a (generator) matrix G_C and forms $r \times k'$-matrix $F = H_V(J)(AG_C)^T$ (i.e., AG_C is an arbitrary generator matrix of the code C). The public key of this system consists of the matrices H_V and F which requires $r(k+k')$ bits, and the private key is relatively small, namely, less than $nh(\frac{n'}{n})$ bits for describing the set J and $(k')^2$ bits for the matrix A. The signer evaluates the signature e as $f(m)AG_C$ on positions of the set J and $\mathbf{0}$ outside of J. All the previously stated arguments concerning the scheme based on equidistant codes are valid for the considered modification. Let us illustrate this construction by the following example.

Example 2. Choose $l = 10$ and $s = 6$. Then $t_1 = 352, t_2 = 672$ and the number of possible signatures equals $2^{60} - 1$. Choose a *random* binary 2808×192-matrix D and form the above described systematic 2808×3000-matrix H_V. With probability $p \geq 1 - 10^{-9}$ the corresponding code V has minimal distance $d_V > 1024$. Public key of this system consists of 7×10^5 bits and a private key of 6.4×10^3 bits. To solve equation (4) for the number of errors in the range $[352, 672]$ takes roughly $2^{nH(t/n)-rH(t/r)} = 2^{54}$ "trials" for the best known algorithm.

This construction can be further randomized and improved by choosing the code C as a random code. The following proposition similar to Proposition 2 guarantees that this can be done.

Proposition 3. *The probability that the weight of every nonzero codeword of a systematic random binary (n, k)-code C lies in the range $\left[\frac{n}{2}(1 - \delta), \frac{n}{2}(1 + \delta)\right]$ is at least $1 - 2^{-r+nh(\delta)+1}$.*

Example 3. Let a code C be generated by a $k' \times n'$-matrix $G = [E_{k'}|D']$, where D' is $k' \times r'$-random matrix, $k' = 60$ and $n' = 280$. Due to Proposition 3 with probability $p_C \geq 1 - 10^{-9}$ this is a code with $t_1 = 50, t_2 = 230$. Choose a random systematic 990×1250-matrix H_V. With probability $1 - 10^{-9}$ the corresponding code V has minimal distance $d_V > 280$. Public key of this system consists of 3×10^5 bits and private key of 4.6×10^3 bits. The number of possible signatures is 2^{60} and complexity of decoding this random code for the number of errors in the range $[50, 230]$ is 2^{46} "trials" on the average.

In the next section we modify these schemes by using just a little bit more complicated enumeration scheme which allows us to use for C codes with not a good minimal distance.

3 A Modified Scheme

The below suggested improvement of the basic scheme based on codes C whose minimal distance is not large can be described simply. Such codes contain codewords (vectors) of low weight. Therefore we try to avoid these vectors in the enumeration procedure. For example, let C be a direct sum of two codes C_1 and C_2. Then all codewords have enough large weight except of vectors of the form $(c_1, 0)$ or $(0, c_2)$. It is clear how to remove such vectors from an enumeration procedure, but an obvious drawback of such a straightforward application is that a standard generator matrix of the code C contains low weight vectors that can

help the opponent to forge a signature. The following scheme seems to be free
of this shortcoming.

Consider the finite field $\mathbb{F}_{q^{k'}}$ as k'-dimensional vector space over $\mathbb{F}_q$, fix
some basis and denote by $M(\beta)$ the matrix corresponding to a linear map
$z \mapsto \beta z$, where β is an element of $\mathbb{F}_{q^{k'}}$, i.e., $\beta z = M(\beta) z^T$. The signer con-
structs generator matrices $G_1, \ldots, G_P$ of (n', k')-codes $C_1, \ldots, C_P$ with the property that the Hamming weight of any nonzero codeword belongs to the interval
$[t_1, t_2]$. He chooses *randomly*: a systematic $r \times n$-matrix H_V, nonsingular $k' \times k'$-
matrices A_j, non-intersecting n'-subsets J_j and distinct elements $\beta_j \in \mathbb{F}_{q^{k'}} \setminus \{0\}$,
for $j = 1, \ldots P$. He forms a public $r \times Qk'$-matrix $F = [F_1, \ldots, F_Q]$, where
$F_i = \sum_{j=1}^{P} H(J_j)(A_j G_j)^T M(\beta_j^{i-1})$ are $r \times k'$-matrices, $i = 1, \ldots, Q$. The signature
equation has the following form:

$$H_V e^T = F[u_1, f(m)]^T : Pt_1 \le wt(e) \le Pt_2,$$

where $f(m) = (u_2, \ldots, u_Q)$, $f : M = \mathbb{F}_q^K \to \mathbb{F}_q^{(Q-1)k'} \setminus \{0\}$ is a public hash
function and $u_i \in \mathbb{F}_q^{k'} \simeq \mathbb{F}_{q^{k'}}$, $i = 1, \ldots, Q$.

Lemma 4. *If e equals $\mathbf{0}$ on positions $\{1, \ldots, n\} \setminus \bigcup J_j$ and equals $U(\beta_j) A_j G_j$ on
the positions of the set J_j (where $U(z) = u_1 + u_2 z + \ldots + u_Q z^{Q-1} = u_1 + u(z)$),
then*

$$H_V e^T = F[u_1, f(m)]^T.$$

*In addition, $Pt_1 \le wt(e) \le Pt_2$ if $u_1 \notin \{-u(\beta_j) : j = 1, \ldots, P\}$, (it is always
possible if $P < q^{k'}$).*

Proof. Denote $\mathbf{u} = (\mathbf{u_1}, \ldots, \mathbf{u_Q}) = (\mathbf{u_1}, \mathbf{f(m)})$. Then

$$F\mathbf{u}^T = \sum_{i=1}^{Q} F_i u_i^T = \sum_{i=1}^{Q} \sum_{j=1}^{P} H(J_j)(A_j G_j)^T M(\beta_j^{i-1}) u_i^T$$

$$= \sum_{j=1}^{P} H(J_j)(A_j G_j)^T \sum_{i=1}^{Q} M(\beta_j^{i-1}) u_i^T = \sum_{j=1}^{P} H(J_j)(U(\beta_j) A_j G_j)^T = H_V e^T.$$

Since $u_1 \notin \{-u(\beta_j) : j = 1, \ldots, P\}$ one has that all $U(\beta_j) \neq 0$. Hence, $U(\beta_j) A_j G_j$
is a nonzero codeword of the code C_j and $t_1 \le wt(U(\beta_j) A_j G_j) \le t_2$ for
$j = 1, \ldots, P$.

Example 4. Let $Q = 14$, $P = 12$ and $C_1 = \ldots = C_P$ be a binary equidistant
$(15, 4)$-code with $t_1 = t_2 = 8$. Choose randomly a systematic $(1100, 335)$-code
V, which has $d(V) \ge 193$ with probability at least $1 - 10^{-9}$ (or one can choose a
hidden, by a McENi system, Goppa code with $n = 1024$, $k = 280$). The number
of possible signatures is $2^{52} - 1$, and complexity of decoding 96 errors by this
random code is at least 2^{53} "trials".

The size of the public key is 3×10^5 bits (approximately the same as for the
McENi system) and the size of the secret key $= 12 \times 16 + 3 \times 4 + 12 \times 15 \times 10 =$

2004 bits, where the first summand is responsible for the description of the 12 nonsingular matrices A_j, the second - for the description of the 12 nonzero elements of $\mathbb{F}_{16}$, and the last one - for the description of the 12 non-intersecting 15-elements subsets of the set $\{1, \ldots, 1100\}$.

4 Summary

In this paper we presented a new digital signature scheme based on error-correcting codes. Our approach is based on the seemingly unknown fact that the set of correctable syndromes, albeit nonlinear, contains a rather large linear space. Our construction, being strictly more than a one-time sgnature scheme, can be favorably combined with the ideas from [12]. The authors invite the readers to analyse the presented scheme.

References

1. W. Xinmei, "Digital signature scheme based on error-correcting codes", *Electronic letters*, 26 (13), 1990, pp. 898-899.
2. L. Harh and D.-C. Wang, "Cryptoanalysis and modification of digital signature scheme based on error-correcting codes", *Electronic letters*, 28 (2), 1992, pp. 157-159.
3. M. Alabbadi and S.B. Wicker, "Cryptoanalysis of the Ham and Wang modification of the Xinmei digital signature scheme", *Electronic letters*, 28 (18), 1992, pp. 1756-1758.
4. M. Alabbadi and S.B. Wicker, "Digital signature scheme based on error-correcting codes", *Proceedings 1993 IEEE International Symposium on Information Theory*, San Antonio, USA, 1993, pp. 199.
5. M. Alabbadi and S.B. Wicker "Susceptibility of digital signature scheme based on error-correcting codes to universal forgery", *Proceedings 1994 IEEE International Symposium on Information Theory*, Trondheim, Norway, 1994, pp. 494.
6. R.J. McEliece, *A public-key cryptosystem based on algebraic coding theory*, DSN Prog. Report, JPL, 1978, pp. 114-116.
7. H. Niederreiter, "Knapsack-type cryptosystems and algebraic coding theory", *Probl Control and Information Theory*, 1986, pp. 157-166.
8. Y.X. Li, R.H. Deng and X.M. Wang. On the equivalence of McEliece's and Niederreiter's public-key cryptosystems, *IEEE Transactions on Information Theory*, 40(1), 1994, pp. 271-273.
9. E.R. Berlekamp, R.J. McEliece, and H.C.A. van Tilborg, "On the inherent intractability of certain coding problems", *IEEE Transactions on Information Theory*, Vol. 24, 1978, pp.384-386.
10. V.M. Sidelnikov and S.O. Shestakov, "On insecurity of cryptosystem based on generalized Reed-Solomon codes", *Discrete Mathematics*, Vol. 4, 1992, pp. 57-63.
11. Lehmer, *Applied Combinatorial Mathematics*.
12. S. Even, O. Goldreich, and S. Micali, "On-line/off-line digital signatures", *Journal of Cryptology*, 9(1), 1996, pp. 35-67.
13. F.J. MacWilliams and N.J.A. Sloane, *The Theory of Error-Correcting Codes*, Amsterdam: North-Holland, 1977.
14. J. Stern, A new identification scheme based on syndrome decoding, *Proc. Crypto'93*, D. Stinson (Ed.), Springer-Verlag, 1994, pp 13-21.

Variable Rate Adaptive Trellis Coded QAM for High Bandwidth Efficiency Applications under Rayleigh Fading Channel

Vincent K.N. Lau and Malcolm D. Macleod
email: knl22@eng.cam.ac.uk, mdm@eng.cam.ac.uk

Signal Processing and Communication Group,
Department of Engineering, University of Cambridge, CB2 1PZ, UK

Abstract. A high bandwidth efficiency variable rate adaptive channel coding scheme, ATCQAM, is proposed. Known pilot symbols are transmitted periodically to aid demodulation. Past channel states are fed back to the transmitter with delay. Current channel state is then predicted at the transmitter to decide on the appropriate modulation mode for the current symbol. At good channel states, high level modulation is used to boost up the average throughput. At bad channel states, low level modulation is used to increase error protection. By matching the variable modulation level with a variable rate channel coder, the physical bandwidth is maintained constant. Design issues for the ATCQAM are considered. Practical schemes to maintain transmitter-receiver synchronization, namely the *quasi-closed loop* control and the *closed-loop control*, are discussed. The effects of finite feedback delay, finite interleaving depth and mobile speed are investigated.

1 Introduction

Error correction codes have been widely used to combat the effect of Rayleigh fading in mobile radio channels. In traditional FEC schemes [1], fixed-rate codes were used which failed to explore the time varying nature of the channel. To keep the performance at a desirable level, they were designed for the average or worst case situation. In this paper, we propose and study a high bandwidth efficiency variable rate adaptive trellis coded QAM (ATCQAM) which varies the code rate and modulation level according to the channel condition. The receiver estimated the channel states and inform the transmitter through the use of a feedback link. When the channel state is good, the transmitter increases the throughput by using a higher level QAM. On the other hand, when the channel state is bad, the transmitter uses a lower level QAM to improve the error protection.

We use M−ary QAM since it is more energy efficient than the M−ary PSK constellations at large M. Known pilot symbols are periodically inserted at the transmitter [2,3] to aid the demodulation. Two methods to ensure correct transmitter-receiver synchronization on the modulation modes are described. Two operations on the ATCQAM scheme, namely the *constant BER* operation and the *constant throughput* operation are introduced. Furthermore, the performance degradations due to finite feedback delay, finite depth interleaving and mobile speed are investigated.

The paper is organized as follows. In section 2, we described design issues, different system components and the operation of the ATCQAM scheme. Simulation results are presented and discussed in section 4. Finally, in section 5, we conclude with a summary of results.

2 Adaptive Trellis Coded QAM

There are three main criteria leading to a good ATCQAM scheme.

(a) **Decoder Complexity:** It is important that a single decoder is used to decode all the modulation modes.
(b) **Rate Compatibility:** It is important that the error path Hamming distance (M-ary) of the trellis diagram is not reduced due to changing modulation modes.
(c) **Constant Bandwidth:** It is important to maintain the physical occupied bandwidth constant as the instantaneous throughput (bits per symbol) varies.

They will be addressed as follows.

2.1 Design of ATCQAM

The simplified system block diagram of the proposed scheme is shown in fig. 1(a). Information bits are convolutionally encoded and the coded bits are mapped with the appropriate M-ary QAM symbol. Known pilot symbols are inserted at the transmitter periodically to aid the demodulation at the receiver. By means of an interpolation filter at the receiver, channel states in between the pilot positions are interpolated and used to demodulate the received symbols. The estimated channel states at pilot positions are fed back to the transmitter via a low noise (error protected) feedback link with certain delay. By means of an instantaneous linear prediction filter, current channel states are predicted at the transmitter and appropriate modulation modes are used for the current symbol.

By matching the *adaptive channel coder* and the *adaptive modulator*, all M-ary symbols have the same duration and hence the occupied bandwidth is constant. This corresponds to the above point (c). The varying instantaneous throughput is achieved by encoding a varying number of information bits per symbol.

The trellis encoder is based on a modified *pragmatic* TCM design [4] using a core rate 1/6 encoder. Between each trellis transition, a variable number of uncoded bits are concatenated with the coded bits and mapped with the appropriate M-ary QAM symbol as shown in fig. 1(b). Hence, we have a trellis with a fixed number of states but a varying number of parallel branches between each transition step. The same Viterbi decoder can be used at the receiver. This corresponds to point (a).

Suppose the estimated path (state sequence) diverged from the correct transmitted path at node 1 in the trellis, the error path will have its Hamming distance (M−ary) unaltered irrespective of the subsequent modulation modes. This corresponds to the rate compatibility of point (b). Therefore, the proposed adaptive pragmatic TCM satisfied the mentioned design criteria. The only disadvantage of the design is the presence of parallel branches. Although parallel branches are detrimental to the performance of the fixed rate TCM under Rayleigh fading, it is not the performance bottleneck of the ATCQAM by careful mapping of the signal constellation [4, 5].

2.2   Frame Structure and Transmitter-Receiver Synchronization

We shall consider two types of frame structure in bandwidth efficient application, namely the FDMA and the TDMA. In FDMA frame, each user occupied the entire time dimension to transmit. Pilot symbols, are inserted periodically in the frame. N_p is the pilot period in terms of the number of symbols and T_s is the symbol duration. Since the channel state could be different for each symbol, we cannot use a single modulation mode for the whole frame. To maintain transmitter-receiver synchronization, we have to employ a *quasi-closed loop* method [5].

For the TDMA frame, each user just occupied a portion of time to transmit or receive. Symbols are transmitted in burst. There is 1 pilot symbol per user burst. Since the burst duration are very short compared with the fading rate, the fading across the user burst do not change much and the whole burst is assigned with one modulation mode. Transmitter-receiver synchronization can be attained with a *closed loop* method [5].

2.3   Interleaving

Interleaving is used to convert the bursty fading into independent fading. Since a variable number of information bits are carried per trellis transition, the interleaving design is not trivial. The simplified block diagram for the interleaver is shown in fig. 1(b). Information bits are passed to the core encoder and the coded bits are interleaved[1]. At the time of transmission, an appropriate number of uncoded bits are drawn from the buffer. The uncoded bits and the interleaved coded bits are mapped with the appropriate M-ary QAM symbol.

2.4   Channel State Interpolation

To avoid severe degradation in QAM performance caused by fast Rayleigh fading, the transmitter inserted known pilot symbol periodically (1 pilot every N_p symbols) to aid the receiver demodulation. The receiver stored a number of

---

[1] The two coded bits are not broken up in the interleaver.

received pilot symbols and made use of an *interpolation filter* to estimate the fading in between the pilots.

Let $\tilde{c}_{r,i}$ be the complex fading at frame position r of the i-th frame. Using a $2P_1 + 1$-th order linear FIR interpolation filter, the interpolated fading at the r-th frame position (in between the pilot instants) of the i-th frame, $\tilde{z}_{r,i}$, is given by:

$$\tilde{z}_{r,i} = [\tilde{c}_{r,i} + \tilde{\epsilon}_{r,i}] + \tilde{e}_{r,i} \quad r \in [1, N_p - 1] \tag{1}$$

Note that $\tilde{\epsilon}_{r,i}$ is modeled as a Gaussian *residual* noise with variance $\sigma_{\tilde{\epsilon}}^2(r)$ due to imperfect filtering, $\tilde{e}_{r,i}$ is a Gaussian noise with variance σ_r^2. The receiver use $\tilde{z}_{r,i}$ to perform match filtering and to compute the decision metrics to be used in the Viterbi decoder.

2.5 Channel State Prediction

It is essential for the transmitter to know the current fading instantaneously to decide on the appropriate modulation mode. To obtain an instantaneous estimate of the current fading given the past fading, we use a two step process.

Firstly, we use a P_2 order linear predictive filter [6] to obtain a prediction of the fading at the next pilot instant. Based on the past $z_{0,i}$, the predicted fading at the next pilot instant, $\hat{z}_{0,k}$, is given by:

$$\hat{z}_{0,k} = \sum_{p=1}^{P_2} \alpha_p \tilde{z}_{0,k-p} = \tilde{c}_{0,k} + \hat{e}_{0,k} + \sum_{p=1}^{P_2} \alpha_p \tilde{e}_{0,k-p} \tag{2}$$

where α_p is the linear predictive coefficients [6] and $\hat{e}_{0,k}$ is the *prediction noise* which is approximately white Gaussian noise at high P_2.

The 2nd step to obtain the predicted fading in between the pilot instants depends on the frame structure.

TDMA: Consider the i-th frame, due to the short burst duration, the predicted fading for the current symbol (position r) is taken to be the fading at the frame's pilot instant (position 0), $z_{0,i}$. However, due to the feedback delay, Δ, $\tilde{z}_{0,i}$ is not available until $r >> \Delta$. At these frame positions, the predicted fading is taken to be $\hat{z}_{0,i}$. Hence, the predicted fading at frame position r, $\hat{z}_{r,i}$ is given by:

$$\hat{z}_{r,i} = \begin{cases} \hat{z}_{0,i} = \tilde{c}_{0,i} + \hat{e}_{0,i} + \sum_{p=1}^{P_2} \alpha_p \tilde{e}_{0,i-p} & 1 \leq r < \Delta \\ \tilde{z}_{0,i} = \tilde{c}_{0,i} + \tilde{e}_{0,i-p} & \Delta \leq r \leq N_p - 1 \end{cases} \tag{3}$$

FDMA: Since fading may vary within a frame, the predicted fading for the frame position r is obtained by 3-rd order interpolation using the $\tilde{z}_{0,i-1}$, $\tilde{z}_{0,i}$ or $\hat{z}_{0,i}$ and $\hat{z}_{0,i+1}$. Hence, the predicted fading at the frame position r, $\hat{z}_{r,i}$, is given by:

$$\hat{z}_{r,i} = \begin{cases} h_{-1,r} \tilde{z}_{0,i-1} + h_{0,r} \hat{z}_{0,i} + \\ h_{1,r} \hat{z}_{0,i+1} & 0 \leq r < \Delta \\ h_{-1,r} \tilde{z}_{0,i-1} + h_{0,r} \hat{z}_{0,i} + \\ h_{1,r} \hat{z}_{0,i+1} & \Delta \leq r \leq N_p - 1 \end{cases} \tag{4}$$

where

$$h_{p,r} = \begin{cases} \text{sinc}(p + r/N_p) & \text{FDMA frame} \\ \text{sinc}(p + r/(N_p N_u)) & \text{TDMA frame} \end{cases} \qquad (5)$$

2.6 Operation of the ATCQAM

There are 7 modulation modes in the proposed ATCQAM. They are listed as follow.

Mode 0: Throughput-1/3, 6 coded bits carried by 3 QAM symbols.
Mode 1: Throughput-1/2, 4 coded bits carried by 2 QAM symbols.
Mode 2: Throughput-1, 2 coded bits carried by 1 QAM symbol.
Mode 3: Throughput-2, 1 uncoded bits + 2 coded bits carried by 1 8PSK symbol.
Mode 4: Throughput-3, 2 uncoded bits + 2 coded bits carried by 1 16QAM symbol.
Mode 5: Throughput-4, 3 uncoded bits + 2 coded bits carried by 1 32QAM symbol.
Mode 6: Throughput-5, 4 uncoded bits + 2 coded bits carried by 1 64QAM symbol.

There are two different ways to operate the ATCQAM, namely the *constant BER* operation and the *constant throughput* operation [5]. The predicted channel state, $\hat{z}_{r,k}$, is partitioned into 7 segments with each segment corresponds to one of the above modes. Let $\frac{E_s}{\eta_0}$ be the average symbol energy to noise ratio. Mode m is chosen if $\frac{E_s}{\eta_0}|\hat{z}_{r,k}|^2 \in [\zeta_m, \zeta_{m+1}]$. Note that $\zeta_0 = 0$ and $\zeta_7 = \infty$.

3 Equivalent Channel Model and Decision Metrics

We assume that all encoded symbols on the length-N error path belong to frame position r from different frames after de-interleaving. Fading for all encoded symbols on the error path is independent after interleaving. The k-th received symbol corresponding to the r-th frame position, $\tilde{y}_k(r)$, is given by:

$$\tilde{y}_k(r) = \sqrt{2E_s}\tilde{c}_{r,k}\tilde{x}_k + \tilde{n}_{r,k} \qquad (6)$$

where $\tilde{x}_k$ is the transmitted symbol normalized to unit variance and $\tilde{n}_{r,k}$ is a complex white Gaussian channel noise. Since the receiver only have knowledge of the interpolated fading, $\tilde{z}_{r,k}$, we express $\tilde{y}_k(r)$ in terms of $\tilde{z}_{r,k}$, E_s and normalize w.r.t. $2\eta_0$. The equivalent channel model is thus described by:

$$\tilde{y}_k'(r) = \sqrt{\frac{E_s}{\eta_0}}\tilde{z}_{r,k}\tilde{x}_k + \tilde{n}_{r,k}' \qquad (7)$$

where $\tilde{n}'_{r,k} = \frac{\tilde{n}_{r,k}}{2\eta_0} - \sqrt{\frac{E_s}{\eta_0}}\tilde{x}_{r,k}(\tilde{e}_{r,k} + \tilde{\epsilon}_{r,k})$ is a Gaussian noise. The maximum likelihood branch decision metric, $\mu_{ML}(r)$, is given by:

$$\mu_{ML}(r) = p(\tilde{y}_k(r)|\tilde{x}_{r,k}\tilde{z}_{r,k}) = \frac{|\tilde{y}_k(r) - \sqrt{\frac{E_s}{\eta_0}}\tilde{z}_{r,k}\tilde{x}_{r,k}|^2}{\sigma^2_{\tilde{n}'}(r)}$$

For simplicity, we use the sub-optimal metric, $\mu_k(r)$, given by:

$$\mu_k(r) = |\tilde{y}_k(r) - \sqrt{\frac{E_s}{\eta_0}}\tilde{z}_{r,k}\tilde{x}_{r,k}|^2 \qquad (8)$$

and we choose $\tilde{x}_{r,k}$ so that $\mu_k(r)$ is minimum.

4 Results and Discussion

We assumed $N_p = 91$ (pilot period), $P_1 = 21$ (interpolation filter order), $P_2 = 16$ (prediction filter order) and $N_u = 8$ (user per TDMA frame). The overhead due to pilot symbols is about 1%. Convolutional code of constraint length 5 is used to construct the ATCQAM. Its performance is compared with that of the optimal fixed rate TCM[2] Hence, the comparison is fair.

4.1 Performance of ATCQAM in TDMA Frame

Zero Feedback Delay ($\Delta = 0$): As an illustration, we assumed zero feedback delay, 100×500 block interleaving and $f_d T_s = 1 \times 10^{-3}$ [3]. The BER and the throughput of the ATCQAM-TDMA scheme using *constant BER* controls are shown in fig. 2(a) and (b). Under the *constant BER* control, the BER remains approximately constant when $\frac{E_s}{\eta_0}$ is within the range of adaptation as shown in fig. 2(a). Along the BER curves of the ATCQAM, the throughput varied as $\frac{E_s}{\eta_0}$ as illustrated in fig. 2(b). At high $\frac{E_s}{\eta_0}$, we trade BER with a higher throughput. To compare with the performance of the fixed rate TCM, we have to consider the relative throughput gain of the ATCQAM at the same $\frac{E_s}{\eta_0}$ and BER. The BER against the relative throughput gains of the ATCQAM w.r.t. the fixed rate 8PSK-TCM and 16QAM-TCM are plotted in fig. 3(a). For example, at $\bar{P}_b = 10^{-4}$, the throughput gains relative to the 8PSK-TCM and the 16QAM are 1.95 and 1.54 times respectively.

On the other hand, when operating under the *constant throughput* control, the throughput of the ATCQAM is maintained at 2 and 3 for comparison with fixed rate 8PSK-TCM and 16QAM-TCM. The $\frac{E_s}{\eta_0}$ gain of the ATCQAM relative to 8PSK-TCM and 16QAM-TCM is plotted in fig. 3(b). For example, there are 7.1dB and 9.3dB gains in $\frac{E_s}{\eta_0}$ at $\bar{P}_b = 10^{-4}$ relative to 8PSK-TCM and 16QAM-TCM respectively.

---

[2] Optimal in the sense of the maximal diversity order that could be achieved. For example, the fixed rate 8PSK-TCM has a diversity order of 3 which is the best out of constraint length 5 convolutional code.

[3] This corresponds to a mobile speed of $24km/hr$ and baud rate of 40kbaud.

174

Effects of Finite Feedback Delay: The $\bar{P}_b$ of the ATCQAM under finite feedback delay at $f_d T_s = 1 \times 10^{-3}$ and $f_d T_s = 6 \times 10^{-3}$ are shown in in fig. 4(a) and (b) respectively with $\frac{E_s}{\eta_0} = 60$ and $\bar{\eta} = 3$. The ATCQAM schemes (FDMA and TDMA), are robust to the feedback delay at slow fading (see fig. 4(a)). For example, at $\Delta = 40$, $\bar{P}_b$ of the ATCQAM schemes are approximately 30 times smaller than the BER of the fixed rate 16QAM. At higher fading rate (see fig. 4(b)), the ATCQAM-TDMA become more sensitive to the feedback delay. At $\Delta = 40$, $\bar{P}_b$ for the ATCQAM-TDMA is the same as the fixed rate BER.

Effects of Block Interleaving: Fig. 5(a) and (b) shows $\bar{P}_b$ and $\bar{\eta}$ against $\frac{E_s}{\eta_0}$ for the ATCQAM-TDMA scheme and the fixed rate 8PSK-TCM at various interleaving depths. $\bar{P}_b$ increases gradually as the interleaving depth decreases. However, at all interleaving depths, although the absolute $\bar{P}_b$ degrades, the ATCQAM-TDMA scheme always out-performs the corresponding fixed rate codes. For example, at $\bar{P}_b = 2^{-3}$ with 100x100 interleaving, the gain in $\frac{E_s}{\eta_0}$ is 8.7dB w.r.t. 8PSK-TCM. At $\bar{P}_b = 4 \times 10^{-3}$ with 35x35 interleaving, the gain in $\frac{E_s}{\eta_0}$ is 10dB. Hence, the proposed ATCQAM also shows significant gains in systems with small interleaving depth.

Effects of Mobile Speed: Fig. 6(a) shows $\bar{P}_b$ against $\frac{E_s}{\eta_0}$ for the ATCQAM-TDMA scheme and the fixed rate 16QAM-TCM at $f_d T_s = 6 \times 10^{-3}$ It is apparent that an *irreducible* error floor at $\bar{P}_b = 3 \times 10^{-4}$ appears (see section 3). Note that above the error floor, ATCQAM always out-performs the fixed rate codes.

The reason of the presence of the error floor is because of aliasing. At this high fading rate, the Nyquist sampling criteria is exceeded and aliasing causes severe degradation in the interpolation process at the receiver. This is particularly serious for high density signal constellations like 64QAM. Hence, σ_0, is non-zero even in the absence of channel noise and this causes the *irreducible* error floor. This can be avoided by reducing the pilot period at the expense of increased pilot symbol overheads.

4.2 Performance of ATCQAM in FDMA Frame

The ATCQAM-FDMA behaves in a similar way as the ATCQAM-TDMA. However, they do have some differences in terms of performance.

Feedback Delays and BER Performance: Under most circumstances, BER for the ATCQAM-TDMA is always lower than the FDMA counterpart. Their difference exagerates at high fading rate (see fig. 4(a) and (b)). At $f_d T_s = 6 \times 10^{-3}$, BER of the ATCQAM-FDMA is even higher than that of the fixed rate 16QAM at all Δ. Hence, the ATCQAM-FDMA is totally not effective at $f_d T_s = 6 \times 10^{-3}$.

This is due to the difference in their prediction errors. The prediction filter for the TDMA scheme only has to predict the channel states for part of the physical frame (assigned timeslot) at the vicinity of the pilot symbol. This is more accurate compared with the FDMA system where the channel states across the whole physical frame have to be predicted.

Irreducible Error Floor: At high fading rate, *irreducible* error floor appears in the BER curves. The error floor depends on the fading rate. However, at the same fading rate, we found that ATCQAM-TDMA always has a lower error floor than ATCQAM-FDMA. This is illustrated in fig. 6(b). At $f_d T_s = 6 \times 10^{-3}$, the error floors of the ATCQAM-TDMA and the ATCQAM-FDMA are 3×10^{-4} and 2×10^{-3} respectively. This means that ATCQAM-TDMA is more robust to high fading rate as well.

5 Conclusions

Seven modes ATCQAM scheme is proposed to exploit the time varying nature of the mobile radio channel. By means of an instantaneous linear prediction filter, current channel states are predicted at the transmitter and appropriate modulation modes are used for the current symbol. Two ways of operation of the ATCQAM, namely the *constant BER* operation and the *constant throughput* operation, are introduced.

Under normal operating conditions, the ATCQAM has relative throughput gains around $1.5 - 1.9$ times and relative $\frac{E_s}{\eta_0}$ gains around $7 - 9$dB w.r.t. the fixed rate codes at $\bar{P}_b = 10^{-4}$. The effects of finite feedback link delay, finite depth interleaver and mobile speed are considered. It is found that ATCQAM-TDMA is in general more robust than ATCQAM-FDMA. At $f_d T_s = 6 \times 10^{-3}$, irreducible error floor occurs but the performance of the ATCQAM-TDMA is degraded less than the ATCMQAM-FDMA.

References

1. J. G. Proakis, *Digital Communications*. McGraw Hill International Editions, NY, third ed., 1995.
2. J. K. Cavers, "An Analysis of Pilot Symbol Assisted Modulation for Rayleigh Fading Channels," *IEEE Trans. on Vehicular Tech.*, vol. 40, pp. 686–693, Nov. 1991.
3. S. Sampei and T. Sunaga, "Rayleigh Fading Compensation for QAM in Land Mobile Radio Communications," *IEEE Trans. on Vehicular Tech.*, vol. 42, pp. 137–147, May. 1993.
4. A. J. Viterbi, J. K. Wolf, E. Zehavi, and R. Padovani, "A pragmatic approach to trellis-coded modulation," *IEEE Communs. Magazine*, pp. 11–19, July 1989.
5. K. N. Lau and M. D. Malcolm, "Variable Rate Adaptive Trellis Coded QAM for High Bandwidth Efficiency Applications under Rayleigh Fading Channel," *Submitted to IEEE Trans. on Communs.*
6. C. W. Therrien, *Discrete Random Signals and Statistical Signal Processing*. Prentice Hall, NJ, first ed., 1992.

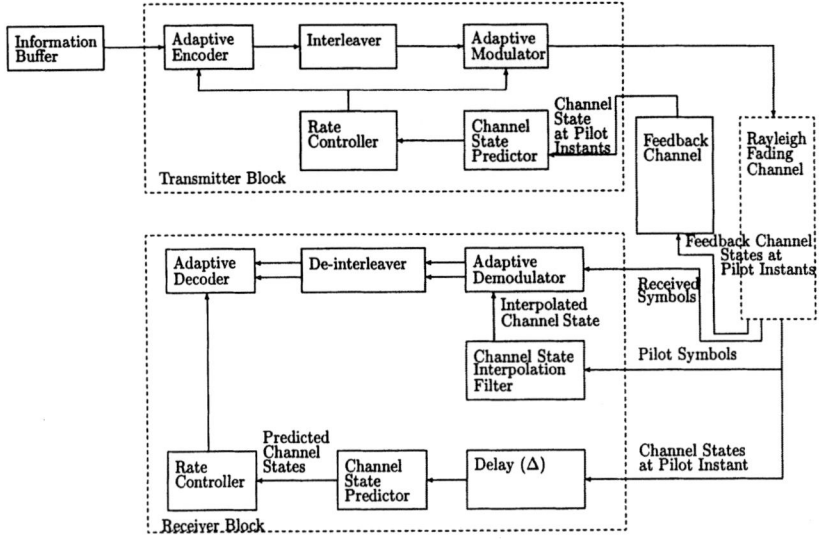

(a) Simplified Block Diagram of the ATCQAM Scheme.

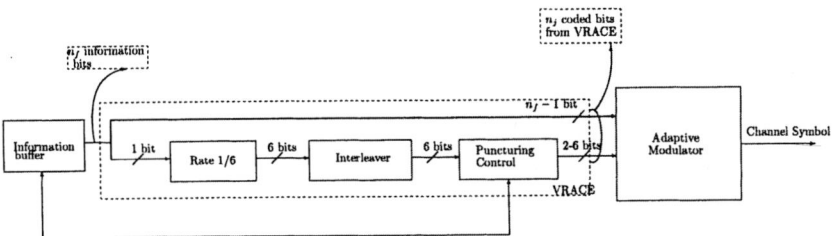

(b) Block Diagram for Variable Rate Interleaver.

Fig. 1. Overall Block Diagrams of the ATCQAM Scheme.

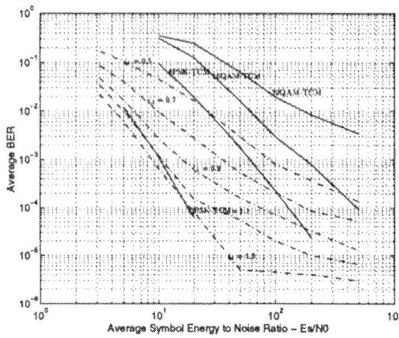 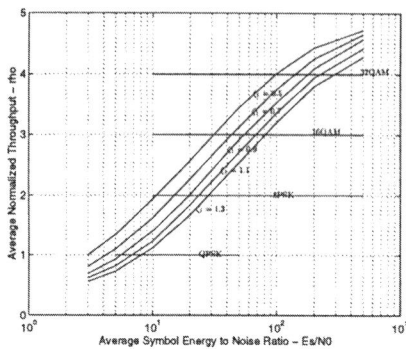

(a) Average BER ($\bar{P}_b$) against $\frac{E_s}{\eta_0}$ for the ATCQAM at various ζ_1 under the *constant BER* operation. The BER curves for the fixed rate codes are shown as solid line. Note that along the BER curves of the ATCQAM (dotted lines), the average throughputs are varying.

(b) Average Normalized Throughputs against $\frac{E_s}{\eta_0}$ for the ATCQAM (dotted lines) at various ζ_1. Note that the throughputs of the QPSK-TCM, 8PSK-TCM, 16QAM-TCM and 32QAM-TCM are 1,2,3 and 4 respectively.

Fig. 2. Performance of ATCQAM-TDMA relative to 8PSK-TCM and 16QAM-TCM. Dotted lines represent ATCQAM. Solid lines represent fixed rate codes.

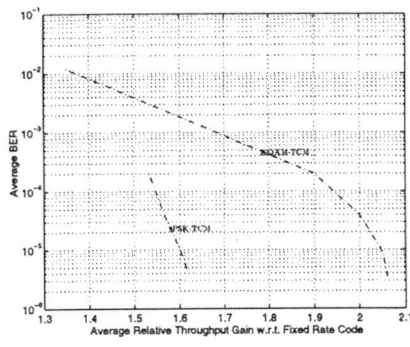

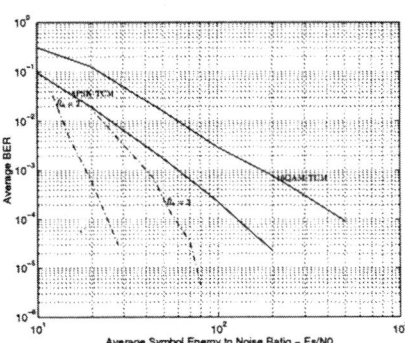

(a) Relative throughput gains of the ACTQAM w.r.t. 8PSK-TCM and 16QAM-TCM.

(b) Relative $\frac{E_s}{\eta_0}$ gains of the ATCQAM w.r.t. 8PSK-TCM and 16QAM-TCM. The ATCQAM is operating in the *constant throughput control*. $\bar{R}_b$ is the average normalized throughput of the ATCQAM.

Fig. 3. Normalized performance of ATCQAM-TDMA relative to 8PSK-TCM and 16QAM-TCM. Dotted lines represent ATCQAM. Solid lines represent fixed rate codes.

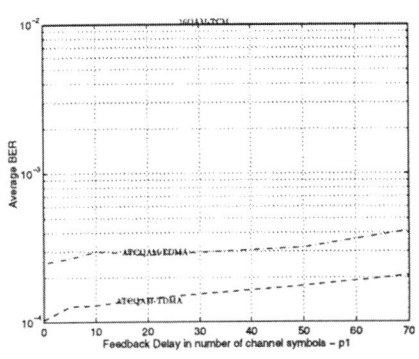

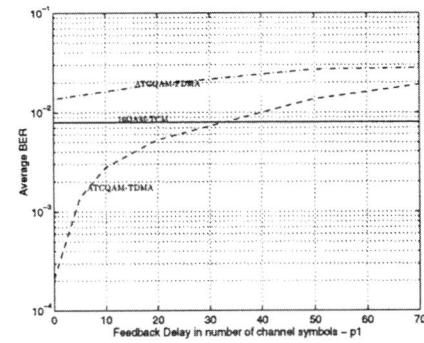

(a) Effects of feedback delay, Δ, on the BER of the ATCQAM-FDMA and ATCQAM-TDMA at $f_d T_s = 10^{-3}$.

(b) Effects of feedback delay, Δ, on the BER of the ATCQAM-FDMA and ATCQAM-TDMA at $f_d T_s = 6 \times 10^{-3}$.

Fig. 4. Effects of Feedback Delay on the performance of the ATCQAM. $\frac{E_s}{\eta_0}$ has excluded the power expenditure due to pilot symbol transmission.

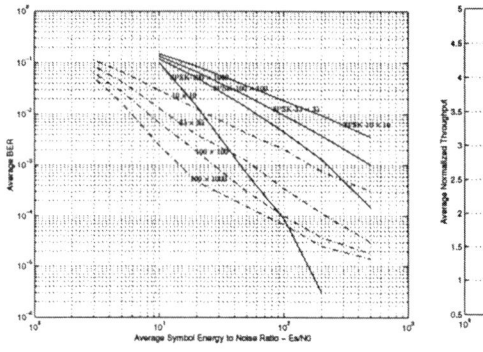 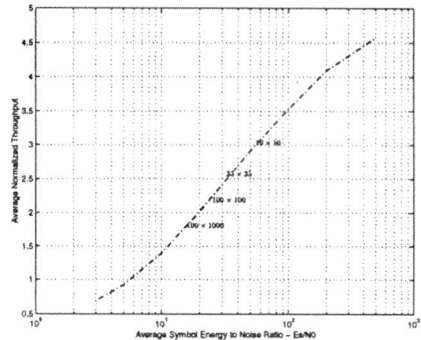

(a) $\bar{P}_b$ against $\frac{E_s}{\eta_0}$ at various inter-leaving depths for the ATCQAM-TDMA are shown in dotted line. The solid line shows the BER curve for 8PSK-TCM fixed rate code.

(b) $\bar{\eta}$ against $\frac{E_s}{\eta_0}$ at various interleaving depths for ATCQAM-TDMA.

Fig. 5. Effects of the Interleaving Depths on the performance of the ATCQAM. $\frac{E_s}{\eta_0}$ has excluded the power expenditure due to pilot symbol transmission.

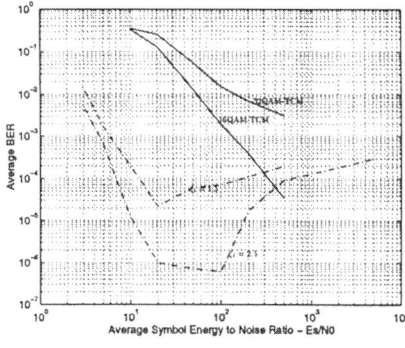 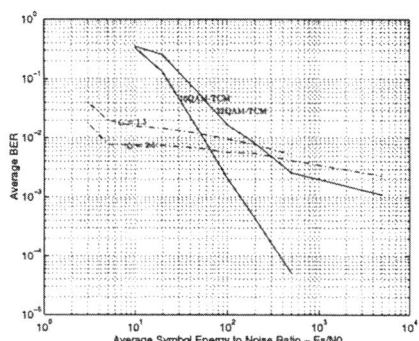

(a) BER of the ATCQAM-TDMA at $f_d T_s = 6 \times 10^{-3}$.

(b) BER of the ATCQAM-FDMA at $f_d T_s = 6 \times 10^{-3}$.

Fig. 6. Effects of mobile speed and the fading rate on the BER performance of the ATCQAM-FDMA and ATCQAM-TDMA schemes. $\frac{E_s}{\eta_0}$ has excluded the power expenditure due to pilot symbol transmission.

Variable Rate Adaptive Channel Coding for Coherent and Non-coherent Rayleigh Fading Channel

Vincent K.N. Lau and S. V. Maric
email: knl22@eng.cam.ac.uk, svm@eng.cam.ac.uk

Signal Processing and Communication Group
Department of Engineering
University of Cambridge
CB2 1PZ
UK

Abstract. We have evaluated the information theoretical performance of *variable rate* adaptive channel coding for the Rayleigh fading channel. The channel states are detected at the receiver and fed back to the transmitter by means of a noiseless feedback link. Based on the channel state informations, the transmitter can adjust the channel coding scheme accordingly. The *channel capacity* and the *error exponent* of are evaluated and the optimal control rules are found for coherent and non-coherent Rayleigh fading channel with feedback of channel states. It is shown that the variable rate scheme can only increase the channel error exponent. The effects of peak time constraint and the finite feedback delays are also considered. Finally, we compare the performance of the variable rate adaptive channel coding in high bandwidth-expansion systems (CDMA) and high bandwidth-efficiency systems (TDMA).

1 Introduction

Error correction codes have been widely used to combat the effect of Rayleigh fading in mobile radio channels. In traditional FEC schemes [8, 3], fixed rate codes were used which failed to explore the time varying nature of the channel. To keep the performance at a desirable level, they were designed for the average or worst case situation. In this paper, we propose and study the general scheme of variable rate adaptive channel coding which varies the code rate according to the channel condition.

In this paper, we evaluated the channel capacity and the error exponent of variable rate Rayleigh fading channel in coherent and non-coherent detection. In section 2, we developed two discrete-time equivalent channel models for coherent and non-coherent detections. In section 3, the expressions for the channel capacity and the error exponent with coherent detection are formulated, taking into account of the effects of feedback delay. In section 4, similar results for non-coherent detection are derived. Numerical results for Rayleigh fading channel are summarized in section 5. Finally, in section 6, we conclude with a summary of results.

2 Variable Dimension Adaptive Channel Coding

In this section, we shall consider two hypothetical channel models, namely the *M1* and *M2* models, and derive the general expressions for the channel capacity and the error exponent. These models will be applied in section 3 and 4 to calculate the channel capacity and the error exponent of the variable rate channel coding scheme for both the coherent and noncoherent detection.

2.1 Equivalent Channel Models

M1 **Channel Model:** The physical channel is a bandlimited (W) continuous time channel in which the channel input can be modelled by a bandlimited random process. We segment the random process into a number of channel symbols with the *i-th* channel symbol, $X_i(t)$, having a duration T_i. It is shown that a continuous time signal which is *approximately* time limited to T and band limited to W can be represented by a $2WT$ dimension vector in the signal space spanned by the Prolate spheroidal wave functions [4, 5, 6]. Hence, the *i-th* channel symbol can be represented by a $2WT_i$ dimension vector, $\mathbf{X}_i$.

There is a channel state A_i associated with the *i-th* channel symbol. Assume that T_i and W are both much smaller than the coherence time and the coherence bandwidth[1], $A(t)$ can be considered as a constant in every dimension of the signal space. Hence, the continuous time model is reduced to the *M1* discrete time model. The channel state is available at the receiver and known to the transmitter (at a delay) through the feedback channel. For each A_i, there is a corresponding prediction on it denoted by $\hat{A}_i$. The channel states A_i (and hence $\hat{A}_i$) are corelated but through ideal interleaving, they becomes i.i.d. and the channel becomes a memoryless channel.

The *i-th* channel symbol, $\mathbf{X}_i$, is a $2WT_i$ dimension vector where T_i is a function of the predicted channel state $\hat{A}_i$ for the current symbol. The output of the channel $\mathbf{Y}_i$ is given by:

$$\mathbf{Y}_i = A_i \mathbf{X}_i + \mathbf{Z}_i \qquad (1)$$

where $\mathbf{Z}_i$ is the uncorrelated Gaussian noise. We assume $\mathcal{E}(A_i^2) = 1$ where $\mathcal{E}(\cdot)$ denotes the expectation. The channel can be described by the channel transition density, $p_{c1}(\mathbf{y}|\mathbf{x}a)$. Note that each symbol in a block of N channel symbols have different dimension. The block diagram of the system with coherent detection is shown in fig. 1.

M2 **Channel Model:** The transmitted physical channel signal[2], $\tilde{s}(t)$, takes one out of M orthonormal signals in the set $\mathcal{S} = \{\tilde{\phi}_0(t), ..., \tilde{\phi}_{M-1}(t)\}$ depending on the input *index* to the orthogonal modulator. Each of the signals in the set $\mathcal{S}$ has a duration $T(\hat{a})$ (varying as a function of $\hat{a}$). Due to M-ary

---

[1] We have a flat fading channel.

[2] We express the physical channel signals in complex envelop.

orthogonal modulation, the channel output, $\tilde{y}(t)$, can be expressed as a M-dimensional vector [3] given by: $\mathbf{Y} = [Y_0, .., Y_{M-1}]$ and $Y_k = |\int_0^{T(\hat{a})} \tilde{y}(t)\tilde{\phi}_k^*(t)dt|^2$. We have a discrete time, discrete-input and continuous-output channel with feedback. The i-th channel symbol, X_i, is a discrete random variable which takes the values in $[0, M_i - 1]$ (*cardinality* of M_i) where M_i is a function of the predicted channel state $\hat{A}_i$ for the current symbol. The output of the i-th channel symbol, $\mathbf{Y}_i$, is a M_i dimensional vector. The channel can be described by a channel transition probability, $p_{c2}(\mathbf{y}|xa)$. To keep the physical bandwidth constant, we have to match $M(\hat{a})$ with $T(\hat{a})$ as follows.

$$\frac{M(\hat{a})}{T(\hat{a})} = \alpha_0 \tag{2}$$

where α_0 is a constant proportional to the bandwidth W. Hence, the effective code rate and the cardinality of channel input is varied symbol by symbol while keeping the bandwidth constant. The system block diagram for the noncoherent detection variable rate channel coding is shown in fig. 2.

2.2 Induced State Distribution

Since the sequence of symbol duration $[T_1(\hat{a}_1), ..., T_N(\hat{a}_N)]$ is varying according to the predicted channel state $\hat{\mathbf{A}}$, it induces a probability density on $\hat{\mathbf{A}}$ which is different from the fading p.d.f. in general. It is shown [2] that the induced probability density, denoted by $\mathcal{P}(\hat{a})$, on $\hat{A}$ is given by:

$$\mathcal{P}(\hat{a}) \overset{def}{=} \lim_{n \to \infty} \frac{n_{\hat{a}}}{n\delta_{\hat{a}}} = \frac{f(\hat{a})}{E_f T(\hat{a})} \tag{3}$$

and

$$\mathcal{P}_N(\hat{a}_1, \hat{a}_2, .., \hat{a}_N) = \frac{1}{E_f^N} \frac{f(\hat{a}_1)f(\hat{a}_2)\cdots f(\hat{a}_N)}{T(\hat{a}_1)T(\hat{a}_2)\cdots T(\hat{a}_N)} \tag{4}$$

where $n_{\hat{a}}$ is the number symbols with $\hat{A} \in [\hat{a}, \hat{a}+\delta_{\hat{a}}]$ in the sequence of n symbols, $f(.)$ is the fading density and $E_f = \int_{\hat{a}} \frac{1}{T(\hat{a})} f(\hat{a}) d\hat{a}$. Furthermore, E_f is shown to be the average symbol rate. Since given $\hat{A}$, the symbol duration, T_i, is constant. The conditional density $\mathcal{P}(a|\hat{a})$ is thus equal to the conditional fading density and is given by [1]:

$$\mathcal{P}(a|\hat{a}) = f(a|\hat{a}) = \frac{2a}{(1-\lambda^2)} e^{-[\frac{a^2+\lambda^2\hat{a}^2}{(1-\lambda^2)}]} I_0 \left(2a\hat{a}\frac{\lambda}{1-\lambda^2} \right) \tag{5}$$

where $\lambda^2 = J_0(\omega_m \Delta)$, ω_m is the Doppler spread, J_0 is the zeroth order Bessel function and I_0 is the zeroth order modified Bessel function.

3 Coherent Detection

3.1 Feedback Channel Capacity with Coherent Detection

Given A and $\hat{A}$, the channel is memoryless and AWGN and by symmetry, the capacity achieving distribution would be a zero mean Gaussian density with variance $\sigma_X^2 = \frac{P_0}{2W}$ so that $\mathbf{X}_i$ has $2WT_i$ i.i.d. components, $[X_{i,1}, .., X_{i,2WT_i}]$. The feedback channel capacity is given by:

$$
\begin{aligned}
C_1^{(fb)} &= I(X; Y|A, \hat{A}) \\
&= WE_f \int_a \int_{\hat{a}} T(\hat{a}) \log_2(1 + \frac{a^2 P_0}{W \eta_0}) f(a|\hat{a}) [\frac{1}{E_f} \frac{f(\hat{a})}{T(\hat{a})}] d\hat{a}\, da \\
&= W \int_a f(a) \log_2(1 + \frac{a^2 P_0}{W \eta_0}) \int_{\hat{a}} f(\hat{a}|a) d\hat{a}\, da \\
&= W \int_a f(a) \log_2(1 + \frac{a^2 P_0}{W \eta_0}) da \\
&= C^{(nfb)}
\end{aligned}
\tag{6}
$$

where $C^{(nfb)}$ is the channel capacity without feedback. Hence, the coherent variable rate scheme cannot increase the channel capacity.

3.2 Error Exponent with Coherent Detection

It is shown that the average error probability, $\bar{P}_{e1}$, is bounded by:

$$
\bar{P}_{e1} \le 2^{-\frac{N}{E_f}[\bar{E}_1^{(fb)}(\rho, Q_1) - \rho \bar{R}_b]}
\tag{7}
$$

for all $\rho \in [0, 1]$ and $Q_1(x)$ where $\bar{E}_1^{(fb)}(\rho, Q_1)$ is the *average error exponent* given by:

$$
\begin{aligned}
\bar{E}_1^{(fb)}(\rho, Q_1) &= E_f \bar{E}_1(\rho, Q_1) \\
&= -E_f \log_2 \left\{ \frac{1}{E_f} \int_a \int_{\hat{a}} \frac{f(a|\hat{a}) f(\hat{a})}{T(\hat{a})} \right. \\
&\quad \left. \left[\int_y [\int_x Q_1(x) p_{c1}(y|xa)^{\frac{1}{\rho+1}} dx]^{\rho+1} dy \right]^{2WT(\hat{a})} da d\hat{a} \right\}
\end{aligned}
\tag{8}
$$

By calculus of Variations, the optimal symbol duration control that would minimize the error probability in (7) is given by:

$$
T(\hat{a}, \bar{R}_b) = \begin{cases} \frac{K}{G(\hat{a}, \rho_0)} & \hat{a}_1 \ge \hat{a} \ge \hat{a}_0 \\ T_p & \hat{a} \in [0, \hat{a}_0] \\ T_l & \hat{a} \ge \hat{a}_1] \end{cases}
\tag{9}
$$

where $G(\hat{a}, \rho_0)$ is given by:

$$
G(\hat{a}, \rho_0) = \log_2 \left[\int_y \left[\int_x Q_1(x) p_{c1}(y|x\hat{a})^{\frac{1}{\rho_0+1}} dx \right]^{\rho_0+1} dy \right],
$$

T_p is given by:

$$T_p = \frac{K}{G(\hat{a}_0, \rho_0)}$$

T_l is given by:

$$T_l = \frac{K}{G(\hat{a}_1, \rho_0)}$$

and K is given by:

$$K = \frac{1}{E_f}\left\{ \int_{\hat{a}_0}^{\hat{a}_1} G(\hat{a}, \rho_0) f(\hat{a}) d\hat{a} \right.$$

$$+ G(\hat{a}_0, \rho_0) \int_0^{\hat{a}_0} f(\hat{a}) d\hat{a}$$

$$\left. + G(\hat{a}_1, \rho_0) \int_{\hat{a}_1}^{\infty} f(\hat{a}) d\hat{a} \right\} \tag{10}$$

4 Non-coherent Channel

4.1 Noncoherent Feedback Channel Capacity

The channel capacity, C_2, in bits per symbol is given by[3]:

$$C_2 = \frac{E_{f,0}}{E_f} \int_a \int_{\hat{a}} \frac{f(a,\hat{a}) M_0}{M(\hat{a})} \int_{\mathbf{y}} \sum_{m=0}^{M(\hat{a})-1} p_{c2}(\mathbf{y}|x = m, a)$$

$$\log_2 \left[\frac{p_{c2}(\mathbf{y}|x = m, a)}{\sum_{m'=0}^{M(\hat{a})-1} p_{c2}(\mathbf{y}|x = m', a)/M(\hat{a})} \right] d\mathbf{y}\,da\,d\hat{a} \tag{11}$$

As a simple illustration, we shall consider a 3-mode variable rate scheme with $M = 4, 8$ or $M = 16$. The optimal control law for the 3-mode scheme is given by:

$$M(\hat{a}) = \begin{cases} 16 & \hat{a} \in [0, \hat{a}_0) \\ 8 & \hat{a} \in [\hat{a}_0, \hat{a}_1) \\ 4 & \hat{a} \in [\hat{a}_1, \infty) \end{cases} \tag{12}$$

where

$$\hat{a}_0 = \sqrt{\left(\frac{\gamma_0}{\bar{\gamma}}\right)},$$

$$\hat{a}_1 = \sqrt{\left(\frac{\gamma_1}{\bar{\gamma}}\right)},$$

and γ_i is the SNR threshold. The corresponding channel capacity is given by:

$$C_2(\Delta = 0) = \frac{E_{f,0}}{E_f} \left[\int_0^{\hat{a}_0} f(\hat{a}) C(16, \hat{a})/2 d\hat{a} \right.$$

$$\left. + \int_{\hat{a}_0}^{\hat{a}_1} f(\hat{a}) C(8, \hat{a}) d\hat{a} + \int_{\hat{a}_1}^{\infty} 2f(\hat{a}) C(4, \hat{a}) d\hat{a} \right] \tag{13}$$

---

[3] By symmetry [7], the capacity achieving distribution is given by $Q_2^*(X|\hat{A}) = \frac{1}{M(\hat{A})}$.

4.2 Evaluation of Variable Rate Error Exponent

Similar to section 3, the average error exponent is given by:

$$\bar{E}_2(\rho, Q_2^*) = -E_f \log_2 \left\{ \frac{E_{f,0}}{E_f} \int_{\hat{a}} \int_a \frac{f(a|\hat{a}) f(\hat{a}) M_0}{M(\hat{a})} \right.$$

$$\int_{\mathbf{y}} \frac{2^{-M} e^{-1/2(\sum_j y_j + a^2 \beta^2)}}{M(\hat{a})^{\rho+1}}$$

$$\left. \left[\sum_{x=0}^{M(\hat{a})-1} I_0^{\frac{1}{\rho+1}} \left(\sqrt{a^2 \beta^2 y_x} \right) \right]^{\rho+1} dy \, da \, d\hat{a} \right\} \tag{14}$$

To simplify the analysis, we shall use the control rule (12) to evaluate the error exponent of the 3-mode variable rate scheme.

5 Results and Discussions

In a microcellular environment at 2GHz with mobiles moving at a maximum speed of 75km/hr, the coherence time is around 1ms and the coherence bandwidth is around 2MHz. We choose the symbol rate, E_f, to be 40ksym/sec and the system bandwidth to be 800kHz. These justified the flat fading assumption made in the channel model. For fix-duration system, the symbol duration, $T(\hat{c})$, is constant and is equal to $1/E_f$. Hence, $T_s = 1/E_f$ is taken to be the reference symbol duration. The channel fading rate is $f_d T_s = 2.5 \times 10^{-3}$.

5.1 Coherent Variable Rate Channel Coding

Coherent Variable Rate Channel Capacity: It is shown in section 3.1 that channel capacity is not increased by the variable rate control. The channel capacity in the example is equal to 616kb/s at reference SNR, $E_b/\eta_0 = 3dB$.

Coherent Variable Rate Error Exponent: Although channel capacity is not increased by variable rate coding, the error exponent is increased significantly compared with fix-duration error exponent. From fig. 3 (a), the improvement in error exponent is 3 times compared with fix-duration case at $\mathcal{R}_b = 0.5\mathcal{C} = 308kb/s$ under ideal situations ($\Delta = 0$, $T_p = \infty$, $T_l = 5\mu s$). The performance is degraded to 2.1 times and 1.13 times if the feedback delay is 18 symbols and 25 symbols respectively.

The effect of peak time constraint is shown in fig. 3 (b). Define the ratio of peak to average symbol duration, ξ, as $\xi = \frac{T_p}{\mathcal{E}[T]} = E_f T_p$. With $\xi = 2.98$ and $\xi = 1.21$, the improvements in the error exponent are 2.46 and 1.40 times respectively. The bandwidth expansion used in the above calculation is $W/\bar{R}_b = 1.3$.

Bandwidth Expansion Consideration: We consider two extreme cases, a bandwidth expansion of 0.25 which models TDMA systems and a bandwidth expansion of 20 which models CDMA systems. The error exponents for small and large bandwidth expansion systems are shown in fig. 4(a) and (b). For the system with small bandwidth expansion (TDMA), there is a significant 5.62 times increase in error exponent at $\bar{\mathcal{R}}_b = 0.8\mathcal{C}$. For the system with large bandwidth expansion (CDMA), there is only 1.5 times improvement in error exponent relative to fix-duration scheme at the same $\bar{\mathcal{R}}_b$.

5.2 Noncoherent Variable Rate Channel Coding

Because of the use of M-ary orthogonal modulation, the bandwidth expansion considered is large. Performances are relative to a reference fix-rate code with $M_0 = 8$.

Noncoherent Variable Rate Channel Capacity: Unlike the coherent channel with average power constraint, the noncoherent M-ary orthogonal variable rate coding increases the channel capacity relative to the non-feedback case. With zero feedback delay, there is 22% increase in the channel capacity for the variable rate system. at $\bar{\gamma} = 7.5$. The results of the capacity enhancements against $\bar{\gamma}$ for the variable rate scheme at various feedback delays are shown in fig. 5. The capacity gain is reduced to 10% under the feedback delay of 18 reference symbols at the same $\bar{\gamma}$.

Noncoherent Variable Rate Error Exponent: The variable rate scheme can increase the error exponent in the ideal conditions. For example, the error exponent for the variable rate scheme with feedback is 2.1 times larger than that without feedback at $\bar{R}_b = 0.7C_{nfb}$. The performance with feedback delay is shown in fig. 6 at $\bar{\gamma} = 7.5$. The improvement increases as $\bar{R}_b$ decreases. The improvement is degraded to 1.55 times at a feedback delay of 18 symbols.

6 Conclusion

In this paper, we have evaluated the channel capacities and the error exponents of variable rate Rayleigh fading channel with feedback of channel states in coherent and non-coherent detections. For the coherent case, the channel capacity with feedback is the same as the non-feedback channel capacity. However, there is a significant increase in the error exponent. This means that less complex codes can be found to achieve the same $\bar{P}_e$ using the variable rate adaptive coding. On the other hand, both the channel capacity and the error exponent are increased by the variable rate scheme for the noncoherent case. This is because we have fixed the modulation format in the latter case and the variable rate scheme can make use of more effective modulation level M according to the current channel state.

For the dependence of the improvement on the bandwidth expansion, we found that when the bandwidth expansion is large (CDMA), the improvements in both the coherent and the noncoherent systems are limited. On the contrary, when the bandwidth expansion is small (TDMA), we have a much more significant gain. This suggests that the variable rate schemes have limited instrinsic gain in the CDMA system compared with the TDMA system.

References

1. Wm. C. Jakes. *Mobile Radio Systems*. McGraw Hill, NY, first edition, 1974.
2. K. N. Lau and S. V. Maric. Variable Rate Adaptive Channel Coding for Coherent and Noncoherent Rayleigh Fading Channel. *Submitted to IEEE Trans. on Communs.*
3. John G. Proakis. *Digital Communications*. McGraw Hill International Editions, NY, third edition, 1995.
4. D. Slepian and H.O.Pollak. Prolate Spheriodal Wave Functions, Fourier Analysis and Uncertainty-I. *The Bell Syst. Technical Journal*, 40(1):43–64, Jan. 1961.
5. D. Slepian and H.O.Pollak. Prolate Spheriodal Wave Functions, Fourier Analysis and Uncertainty-II. *The Bell Syst. Technical Journal*, 40(1):65–84, Jan. 1961.
6. D. Slepian and H.O.Pollak. Prolate Spheriodal Wave Functions, Fourier Analysis and Uncertainty-III. *The Bell Syst. Technical Journal*, 41(4):1295–1336, July 1962.
7. W. E. Stark. Capacity and Cutoff Rate of Noncoherent FSK with Nonselective Rician Fading. *IEEE Trans. on Communs.*, 33:1153–1159, Nov. 1985.
8. S. G. Wilson. *Digital Modulation and Coding*. Prentice Hall, NJ, first edition, 1996.

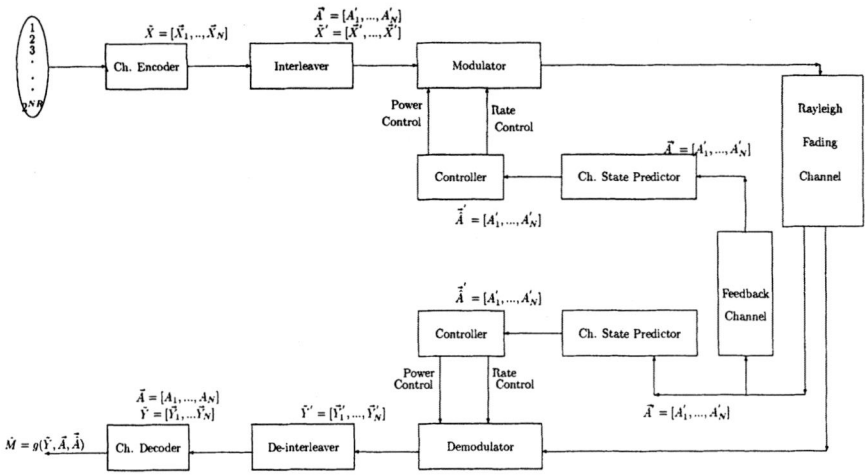

Fig. 1. Block Diagram of the Variable Rate Adaptive Channel Codec - Ideal Interleaved Coherent Channel

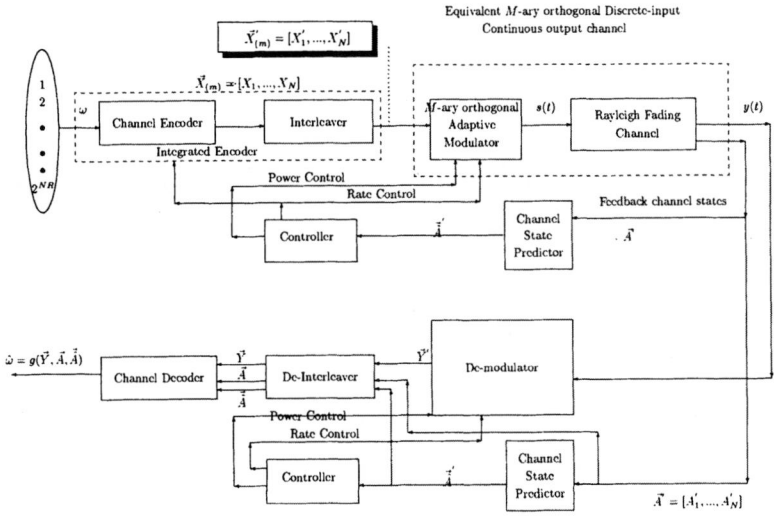

Fig. 2. Block Diagram of the Variable Rate Adaptive Channel Codec - Ideal Interleaved Noncoherent Channel using M-ary orthogonal modulation

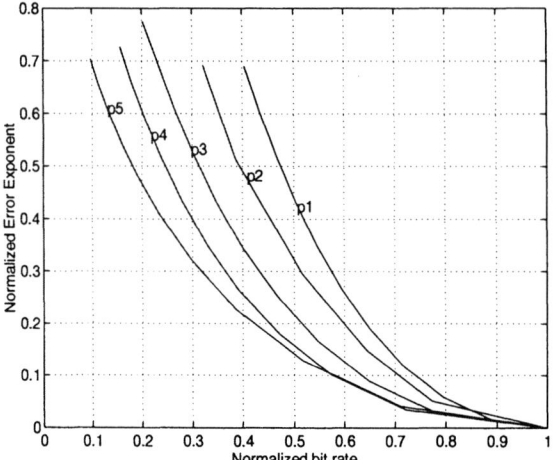

a) Error exponent at various feedback delays. Note : $p_1 = \Delta = 0$, $p_2 = \Delta = 18$, $p_3 = \Delta = 21$, $p_4 = \Delta = 25$ and $p_5 = No feedback$.

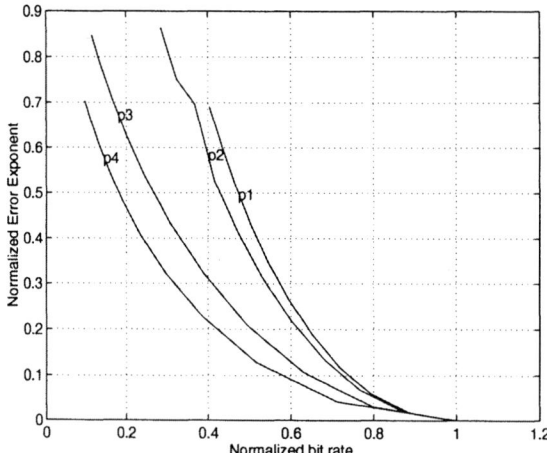

b) Error exponent at various peak time constraints. Note : $p_1 = \xi = \infty$, $p_2 = \xi = 2.98$, $p_3 = \xi = 1.21$, $p_4 = No feedback$.

Fig. 3. Error Exponents of Coherent Variable Rate Adaptive Channel Coding at Various Feedback Delays and Peak Time Constraints. $\xi \overset{def}{=} E_f T_p$. Bit rate is normalized w.r.t. the channel capacity without feedback, C_{nfb}. Error exponent is normalized against $\bar{E}_1^{nfb}(1, Q)$. Delays are expressed as number of symbols in the figure.

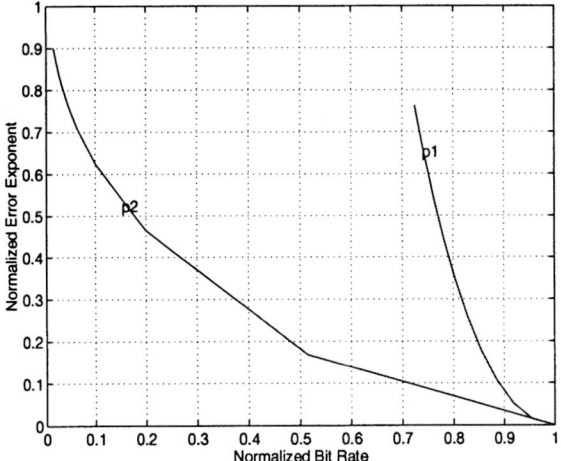

a) Small Bandwidth Expansion (TDMA) System. Note :
$p_1 = VariableRate$ and $p_2 = Nofeedback$.

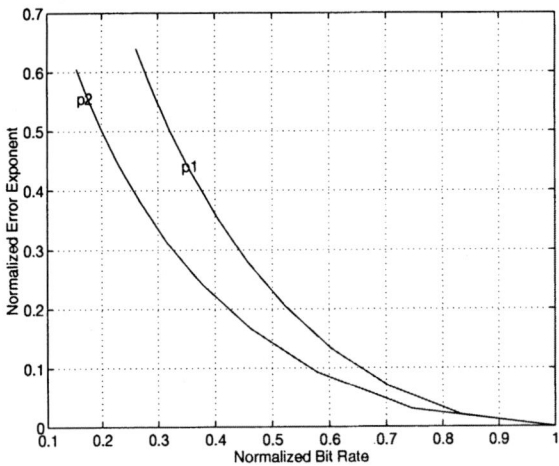

b) Large Bandwidth Expansion (CDMA) System. Note :
$p_1 = VariableRate$ and $p_2 = Nofeedback$.

Fig. 4. The Error Exponent against Bit Rate for Coherent Variable Rate Adaptive Coding at Small and Large Bandwidth Expansion. Bit rate is normalized against the non-feedback channel capacity, C_{nfb}. Error exponent is normalized against $\bar{E}_1^{nfb}(1, Q)$.

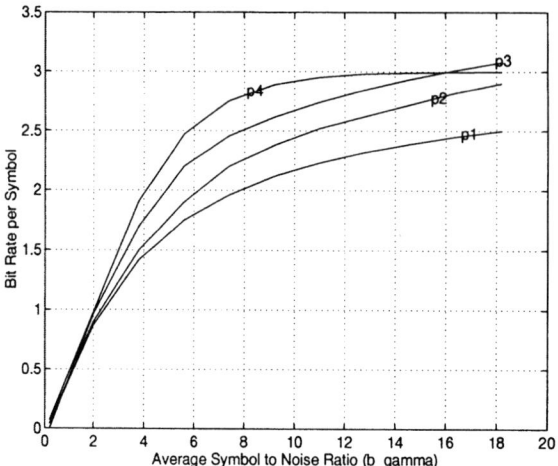

Fig. 5. Average Channel Capacity (bits/ref symbols) for Noncoherent Variable Rate Adaptive Coding Scheme against $\bar{\gamma}$ at Various Feedback Delays. Delays are expressed as the number of reference symbols. Both the channel capacity and the $\bar{\gamma}$ are relative to $M_0 = 8$. Note : $p1 = No feedback$, $p2 = \Delta = 18$, $p3 = \Delta = 0$ and $p4 = AWGN$.

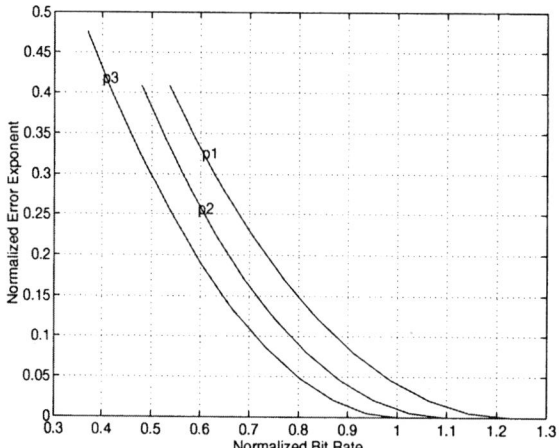

Fig. 6. Error Exponents of the Noncoherent Variable Rate Adaptive Coding Schemes at various Feedback Delays. Bit rate is normalized w.r.t. the channel capacity without feedback, C_{nfb}. Error exponent is normalized against $\bar{E}_2^{nfb}(1, Q^*)$. Delays are expressed as the number of reference symbols. $M_0 = 8$ and $\bar{\gamma} = 7.0$. Note : $p_1 = \Delta = 0$, $p_2 = \Delta = 18$ and $p_3 = No feedback$.

LABYRINTH: A New Ultra High Speed Stream Cipher

Bo Lin * and Simon Shepherd **

Abstract. We present a new high speed, high security, 32-bit oriented stream cipher. The algorithm was developed for use in very high speed systems, in particular, ATM fibre networks. The design is based on simple, proven components to minimise complexity and allow detailed analysis of the algorithmic security. A complete prototype implementation has been built to high grade equipment standards for use in ATM circuits at data rates up to 1.6 Gbps. We have called the new algorithm "Labyrinth".

1 Introduction

The advent of very high speed communications, in particular ATM, has meant that conventional cryptographic mechanisms cannot be used due to their inability to operate at very high data rates. For example, the cheaper commercial grade DES [1] chips available from Lintel nv/sa [2] have a maximum throughput of about 30 Mbps while the highest specification devices are limited to around 100 Mbps. Although end-to-end cryptographic protection can be achieved on a per-application basis, the onus placed on (generally non-technical) users is intollerable.

As a result, we introduce Labyrinth, an ultra high speed, high security stream cipher designed specifically for use in high speed networks for bulk traffic encryption. The prototype implementation is fully key-agile and can be installed either side of the ATM switch depending on requirements. One unit can handle the complete traffic of a medium to large scale enterprise at a level of security hitherto unavailable commercially.

Some of the basic ideas behind Labyrinth draw partial inspiration from existing algorithms, in particular, A5 [3], Rambutan [4] and SEAL [5]. Although the algorithm operates on "blocks" of text, it is in every formal respect a stream cipher. (A simple criterion for distinguishing between these fundamentally different mechanisms is that, given no key reload between symbols, a block algorithm will encipher two identical consecutive plaintext symbols to identical ciphertexts whereas a stream cipher will produce different ciphertexts under these conditions.)

---

* Cryptographic Technologies Group, Motorola Ltd, East Kilbride, G75 0TG, UK.
** C^3SG, Telecommunications Research Centre, Bradford University, BD7 1DP, UK.

Unlike conventional stream cipher designs where the output of a number of bitstream generators are combined in some (complex) way and then added bit by bit in GF(2) to the data stream, Labyrinth uses three medium length shift registers and a lookup table to produce a keystream in 32-bit blocks which are used to encrypt or decrypt the data. This arrangement is very efficient while providing a high degree of resistance against entropy leakage and correlation attack. The current version of the algorithm uses a 128-bit key, although a 256-bit version is available.

2   Detailed Design

2.1   General Description

Labyrinth consists of three main components. The first is a set of three linear feedback shift regsiters (LFSR), designated **A**, **B** and **C** with lengths 43, 32 and 32 respectively. The three shift registers possess different mathematical characteristics and are carefully designed to contribute various properties from orthogonal algebraic objects. Register **A** is a completely standard 43-stage LFSR with a primitive feedback polynomial to yield a sequence of maximum period. Register **B** is a shift register based on the design in [6]. Register **C** implements a hash function suggested by [7] whose period is variable and dependent on its initial value.

The use of these three registers ensures the two important properties necessary for high security. Firstly, the combination of orthogonal algebraic properties makes cryptanalysis based on statistical models extremely difficult. Secondly, the period is very high but variable and the linear complexity is provably equal to the period, which offers extremely high resistance to correlation and spectral attacks. The period and linear complexity is at least 2^{97} which at 1 Gbps represents a cycle period of about 5×10^{12} years.

The second main component of Labyrinth is the lookup table. We have unashamedly used the eight S-boxes from the DES [1] due their excellent design and suitable non-linear properties [8]. A total of 48 bits are tapped from the registers described above (16 bits from each) and used to address the lookup table to produce a 32-bit output.

The third component is a very high entropy, key controlled hash function which processes the output of the lookup table to introduce a high level of confusion.

The key is split into a number of subkeys and used to initialise the three shift registers and variables within the key controlled hash function. The complete system is shown in Figure 1.

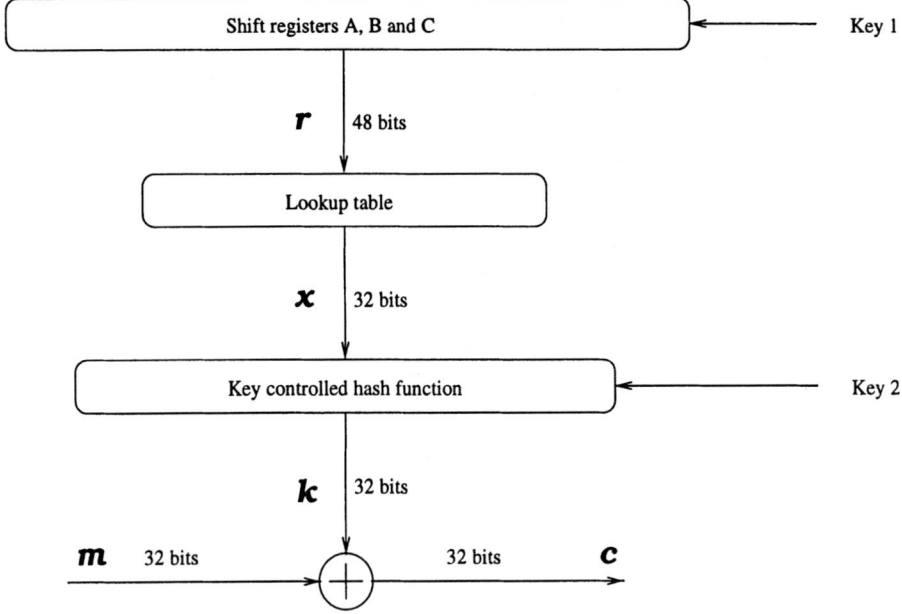

Fig. 1. The components of the Labyrinth algorithm

2.2 The Three Registers

The three shift registers are the main driving component of the Labyrinth engine. Their orthogonal algebraic properties make it very difficult to find any affine approximation to the sequence tapped from them.

Register **A** is a 43-bit LFSR with primitive feedback polynomial $x^{43} + x^6 + x^4 + 1$ and period $2^{43} - 1$. A 43 stage register was chosen specifically because of its very large *witness* which increases its resistance to the powerful spectral redundancy attack described in [9].

Register **B** is a 32-bit register based on the linear transforms $y_i = x_{i-1} \oplus (x_{i-1} << 18)$ and $x_i = y_i \oplus (y_i >> 13)$. It provides a sequence with period $2^{32} - 1$.

Register **C** is based on the design in [7]. It provides a sequence with variable period and its operation may be summarised as:

– $(Y, Z) = (x_{i-1}$ rotate right 1 bit$)$
– Z rotate right 1 bit
– $x_i = (Y, Y \oplus Z)$ and output Z

Each register contributes two tapped bits to each S-box. The taps to a given box are placed as far apart as possible to thwart correlation attack, although the internal bit stream at this point would not be available to an attacker. The taps are summarised in Table 1 where $i = 0 \ldots 7$. For every i, the six tapped bits are taken to the input of S-box S_i.

Table 1. Tap points on the shift registers

Register	First tap	Second tap
A	$2i + 1$	$2i + 17$
B	$2i$	$2i + 17$
C	$2i$	$2i + 1$

2.3 The Lookup Table

In order to combat correlation attacks, the use of non-linear components ideally with complex null-space is recommended [8] [10] [11] [12]. Careful analysis [8] has shown that it is difficult to do better then the eight excellent S-boxes designed for the DES [1] and so these were adopted in our design.

2.4 The Key-Controlled Hash Function

The key controlled hash function modifies the 32-bit output of the lookup table. It comprises of a bitwise XOR, a right register rotation controlled by part of the key and an addition with another part of the key. If we let (U, V) be the output of the lookup table, the hash function can be summarised as follows:

1. $Y = U \oplus V$
2. $(W, Z) = Y$ rotated left by t bits where $t = 1, 3, \ldots, 31$ specified by part of the key
3. Output $(A, B) = (W, Z) + (K, W)$ where K is the continous p bit left rotation of $K_{15}, K_{14}, \ldots, k_0$ and $p = 1, 3, \ldots, 15$.

The odd number rotation in the above is chosen due to its superior security properties [7].

2.5 Key Arrangement

The key used in Labyrinth is 128 bits designated $k_{127}, \ldots, k_0$. The 16 bits $k_{15}, \ldots, k_0$ are used for the initial value of K at step (3) in the above description of the hash function. The next 4 bits $k_{19}, \ldots, k_{16}$ are used to control the t bit left rotation at step (2) above. The following three bits $k_{22}, \ldots, k_{20}$ are used to control the left rotation of K at step (3). The other 96 bits are used to initialise the three shift registers, 32 bits in each. The remaining 11 bits for the 43-stage register are simply padded with a fixed number as this register contributes only 32-bits worth of entropy to the overall algorithm and thus its key need not exceed this size.

3 Implementation

Although the design can be implemeted easily in fast TTL, we chose to use four
Altera FPGA (*Field Programmable Gate Array*) devices for simplicity, reliability
and compactness. The three shift registers together with the control logic are in
one device, the key controlled hash function is in another device and the remaining two devices house the S-boxes, four in each. Although each S-box represents
only a 64 × 4 bit RAM, the FPGA compiler reduces lookup tables to combinatorial logic for speed. If a table is quite linear, then the logical reduction can be
very compact. However, since the S-boxes by their very purpose are designed to
be non-linear, little reduction is possible! (The ease or otherwise with which a
lookup table can be reduced in this way is a good measure of its effectiveness as
a non-linear function.)

For use in the ATM environment, the cipher engine has been provided with
64 Kbytes of ultra fast RAM to store keys and cipher state. As the ATM cells
arrive at the device for processing, pre-fetch and lookahead buffer logic ensure
that the correct key and cipher state are ready in advance for a single cycle load
operation.

The hardware has been built to commercial high–grade equipment standards.
The "red" and "black" data are physically and electrically isolated by encapsula-
tion in screened and sealed compartments designed to meet the EMC/TEMPEST
specifications of MIL-STD-462 for conducted and radiated emissions and susceptibility. All data paths external to the TEMPEST protection are optical. Separate power supplies with active filtering are used to supply the two halves of
the equipment. The operation of the crypto engine is monitored constantly by
an optically-isolated comparitor which compares the red and black data. In the
(extremely unlikely) event of a crypto engine or keystream generator failure, the
red data is isolated immediately and the output buffer flushed before data can
leave the device.

Our prototype shows that operation is possible with clock speeds up to 58
MHz. Since the device enciphers 32 bits of data on each clock cycle, this represents a peak data rate of 1.856 Gbps. However, due to overhead, key fetch and
loading and statistical queueing peculiarities in ATM systems, the maximum
usable throughput is about 1.2 Gbps. This is more than adequate for the OC-12
defined rate of 622.56 Mbps in the SONET Optical Signal hierarchy, see Table
2.

4 Security Analysis

We will not attempt to give a detailed security analysis in the limited space
available in this paper, but will only summarise the security features of the various design choices.

Table 2. The SONET Optical Signal Hierarchy

Level	Line rate (Mbps)
OC-1	51.84
OC-3	155.52
OC-9	466.56
OC-12	622.56
OC-18	933.12
OC-24	1244.16
OC-36	1866.24
OC-48	2488.32

Firstly, a high speed bitstream generator is required and the LFSR is the ideal candidate for this task due to its simplicity and ease of implementation in hardware. A simple LFSR, however, offers little or no security as it is a linear device and is easily analysed [13]. We have used a combination of three shift registers but which produce bitstreams based on different algebraic groups. The result is a generator that is as easy and fast to implement as a simple LFSR but with much stronger resistance to statistical and correlation analyses.

The use of a non-linear lookup table completes the disguise of the underlying generator by adding significant *confusion* to the bitstream. Confusion has the effect of making the plaintext/ciphertext transformation depend on the key in a highly complex way. The aim here is to make it as difficult as possible for an attacker to identify from a plaintext/ciphertext pair (or a number of such pairs) a subset of the key space where the key is located, in order to force him to search such a large portion of the keyspace as to be infeasible.

Finally, the use of a key-controlled hash function introduces *diffusion*, the effect of which is to mask in the ciphertext those statistical characteristics in the plaintext which, in the absence of diffusion, would render the cipher susceptible to statistical cryptanalysis.

Extensive tests using the proprietary Motorola Cryptographic Test Suite has shown that the algorithm exhibits the other desirable overall properties required of a good cipher. One of the most important is high *dispersion*, meaning that a small change in the key or input data produces a large change in the output data, for example, a 1-bit change in the plaintext will produce changes in many bits of the resulting ciphertext.

5 Conclusions

We have described a new, ultra high speed stream cipher algorithm for use in very high speed communications networks. The cipher can be implemented easily in standard logic components and can be incorporated into network switches, routers, bridges and other components without difficulty.

We have called the new algorithm "Labyrinth" (or *mi gong* in Mandarin – literally *mysterious palace*) in the sincere hope that its security will prove likewise. However, as is customary with the introduction of a new cryptographic primitive, we invite and welcome analyses, attacks and comments on our design.

References

1. NBS FIPS Publication 46, *Data Encryption Standard*, US National Bureau of Standards, US Dept. of Commerce, January 1977.
2. *"CRY12C102 DES CHIP"*, Data sheet, Lintel NV/SA, Belgium.
3. *"The French Proposal for the Cipher"*, Racal Internal Document No. 10-1617-01, 10 June 1988, (private communication).
4. *"Secure Rambutan Kernel"*, Marconi Electronic Devices Ltd, 1989, (private communication).
5. P. Rogaway and D. Coppersmith, *"A Software-Optimised Encryption Algorithm"*, Proc. Cambridge Security Workshop on Fast Software Encryption, *Lecture Notes in Computer Science 809*, Springer Verlag, Berlin, 1994, ISBN 0-387-58108-1, 56-63.
6. G. Marsaglia, *"A current view of random number generators"*, Keynote address in *Proceedings of 16th Symposium on Computer and Statistics*, Elsevier, 1985.
7. D. E. Knuth, *"The Art of Computer Programming"*, Volume 3, Section 6.7, Addison-Wesley, Reading, MA, 1981.
8. S. N. Blackler and S. J. Shepherd, *"Characterization of Non-linearity in Block Ciphers"*, selected paper in *Communications Coding and Signal Processing*, eds. Honary, Darnell and Farrell, John Wiley & Sons, Chichester, 1997, ISBN 0-471-97821-3, 233-247.
9. S. J. Shepherd and S. K. Barton, *"On the Decomposition of Mixed Shift Register Sequences"*, to appear.
10. T. Siegenthaler, *"Decrypting a Class of Stream Ciphers Using Ciphertext Only"*, IEEE Transactions on Computers, 34 (1), January 1985, 81-85.
11. R. Rueppel, *"Design and Analysis of Stream Ciphers"*, Springer Verlag, 1986, ISBN 0-387-16870-2.
12. J. Golic, *"Linear Models for Keystream Generators"*, IEEE Transactions on Computers, 45 (1), January 1996, 41-49.
13. J. L. Massey, *"Shift Register Synthesis and BCH Decoding"*, IEEE Transactions on Information Theory, 15, January 1969, 122-127.

Resisting the Bergen-Hogan Attack on Adaptive Arithmetic Coding

Xian Liu, Patrick G. Farrell, and Colin A. Boyd*

Communications Research Group, School of Engineering, University of Manchester,
Manchester M13 9PL, UK
liu@comms.ee.man.ac.uk, farrell@man.ac.uk
*School of Data Communications, Queensland University of Technology, Brisbane Q4001,
Australia
boyd@fit.qut.edu.au

Abstract. Arithmetic coding is an optimal data compression algorithm in terms of entropy. As it is a shortest length coding, it is well recognized that the randomness of the output of arithmetic coding is very good. Based on arithmetic coding, we propose a novel encryption algorithm. The algorithm can resist existing attacks on arithmetic coding encryption algorithms. The statistical properties of the system are very good. General approach to attack this algorithm is difficult. The algorithm is easily programmed. The algorithm on its own is also an effective data compression algorithm. The compression ratio is only 2% worse than the original arithmetic coding algorithm.

1 Introduction

Arithmetic coding [1] is an optimal data compression algorithm in terms of entropy, recently developed. The first practical implementation of arithmetic coding with either a static model or a first-order adaptive model was published in 1987 by Witten et al [9] (which we call the WNC implementation). Since then there have been a number of different implementations for arithmetic coding with adaptive models in which the symbol probabilities are evolving on the fly. Principally, the better the compression is, the less redundancy in the output. On the other hand, for cryptography, redundancy contained in the output of a cryptosystem is usually one of the main resources to be used by the cryptanalyst. Based on these facts Witten et al [8] suggested that a higher-order adaptive arithmetic coding algorithm may provide high level security. They also indicated that to use an adaptive model compression algorithm as an encryption algorithm, it was enough to transmit the initial model as a key over a secure channel, or that a constant initial model could be used and a short message which was sent by a secure channel and provided to the model by both encoder and decoder before transmission could be applied. Aiming at the WNC implementation, Bergen and Hogan suggested a chosen plaintext attack on first-order adaptive arithmetic coding in 1993 [3]. Instead of trying to recover the initial model the Bergen-Hogan attack tries to take control of the model and reduce it to a manageable form. If the encoder does not initialize its model, the attacker can decrypt any message transmitted after the attack is done. To be successful, in the Bergen-Hogan attack an associate as well as an attacker are necessary. The associate needs to send 2^{18} symbols and the attacker needs to try decoding the test string 2^{14} times. Up until now the Bergen-Hogan attack is the only feasible attack on the adaptive arithmetic coding encryption algorithm. It is an open question whether or not a modified Bergen-Hogan attack will succeed in breaking the arithmetic coding algorithm with an advanced adaptive model, such as higher-order adaptive models as well as Prediction by Partial Match Model (PPM) [1]. In this paper we propose a novel encryption scheme based on arithmetic coding. We do not rely on complicated advanced adaptive models. Instead, we work on the WNC implementation. The resulted algorithm can resist the Bergen-Hogan attack. The techniques are easily programmed and there is no significant impact on compression.

2 Arithmetic Coding

Arithmetic coding is based on the fact that the cumulative probability of a sequence of statistically independent source symbols equals the product of the source symbol probabilities. In arithmetic coding each symbol in the message is assigned a distinct subinterval of the unit interval of length equal to its probability. This is the encoding interval for that symbol. As encoding proceeds, a nest of subintervals is defined. Each successive subinterval is defined by reducing the previous subinterval in proportion to the current symbol's probability. When the message becomes longer, the subinterval needed to represent it becomes smaller, and the number of bits needed to indicate that subinterval grows. The more likely symbols reduce the subinterval by less than the unlikely symbols and thus add fewer bits to the message. This results in data compression. When all symbols have been encoded, the final interval has length equal to the product of all the symbol probabilities and can be transmitted by sending any number belonging to the final interval. That means if the probability of the occurrence of a message is p, arithmetic coding can encode that message in $-\log_2 p$ bits, which is optimal in the sense of the shortest length encoding. The following pseudo code [1] illustrates the encoding and decoding procedure of arithmetic coding:

```
/* In the model, symbols are numbered 1, 2, 3, ...The cum_prob[ ] stores the    */
/* cumulative probabilities of symbols with cum_prob[i] increasing as i         */
/* decreases and cum_prob[0]=1. The encoding                                    */
/* transmits any value in the final [low, high) when it is finished.            */

        Encoding: The initial encoding interval [low, high) = [0, 1)
        EncodeSymbol (symbol, cum_prob)
        range = high - low;
        high = low + range * cum_prob[symbol-1];
        low = low + range * cum_prob[symbol];
```

```
Decoding:  The initial decoding interval [low, high) = [0, 1)
DecodeSymbol (cum_prob)
find symbol such that cum_prob[symbol]
    <= (value - low)/(high - low)
    < cum_prob[symbol - 1];
range = high - low;
high = low + range * cum_prob[symbol - 1];
low = low + range * cum_prob[symbol];
return symbol;
```

The WNC implementation for arithmetic coding [9] is the first practical algorithm and is widely accepted. The algorithm is provided with either a static model or a first-order adaptive model. The algorithm realizes integer arithmetic and incremental transmission. The arithmetic precision is 16-bit. In their first-order adaptive model, all the frequencies are initialized to 1. If the current model exceeds the maximum cumulative frequency, the model reduces all frequencies by half and recalculates cumulative frequencies. If necessary the model reorders the symbols to always put the current one in its correct rank in the frequency ordering. Adaptation is performed by incrementing the proper frequency count and adjusting cumulative frequencies accordingly. Due to the limit of first-order adaptation, the compression ratio is 50% to 70% according to the size and type of the file. However it can be greatly improved by using a higher-order adaptive model.

3 Witten-Cleary Proposal

The Witten-Cleary proposal [8] introduces a secret key in the model and the coding procedure is completely unchanged. The reason for their doing so is based on the specific characteristics of data compression:

• Models for data compression are often very large, so the number of the keys is large.

• The model acts as an enormous key, since without it decoding is impossible. Partial knowledge of the model is of little use.

• The model is never transmitted explicitly, and is observable only through experiment, by causing a particular message to be sent and examining the compressed version; such experiments change the model in ways which can be very difficult to predict.

• Adaptive modeling means that the key depends on the entire text which has been transmitted so far since the encoder/decoder system was initialized.

They suggested two ways to insert the key:

Method 1: The initial model is used as the key in which an array of single-character frequencies in the range of 1-10 would do.

Method 2: A constant initial model is used and before transmission begins both the encoder and decoder assimilate a short secret message into the model.

Their further suggestion is that the adaptive links should be maintained over long periods of time; i.e. the final model of encoding the current message will become the initial model to encode the next message.

4 Bergen-Hogan Attack on the Adaptive Model

In [3] Bergen and Hogan introduce a chosen plaintext attack against method 1 in Witten-Cleary proposal with WNC implementation which uses first-order adaptive modeling. This attack is unable to recover the initial model being attacked but tries to take control of the model and reduce it to a manageable form. In this attack an associate and an attacker are necessary, in which the associate sends specific messages with the encoder being attacked and the attacker tries to decode the test string. During the whole attack preventing any other messages from being sent is the associate's task. The key point of this attack is that due to the structure of the first-order adaptive model in WNC implementation it is possible to reduce the frequency table from random form to the standard form:

$$m_1, 1, 1, ..., 1; \qquad m_1+255$$

in which m_1 (unknown) is the frequency count of the symbol a and the frequency counts of other symbols are one, by sending a large homogenous string of symbols (say a). At the beginning the frequency table in the model being attacked is of random form. The frequency table in the attacker's model is in standard form and the symbols are in ASCII order. Here is the procedure:

(1). The associate sends a homogenous string of the symbol *0x00* with length 112882.
(2). The associate sends the symbol: *0x01, 0x02, ..., 0xFF*, each once.
(3). The associate sends the test string: *0x000x010x02...0xFF*.
(4). The attacker comes on line to receive the test string; leaves off line and tries 7809 different values for symbol *0x00*'s frequency count to decode the test string. If successful goes to (6).
(5). Repeats (1), (2), (3), and (4).
(6). The attacker comes on line; sets the correct current state in the model and decrypts any text as it is sent.

During the whole attack the associate needs to occupy the encoder being attacked and send 2^{18} symbols; the attacker needs to try to decode the test string 2^{14} times.

5 Our Proposal

The vulnerability of the Witten-Cleary proposal with WNC first-order adaptive implementation results partially from the structure of their first-order model being very simple. This model only maintains one alphabet with 256 symbols together with their frequency counts. Modern adaptive models are often very complicated and large. For example, for an n-order adaptive model, it needs to maintain a^n frequency counts where a is the size of the alphabet. The variable-order adaptive models such as PPM are also very large and with flexible structures. These models can provide more accurate prediction of the occurrences of symbols in the source and therefore greatly increase compression efficiencies if they are used with arithmetic coding.

To enhance the security, we do not rely on complicated adaptive models although they are preferable. Instead, we concentrate on the first-order adaptive WNC implementation. The arithmetic coding procedure is highly sensitive to the end points of the current interval, *high* and *low*, as well as the initial interval. These values are evolving on the fly as the coding is going on. The decoder is unable to work at all if at some place these values maintained by the encoder and decoder have only very slight differences. Together with a random initial state in the model we insert some secret keying materials into the WNC implementation of first-order adaptive arithmetic coding.

5.1 The Proposal

1. Select 256 3-bit random numbers as the initial frequency counts for every symbol, which act as the initial state in the model. The only restriction is that these numbers are all larger than 0.

2. Select the initial interval randomly. The length of the initial interval should not be less than $(2^{16}-1)/4$.

3. Select a secret 16-bit substitution which is used to substitute the first 16-bit output of the encoding.

4. Choose two secret parameter pairs $(\varepsilon_l^0, \varepsilon_h^0)$ and $(\varepsilon_l^1, \varepsilon_h^1)$ which are used to shrink the current interval controlled by a random 128-bit string cyclically, where the four parameters are different.

The keying materials should be inserted in the proper places in both the encoder and the decoder.

5.2 Effectiveness

If we choose the initial symbol frequency counts which are larger than 0 as 3-bit numbers, the number of possibly different initial states in the model is 7^{256}. The only concern here is to prevent the attacker from just trying the initial state, so 3-bit numbers are enough. Normally the unrepresentative initial model only influences the compression at the very beginning and the influence will be rapidly overwhelmed by the adaptivity, so the affect from the model is negligible. The only restriction to the selection of the initial interval is to prevent underflow. An initial interval with length a quarter of the full range is large enough, so there are 2^{30} different selections. As WNC implementation uses 16-bit precision, a secret 16-bit substitution is appropriate and is easily manageable in software. The function of the parameters $(\varepsilon_l^i, \varepsilon_h^i)$ is that whenever the encoder (decoder) finishes encoding (decoding) the current symbol, if the current control bit from the random bit string, which is shifted cyclically, is i, scale the *high* down by multiplying ε_h^i with the *range* and then scale up the *low* by multiplying ε_l^i with the *range*. ε_l^i is chosen to be 4 decimal digits in the form of 0.0*** and ε_h^i is chosen to be 4 decimal digits in the form of 0.9***. Therefore the number of different combinations of both $(\varepsilon_l^0, \varepsilon_h^0)$ and $(\varepsilon_l^1, \varepsilon_h^1)$ are 2^{19}. The experiment shows that all the changes in the coding only increase the output file length by 2% compared with the original WNC implementation.

5.3 Communication Protocol

Before communication begins, all of the keying materials are inserted in the encoder and decoder.

Protocol 1: The final state in the model of encoding current message becomes the initial state of the model to encode the next message and during the lifetime of using the same key the model will be initialized regularly. When encoding the current message is finished, shift the cyclic shift register storing the 128-bit random string one step and the next bit will become the first control bit to control the first shrinking in encoding the next message.

This protocol has the advantage that the initial model to encode the next message is relevant to all of the messages which have been sent since initialization.

Protocol 2: The only change compared with protocol 1 is that whenever encoding is finished, initialize the model.

6 The Strength

6.1 Resisting Bergen-Hogan Attack

In Bergen-Hogan attack the attacker knows he matches the keying materials only when he successfully decodes the test string. To use Bergen-Hogan attack on our proposal, the associate's strategy is the same as that to attack with the Witten-Cleary proposal, but the attacker's work will be increased dramatically. In order to decode the test string the attacker has to find the first symbol's frequency count in the standard form, the initial interval, the substitution, the pairs $(\varepsilon_l^0, \varepsilon_h^0)$ and $(\varepsilon_l^1, \varepsilon_h^1)$, and the 128-bit control string all together, instead of just trying the first symbol's frequency count in the standard form 2^{14} times in breaking with the Witten-Cleary proposal. As calculated previously, the attacker has to try to decode the test string $2^{14} \times 2^{30} \times (2^{16}!) \times 2^{19} \times 2^{19} \times 2^{128}$ times. Partially finding the keying materials is also very difficult. The reasonably simplest way for the attacker would be to firstly find the first symbol's frequency count in the standard form together with the initial interval. For this purpose the attacker only needs to decode the first symbol in the test string, but he has to try decoding $2^{14} \times 2^{30} \times 2^{16} \times 2^{19} = 2^{79}$ times. The analysis above is based on the assumption that our proposal is with communication protocol 1. Our proposal with the protocol 2 definitely denies the Bergen-Hogan attack because their attack is absolutely unable to recover the initial state in the model.

6.2 Resisting Other Potential Attacks

Irvine et al [6] suggested a way to analyze arithmetic coding encryption algorithms. They show in the binary static model arithmetic coding that if the length of the encoding file exceeds a specific value it is possible to theoretically determine the symbol probabilities. We think their result has the same significance and the same role as the concept of unicity distance in a cipher system. In [7] they suggest another approach in binary static model arithmetic coding with known symbol frequencies to analyze their way to use key bits. If we just extend their approach in [7] to the ASCII set static model arithmetic coding in which the symbol probabilities are unknown, one has to solve two polynomial equations which have 383 variables and have the degree 128 to find the proper solution.

6.3 Some Further Results

Our proposed scheme has been implemented in C programming. In this implementation the initial state of the adaptive model, the initial interval, the two shrinking parameter pairs, and the 128-bit control string were generated randomly. We used a randomly generated 16-bit substitution. Here are some further results, in which we firstly report the running time and compression ratio of our algorithm, which are compared with those of the WNC adaptive arithmetic coding algorithm, to four encoding files with different sizes and different types, and secondly the experimental results of related statistical properties, and thirdly the theoretical approach to the strength.

1. File size: 2785799 bytes.   Type: Doc. (binary).

	output size (bytes)	compression ratio	encoding time	decoding time
WNC adapt	1454519	52%	1'48''	2'33''
ours	1507421	54%	3'01''	4'05''

2. File size: 2832325 bytes.   Type: ASCII.

	output size (bytes)	compression ratio	encoding time	decoding time
WNC adapt	1882514	66%	1'58''	2'43''
ours	1936123	68%	3'47''	5'15''

3. File size: 1870823 bytes.   Type: PostScript.

	output size (bytes)	compression ratio	encoding time	decoding time
WNC adapt	947674	51%	1'03''	1'26''
ours	983217	53%	2'08''	2'55''

4. File size: 2277376 bytes.   Type: Image.

	output size (bytes)	compression ratio	encoding time	decoding time
WNC adapt	1506804	66%	1'34''	2'10''
ours	1549859	68%	3'17''	4'40''

Our algorithm inherits the excellent statistical properties of arithmetic coding, one of the optimal compression algorithms. With WNC adaptive coding, the encoded file has very good randomness. Changing any number of bits in the file to be encoded results in the fact that from the position in the encoded file the first changed bit corresponds to, then in the subsequent output, if this encoded file is compared with the encoded file resulted from the totally unchanged original file to be encoded, the changed bits and unchanged bits take the probabilities 0.5, and distribute uniformly and randomly. Exactly same things are found with our algorithm. So good plaintext diffusion and ciphertext avalanche and good randomness in the output are achieved.

Concerning the keying materials effects, no matter how many different combinations of the keying materials and how many different types of the input we use, the randomness of the output is very good. The diffusion and avalanche properties of the keying materials are also very good. First we tested the diffusion and avalanche properties of changing only a single parameter in the keying materials (in the initial state, the 128-bit control string, the initial interval, and the shrinking parameter pairs). No matter what kind of change was made (from changing one bit or one digit randomly, to several bits or digits and even the total parameter), when we compare the output file from changed single keying materials encoding with the output file from our algorithm with the keying materials unchanged, in the output from the position the parameter effects (the initial state effects the whole output file; the initial interval effects the whole file; the shrinking parameter pairs effect the whole file; the effect of changing the 128-bit cyclic control string depends on the position of the first changed bit) to the end, the changed bits and unchanged bits take the probabilities 0.5 and distribute uniformly and randomly, except for the effect of changing the initial interval. The same things are found when changing several parameters or all of them randomly including the initial interval. That means except for the initial interval, all of the keying materials have very good effects on balance, diffusion, completeness, and avalanche, which are desired in block ciphers and some of them in stream ciphers and are particularly difficult to achieve in all of them.

As to the effect of only changing the initial interval, sometimes in the output file the changed bits and unchanged bits take the probabilities one half and distribute uniformly and randomly but sometimes not. There seems not to be any regularity. But sometimes (not always) when the changed initial interval is very close to the original, for example, from [13456, 45321) to [13455, 45321), in the output file the unchanged bits are more frequent than changed bits. The worst case we have found is that very *occasionally* if the distance of the changed *high* or *low* to the original is 1, 2, or 3 (which has been the largest distance we have found in more than three months of experiments) the output file is totally unchanged. *But* the negative effect of the initial interval is absolutely totally overwhelmed by changing any other parameter(s) in any way, resulting in uniform distributions of the changed bits and unchanged bits in the output file.

As we calculated, the number of different legal initial intervals is 2^{30}. If we add the restriction that the distances of every *high* and *low* are not less than 4, the number of different legal initial intervals is 2^{26}. If they are not less than 10, the number of different legal initial intervals is 2^{23}. Now we recalculate the complexity to find the equivalent state and the initial interval by Bergen-Hogan attack:

The complexity is $2^{14} \times 2^{23} \times 2^{16} \times 2^{19} = 2^{72}$ if the *high* and *low* in any of two initial intervals have the distances at least 10.

208

As its name means, the arithmetic coding encoding procedure can be represented by an arithmetic procedure. If the precision is infinite, in original arithmetic coding, the current *low*, *range*, and *high* can be calculated from the exact knowledge of the symbol probabilities and the encoding symbol string. This technique can be used to obtain a theoretical approach to the strength of the system. For clarity we assume that the system uses a static binary model with the symbol probabilities being kept secret and that the arithmetic precision is infinite. The substitution is neglected. We can get polynomial representation for the current *low*, *range*, and *high*. However, if the attacker mounts a chosen plaintext attack on the keying materials by using this technique, he will find that he will be facing the problem of solving three polynomial equations with degrees of 512 and 510 and with unknown coefficients. What he knows are only the structures of the three polynomials.

7 Conclusion

In this paper we suggest a novel encryption algorithm based on the first-order adaptive model arithmetic coding. The techniques are easily programmed. This scheme can resist the Bergen-Hogan attack on adaptive arithmetic coding. This algorithm in itself is also an effective data compression algorithm. There is not any significant impact on compression compared with the WNC implementation with a first-order adaptive model for arithmetic coding.

References

1. Bell T., Cleary J. and Witten I.: *Text Compression*, Prentice Hall, 1990.
2. Bergen H. and Hogan J.: "Data security in a fixed-model arithmetic coding compression algorithm", Computers and Security, Vol.11, 1992, pp.445-461.
3. Bergen H. and Hogan J.: "A chosen plaintext attack on an adaptive arithmetic coding compression algorithm", Computers and Security, Vol.12, 1993, pp.157-167.
4. Boyd C.: "Enhancing secrecy by data compression : theoretical and practical aspects", Proceedings of Eurocrypt'91, LNCS 547, Springar-Verlag, 1991, pp.266-280.
5. Boyd C., Cleary J., Irvine S., Rinsma-Melchert I. and Witten I.: "Integrating error detection into arithmetic coding", IEEE Trans. COM, Vol.45, No.1, 1997, pp.1-3.
6. Cleary J., Irvine S. and Rinsma-Melchert I.: "On the insecurity of arithmetic coding", Computers and Security, Vol.14, 1995, pp.167-180.
7. Irvine S. and Cleary J.: "The subset sum problem and arithmetic coding", private communication, 1995.
8. Witten I. and Cleary J.: "On the privacy afforded by adaptive text compression", Computers and Security, Vol.7, 1988, pp.397-408.
9. Witten I., Neal R. and Cleary J.: "Arithmetic coding for data compression", Communications of the ACM, Vol.30, No.6, 1987, pp.520-540.

Novel Decoding Technique for the Synchronous and Quasi-synchronous Multiple Access Adder Channel

G. Markarian

P. Benachour

B. Honary

Lancaster Communications Research Centre

SECAMS, Lancaster University, Lancaster, LA1 4YR

E-mail : g.markarian@lancaster.ac.uk

Abstract

A novel technique that allows efficient trellis decoding for the M-choose-T synchronous multiple access (MA) adder channel is investigated. Decoding in the M-choose-T system is performed in two operations, identifying the set of active users in the channel and decoding the received data according to which set of users are active. The results show that a sufficient performance improvement is achieved without the need for additional information regarding the active users. The application of the proposed technique to the quasi-synchronous MA adder channel is also analysed. The modified decoding procedure shows a further improvement when compared to the conventional hard decoding technique.

1. Introduction

A MA adder channel is said to be synchronous, if the encoders and the decoder are in block synchronism. If no block synchronism exists between the encoders and the decoder, the system is said to be asynchronous. However, it is quasi-synchronous if the encoders are not in block synchronism with each other, but the decoder knows the block position of each encoder in the received sequence of symbols, and bit synchronism is maintained amongst the encoders and the decoder. In the MA adder channel, the output symbol value is the arithmetic sum of the input symbol values [1,2]. The original model of such a channel was proposed by Kasami and Lin [2] and represents a uniquely decodable code pair of block length $n=2$, as is shown in Tab. 1.

Tab. 1. Conventional MA Coding Scheme With 2 Users

User$_2$\ User$_1$	(00)	(11)
(00)	(00)	(11)
(01)	(01)	(12)
(10)	(10)	(21)

In this table, User$_1$ has only two codewords, $C_1=(00,11)$, and User$_2$ has three code words, $C_2=(00,01,10)$. Since all the sum codewords are distinct, the decoder can reconstruct the two messages without ambiguity. The overall rate of this MA coding scheme is $R=1.292$ which is better than time sharing [1,2]. Trellis design for the MA adder channel has recently been investigated and implemented [3,4]. The technique is based on the concept of the Shannon product of trellises [5,6] and is applicable to MA coding schemes with M (M $\geq$ 2) users. Based on this technique, trellis diagrams for some known 2-user schemes have been constructed; in addition, a DC-free 2-user binary code together with its trellis structure was introduced and investigated. It was shown that such a technique has allowed the achievement of about 2.2 dB energy gain for the overall coding scheme in comparison with the conventional hard decision case.

The results achieved so far use code constructions that are based on a MA model where all the M-users are simultaneously active and the individual users codewords are combined in the channel to form a composite codeword. However, a more practical channel model describes the situation where there are M potential users in the system of which at most T are active simultaneously [7]. For the case when the set of active users is known, a simple decoding algorithm over the reals which separates the received codeword into the components sent by the individual users has been developed . Here, identification of the set of active users is achieved by using a special signalling codeword at the beginning (and possibly at the end) of transmission. For the case when the set of active users is unknown, it was shown that such a set can be identified uniquely provided that at most T/2 are active simultaneously. However, the need to identify the set of active users using a signalling codeword and the use of hard decision decoding reduces the efficiency of such a scheme.

In this paper, we present a novel decoding technique for the M-choose-T system and we show that a sufficient performance improvement could be achieved without the need for additional information regarding the active users. This approach termed *look-ahead/look-back*, is based on the assumption of finite time transmission by the set of active users and a metric accumulation process. Initially, the system is tested under the assumption of perfect block and bit synchronisation. For the quasi-synchronous system, we use an existing coding scheme from the M-choose-T system and we perform a shifting operation on the user's codewords. The overall trellis structure is a modified version with one sub-trellis corresponding to the synchronised case and a further n sub-trellises were n is the maximum number of possible bit shifts. The modified decoding procedure for the quasi-synchronous case shows a further improvement when compared to the conventional hard decoding techniques.

2 Trellis Decoding in the M-choose-T Synchronous Scheme

The *look-ahead/look-back* technique is based on the codes developed in [7]. In this case, the overall coding scheme is represented by a trellis diagram from which the composite and constituent codewords are decoded and the individual user's information vectors extracted. Furthermore, such a technique attempts to identify the set of active users without the need for a special signalling codeword; in fact no extra information is required on the set of active users at all. This approach is based on the assumption that a set of active users will be active during a finite period (frame) in which they will transmit their information. The system keeps track of several received composite codewords and makes a decision when it is confident that a set of users has been active for some time. The amount of time and the number of codewords required to make a sensible decision may be fixed or variable and will be determined upon the system's requirements.

2.1 Decoding Procedure

The *look-ahead/look-back* decoding steps are as follows :
Step One :
Implement *look-ahead* operation by performing maximum likelihood decoding (MLD) on all received composite codewords. At the same time, keep track of metric values accumulated over the transmission period.
Step Two :
look-back to the metric values and make a decision on the set of active users by identifying the decoder with the least accumulated metric.

The technique is illustrated by means of a simple example in which M=3 and T=2. The number of simple decoders required is 3 and the overall decoder will be as shown in Fig. 1.

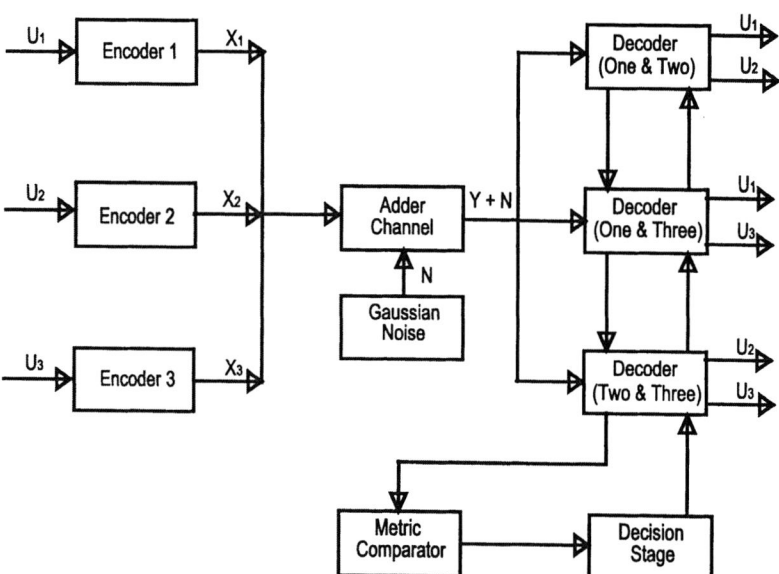

Fig. 1. Block diagram for the M=3-choose-T=2 scheme

In Fig. 1, the system selects a set of users for transmission, since synchronisation is assumed, the decoders know the start and end of each composite codeword block. During transmission, the decoders will attempt to decode the received data according to their structures; and at the end, the identification process and decision making are performed by the *Metric Comparator* and *Decision Stage* blocks respectively.

2.2 Metric Calculation and Accumulation

The overall metric accumulation process is based on calculating the Eucliden distance for each received composite codeword and updating the metric for the next codeword. Each codeword is decoded according to it's own minimum metric, but an overall metric is updated based on the accumulation of all metrics calculated during the transmission period. Since the scheme is implemented in an adder channel, the output symbol with added noise will be :

$$s = \sum_{i=1}^{T} x_i + N \tag{1}$$

where x_i, $i=1, 2, ..., T$, is the *ith* user channel input symbol , s the composite channel output symbol and N the Gaussian noise random variable. In each of the decoders in Fig. 1, the Euclidean distance between each received composite vector and all the possible composite codewords is expressed as :

$$\left\| d_j \right\|^2 = \sum_{i=1}^{n} \left[r_i - \left(\sum_{i=1}^{T} x_i + N \right) \right]^2 \tag{2}$$

where r_i and n are the received data and the number of bits in the sequence (codeword) respectively and $\left\| d_j \right\|$ means that the distance is a metric where $j=1,2,...,$ $^M C_T$ refers to the number of decoders in such a system.

The total metric value accumulated over a transmission period τ can be expressed as follows :

$$\sum_{t=0}^{\tau} \left\| d_j \right\|^2 = \sum_{t=0}^{\tau} \sum_{i=1}^{n} \left[r_i - \left(\sum_{i=1}^{T} x_i + N \right) \right]^2 \tag{3}$$

At the end of transmission, the system makes a decision on which set of active users have combined in the channel by identifying the decoder which corresponds to the smallest accumulated metric over τ .

$$\sum_{t=0}^{\tau} \left\| d_j \right\|_{\min^k}^2 = \min \left\{ \sum_{t=0}^{\tau} \left\| d_1 \right\|_{\min^1}^2, \sum_{t=0}^{\tau} \left\| d_2 \right\|_{\min^2}^2, ..., \sum_{t=0}^{\tau} \left\| d_{^M C_T} \right\|_{\min^{^M C_T}}^2 \right\} \tag{4}$$

where $j=k=1, 2, ..., {}^M C_T$.

3 Trellis Decoding in the T-user Quasi-Synchronous Scheme

It is a well known fact that reliable decoding in the MA adder channel is based on code constructions that satisfy the unique decodability criteria; this is not the case when the users do not maintain block synchronisation. The search for good coding schemes for the asynchronous MA adder channel is limited as previous research has shown that it is much more difficult to design a certain number of asynchronous codes than to design the same number of synchronous ones. The asynchronous MA adder channel can be reduced to the quasi-synchronous by the following scheme [8], [9] : at the beginning of data transmission, each encoder sends a synchronising sequence to the decoder. The synchronising sequence has the property that, upon reception of a synchronising sequence from one encoder, the decoder can establish block synchronism with it, regardless of what codeword the other encoder is transmitting during the same period. After the reception of synchronising sequence from both encoders, the decoder has established synchronism with each individual encoder, and the asynchronous MA adder channel is reduced to the quasi-synchronous MA adder channel.

Quasi-synchronous uniquely decodable codes (QSUD) have been designed and constructed in [10] based on the definition above. In this paper, we define our version of the quasi-synchronous scheme where the decoder does not know the block position for each encoder but the block position of the composite codeword instead without the need for a synchronisation sequence. We shall use this definition of the quasi-synchronous MA adder channel to investigate the performance of this scheme.

3.1 Description of System

In this case, we use an existing 2-user uniquely decodable code in [7] and we shift one of the user's codeword . The original codes for such a scheme are presented below in Tab. 2.

Tab. 2. 2-User Synchronous Scheme

User$_2$\User$_1$	000	001	110
000	000	001	110
010	010	011	120
101	101	102	211

In this table, User$_1$ and User$_2$ have three codewords each, $C_1=(000,001,110)$ and $C_2= (000,010,101)$ and the overall rate is $R=1.057$. If User$_2$ codewords were shifted by 1 bit and assuming the first bit entering the decoder is the leftmost bit, then shifting is performed from right to left and the result for such shifted version is presented in Tab. 3.

Tab. 3. 2-User Quasi-synchronous Scheme

User$_2$\User$_1$	000	001	110
00X	000	001	110
(000 or 001)	001	002	111
10X	101	102	211
(101 or 100)	100	101	210
01X	010	011	120
(010 or 011)	011	012	121

Inspection of Tab. 3 reveals that the 2-user scheme is not uniquely decodable which makes the decoding process unreliable and delivering correct data to its corresponding destination will not be achievable in some cases . Although this can be seen as a drawback, we will use this effect for good cause. It can also be seen in Tab. 3 that User$_2$ constituent codes have doubled due to the shifting operation performed and therefore $C_2= (000,001,101,100,010,011)$ and consequently the overall rate has increased to $R=1.389$.

3.2 Trellis Construction and Representation

The original trellis structures for the User$_1$ and User$_2$ codes in Tab. 2 are illustrated in Fig. 2 and 3 respectively.

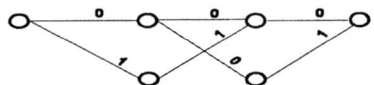

Fig. 2. Trellis Structure of User₁ Code

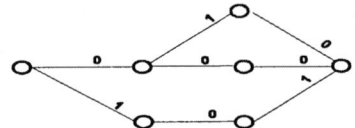

Fig. 3. Trellis Structure of User₂ Code

Since User$_2$ constituent codewords have increased from three to six, the trellis diagram in Fig. 3 has to be modified in order to construct the new structure. Tab. 3 can be seen to have three main rows, each having two sub rows : upper and lower. The upper sub-rows can be seen to represent the original synchronous scheme in Tab. 2 and the lower sub-rows represent the remainder from the effect of the shifting operation. Therefore, trellis construction for the User$_2$ case will be based on upper and lower sub-trellises. The modified version for such a code is shown in Fig. 4.

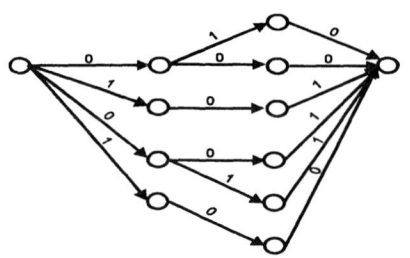

Fig. 4. Modified Trellis Structure of User₂ Code

In order to construct the composite trellis structure of this scheme, the Shannon Product of trellises is applied to the codes in Fig. 2 & 4. Since User₂ trellis has been modified as illustrated in Fig. 4, the composite trellis should also be made up of two sub-trellises. Consequently, the overall composite trellis is illustrated below in Fig. 5.

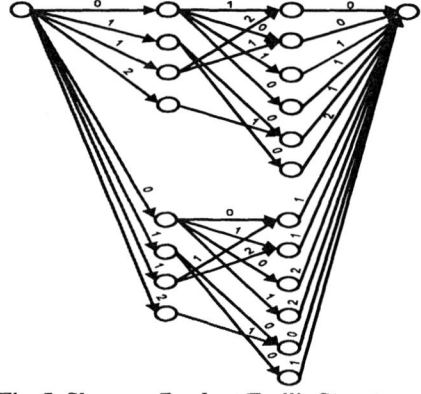

Fig. 5. Shannon Product Trellis Structure of
User₁ + User₂ Code

If a further shifting operation takes place (2 bit shifts), User₂ constituent codewords will be $C_2=(000,001,010,100,101,110)$ and similarly to the 1 bit shift case, the trellis construction for the User₂ and the overall composite scheme will be based on upper and lower sub-trellises. The trellis structure for User₂ and the overall 2 bit shift version are shown in Fig. 6 and 7 respectively.

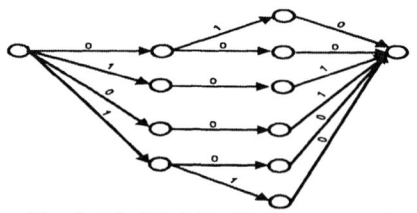

Fig. 6. Modified Trellis Structure of
User₂ Code with 2 bit Shifts

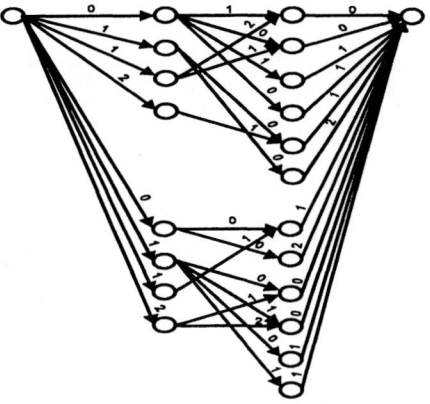

Fig. 7. Shannon Product Trellis Structure of
User₁ + User₂ Code (2 bit shift case)

In a situation where the number of bits in a codeword is known, it is then possible to workout the number of shift operations required. Furthermore, it is possible to construct sub-trellises for each of the shifted versions independently. The overall trellis structure will be based on one sub-trellis corresponding to the perfectly synchronised case and n sub-trellises were n is the maximum number of possible bit shifts (n=b-1 where b is the number of bits in the codeword). In the example above, n=2, hence there are 3 sub-trellises, the synchronous case corresponding to Tab. 2 and two sub-trellises for the 1 and 2 bit shifts respectively. The general decoder structure for the 2-user quasi-synchronous scheme is represented in Fig. 8.

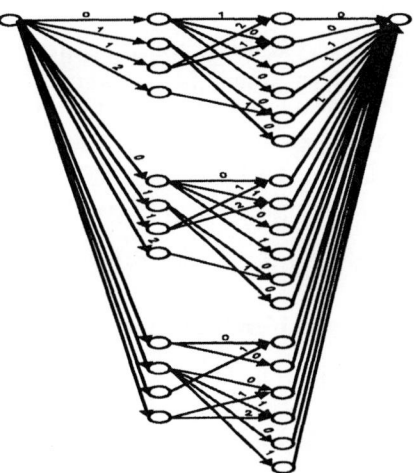

Fig. 8. General Trellis Structures for the
Quasi-Synchronous Scheme

3.3 Combined Maximum Likelihood Trellis and erasure Decoding

The decoding procedure used here is termed combined maximum likelihood (ML) trellis and erasure decoding. The term "Combined" is employed because the decoder performs ML estimation on the received composite data, but in some cases, the ML process can only decode 2 constituent bits out of 3. The last bit for each individual users is erased if the composite codeword is not uniquely decodable. Therefore, to make the decoding process complete, the users have to either request a retransmission of the last bit or make a half chance correct decision. Error control coding can be used to correct erasures or to correct errors and erasures simultaneously. If a code has minimum distance d_{min} , any pattern of p or fewer erasures can be corrected if

$$d_{min} \geq p + 1 \qquad\qquad (5)$$

In our case, $d_{min}=1$ for the individual user's and composite codewords and therefore erasure correction is not possible. Consequently, a bit error in the individual user's codes is inevitable sometimes.

The decoding algorithm for the quasi-synchronous 2-user coding scheme is based on the following steps:

Step One :

Perform MLD on the received composite codeword.

Step Two :

If the received codeword is uniquely decodable, retrieve the individual user's codewords and in the $User_2$ case retrieve the original code.

Step Three :

If the received codeword is a duplicate, decode the first 2 bits of the individual user's codes and erase the last bit.

Step Four :

To complete the decoding process, request a 1 bit transmission from each user (independently) or retransmit a new composite codeword.

Step four is repeated until the composite codeword received is uniquely decodable in order to extract the component codes without ambiguity.

4 Simulation Tests and Results

4.1 Simulations for Active Users Identification

The system was tested under additive white Gaussian noise (AWGN) conditions and graphical results for each of the decoder's accumulated metrics for the example in section 2) are presented in Fig. 9. The results show that for higher ratio values of E/No, where E is the average energy per user, the scheme performs reliably and the system can identify the set of active users successfully by detecting the decoder with the least accumulated metric. In Fig. 9, curves 1, 2 and 3 correspond to decoders for active users 1&2, 1&3 and 2&3 respectively and the set of users activated for transmission is Users 1&2. However, there are some practical limitations with this system, ideally the decoding process should be performed for each codeword without the need for a finite transmission time. There is also the problem of time delay that can be incurred when the system is idle during the active user's identification process, delivering the received data to it's intended destination and transmitting the next frame.

4.2 Simulations for the Synchronous and Quasi-Synchronous Systems

Fig. 10 shows the performances for the synchronous and quasi-synchronous systems under AWGN and different decoding methods and structures. Curves 1 and 2 represent the conventional and trellis decoding characteristics for the synchronous scheme. It can be seen that the soft decision case shows a performance improvement when compared to the conventional hard decoding technique. The modified decoder " curve 4 " for the quasi-synchronous case also shows an improvement and outperforms the conventional decoding technique by achieving a coding gain of about 0.5 dB. Curve 3 represents the quasi-synchronous performance when such a scheme is fed through the synchronised decoder in place of the modified version; the graphical results show that even though the modified technique " curve 4 "does not outperform the original synchronous decoder "curve 2", there is a substantial improvement when compared to curve 3.

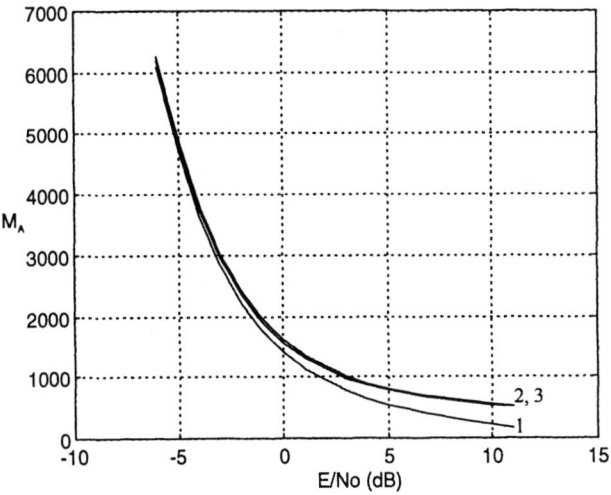

Fig. 9. Graphical plot of accumulated metrics versus E/No

M_A: Metrics Accumulated
Active Users: Users One & two
Number of Transmitted Codewords: 1000

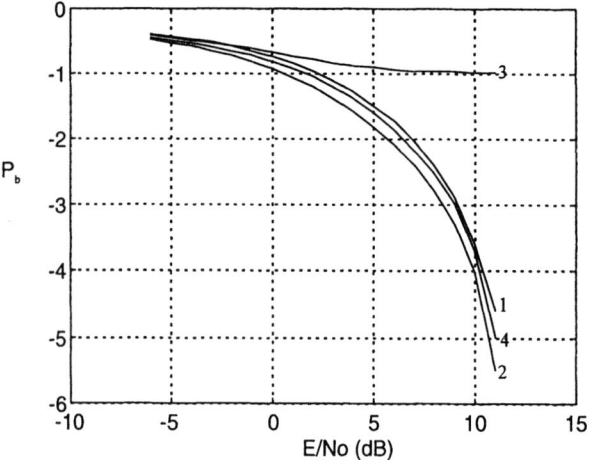

Fig. 10. Performance of Synchronous and Quasi-Synchronous 2-user schemes
P_b : **Probability of Bit Error Rate**

5. Conclusion and Open Problems

The performances of the synchronous and quasi-synchronous MA adder channel schemes have been investigated and tested. For the M-choose-T synchronous case, a sufficient performance improvement has been realised without the need for additional information on the set of active users. In the T-user quasi-synchronous system, it has been demonstrated that it is possible to improve such a scheme by modifying the constituent codes in use with further increase in overall code rate at the expense of decoder complexity.

There are a number of open questions in connection with the quasi-synchronous system. Firstly, decoding in the M-choose-T case and the identification process cannot always be implemented reliably in the quasi-synchronous channel. This is due to the fact that the codes used here do not exhibit the same characteristics when a shifting operation takes place and modified trellis structures for the different set of active users cannot be constructed in an M-choose-T quasi-synchronous environment. The search for efficient coding schemes that can be applied to such an environment is a subject of further investigation. One possible solution that can be implemented is to modify the metric calculation and accumulation process by introducing a new metric other than the Euclidean distance. The idea stems from the fact that when the MA adder channel is synchronous, the Euclidean distance metric is used as the noise in the channel is assumed to be Gaussian. This is not the case when the channel is quasi-synchronous.

6. References

[1] P. G, Farrell "Survey of channel coding for multi-user systems", in *Skwirzynski, J.K.* (Ed.) : 'New concepts in multi-user communications' (Sijthoff and Noordhoff, 1981), pp. 133-159.

[2] T.Kasami and S.Lin, "Coding for a multiple access channel", *IEEE Transactions. on Inform. Theory*, Vol. IT-22, No.2, pp 129-137, March 1976.

[3] G.Markarian, B.Honary, P.Benachour, "Trellis decoding techniques for the binary
adder channel with M users", *IEE Proceedings on Communications,* vol.144, 1997.

[4] P. Benachour. G. Markarian, B. Honary, " Trellis decoding techniques in an M-choose-T multiple access adder channel ", *in Proceedings of the 4th Int. Symp. On Communication Theory and Applications, Ambleside, UK,* pp 84-85, July 1997.

[5] V.Sidorenko, G.Markarian, B.Honary, "Minimal trellis design for linear codes based on the Shannon product ", *IEEE Trans. Inform. Theory,* vol. 42 pp.2048-2053, Nov.1996.

[6] B. Honary, G. Markarian, " Trellis decoding of block codes : A practical approach ", *Kluwer Academic Publishers,* 1997.

[7] P.Mathys, "A class of codes for T active users out of N, for a multiple access communication system," *IEEE Trans. Inform. Theory,* vol. 36 pp.1206-1219, Nov.1990.

[8] T. Kasami, S. Lin, V. Wei, and S. Yamamura, " Graph theoretic approaches to the code construction for the two-user multiple access binary adder channel," *IEEE Trans. Inform. Theory,* vol. IT-29, pp.114-130, 1983.

[9] R. J. McEliece, E. C. Posner, "Multiple access channels without synchronisation," *in Conf. Rec. Int. Conf on Communication,* vol. 2, pp. 246-248, 1977.

[10] S. Lin, V. K. Wei, "Non-homogeneous trellis codes for the quasi-synchronous multiple access binary adder channel with two users", *IEEE Trans. Inform. Theory,* vol. IT.32 pp.787-796, Nov. 1986.

Increasing Efficiency of International Key Escrow in Mutually Mistrusting Domains

Keith M. Martin*

Katholieke Universiteit Leuven, Dept. Elektrotechniek - ESAT
Kardinaal Mercierlaan 94, B-3001 Heverlee, Belgium
keith.martin@esat.kuleuven.ac.be

Abstract. This paper is concerned with key escrow protocols for use in international communications environments, where communication domains do not necessarily trust one another. It is concerned particularly with systems where users place their trust collectively with groups of trusted third parties. We consider two different protocols, discuss and improve their efficiency and generalise the type of key splitting.

1 Introduction

There is a great deal of current interest in *key escrow* protocols, loosely defined as systems that protect data using conventional cryptographic methods but, under special circumstances, make it possible for the cryptographic protection to be circumvented allowing access to either the data itself, or some cryptographic key that protects it. Such a circumstance is when law enforcement agencies have a warrant to obtain access to certain specified communications. For an introduction to many of the existing key escrow systems see [5].

Most proposed key escrow schemes rely on the use of *trusted third parties* (TTPs). These TTPs are trusted organisations who provide network services which include acting as a trusted itermediary between network users and law enforcement agencies. In most key escrow schemes a user lodges some secret information with a TTP. This information is kept confidential during normal network use, however under special circumstances the information can either be released, or used, in order to permit access to the user's communications.

Providing a key escrow service within one *domain* (perhaps a nation state) is likely to require a network of TTPs. Each user identifies with one (or more) of these TTPs (their *home* TTPs). Within a single domain it may be possible that different TTPs have established working relationships that allow secure information to be exchanged under suitable circumstances. Key escrow in a network spanning several domains (international telecommunications) is considerably more challenging. The main problem is that it is likely that law enforcement agencies belonging to one domain will only legally be able to approach TTPs

---

* Supported by the European Commission under ACTS project AC095 (ASPeCT)

that also belong to that domain. This has the potential to limit the scope and efficiency of interception. Further, it may be more difficult, or indeed impossible, to establish trust relationships between TTPs belonging to different domains. In this paper we are interested in key escrow within a multiple domain environment.

For application within international telecommunications networks, one of the most studied key escrow proposals is the *JMW Protocol* [12, 13]. In the basic version of the protocol each user has one home TTP. It is assumed that this is the only TTP with which the user can communicate. The JMW Protocol can be used to establish a session key between the two users (see Sect. 2.1).

One, of several, generalisations of the JMW Protocol is to weaken the requirement that a user fully trusts their single home TTP. Instead, the trust is spread among a *group* of home TTPs in such a way that the user need only *jointly trust* certain sets of TTPs [4]. We recall two of the main protocols from [4] (Sect. 2) and generalise the key splitting techniques used in each variant (Sects. 3 and 4). We then reduce several different communication costs for the second variant (Sects. 5 and 6) and consider in detail the case of threshold splitting (Sect. 7).

2 The JMW Protocol and Splitting Variants

2.1 The JMW Protocol

We describe the basic protocol proposed in [12]. We assume that user A wants to establish a session key with user B. Let the home TTP of A be TA, and the home TTP of B be TB. Let p be a prime and g be a primitive element modulo p. Each TTP is equipped with a signature and verification algorithm pair, which we denote by $\text{sig}_{TA}(\cdot)$, $\text{ver}_{TA}(\cdot)$ and $\text{sig}_{TB}(\cdot)$, $\text{ver}_{TB}(\cdot)$. The TTPs TA and TB share a secret key K_{TAB} and a key generation function f, which takes as input a user identification number and K_{TAB}, and outputs a *private receive key* for that user. We assume that prior to communication, B has received a copy of their *private receive key* $b = f(B, K_{TAB})$. The protocol runs as follows:

1. User A sends a message to TA that they want to communicate with user B.
2. TA chooses a *private send key* a for A and computes $b = f(B, K_{TAB})$. TA then sends the values a, $\text{sig}_{TA}(g^a)$, and g^b, to A.
3. A computes $K_{AB} = g^{ba} \bmod p$ and sends $\text{sig}_{TA}(g^a)$ to B.
4. B verifies A's public send key g^a and computes $K_{AB} = g^{ab} \bmod p$.

Thus users A and B can now communicate using session key K_{AB}. Further, both TA and TB can provided interception authorities with either a private send or a private receive key for A or B, or a copy of session key K_{AB}.

2.2 JMW Protocol Variants Permitting Key Splitting

We now consider weakening the requirement that a user has just one home TTP in which they must place full trust. The formulation of a system by means of which trust can be spread across a group of TTPs is achieved using *key splitting* techniques (more commonly known as *secret sharing*). Secret sharing was first suggested for use in key escrow systems in [14].

A *secret sharing scheme* is a protocol for protecting a secret among a group of entities (*participants*) in such a way that only certain sets of participants can compute the secret. Each participant is issued with a *share* of the secret. The *access structure* Γ is the collection of participants desired to be able to compute the secret. We assume that Γ is *monotone* (given a set of participants in the access structure, all larger sets that include the given set are also in the access structure). An access structure Γ is uniquely determined by its *minimal sets* (those sets that do not strictly contain another set in the access structure).

The most well-known access structures are the (k, n)-*threshold* access structures, which consist of all subsets of n participants of size at least k [2, 15]. It is possible to construct secret sharing schemes for *any* access structure.

We now describe two JMW Protocol variants that permit key splitting. We assume that each user has a collection of home TTPs. Let the home TTPs of A be $T_1, \ldots, T_m$ and the home TTPs of B be $U_1, \ldots, U_n$.

First Key Splitting Variant [13]. In this scheme, A has a distinct modulus p_A and base g_A. We assume that every pair of TTPs (T_i, U_j) share a secret key K_{ij} and assume the existence of a function f as before. This f can be used by any pair (T_i, U_j) to compute $b_{ij} = f(B, K_{ij})$. Prior to communication, B is issued with a copy of their private receive key b as follows: each U_j sends b_{ij} $(i = 1, \ldots, m)$ to B; user B computes $b_i = \sum_{j=1}^{n} b_{ij}$; user B computes $b = \sum_{i=1}^{m} b_i$.

Protocol 1. The protocol runs as follows:

1. User A sends a message to each T_i that they want to communicate with B.
2. Each T_i chooses a_i for A, computes b_{ij} $(j = 1, \ldots, n)$ and then $g_A^{b_i} = \prod_{j=1}^{n} g_A^{b_{ij}}$. Each T_i then sends back to A the values a_i, $\mathrm{sig}_{T_i}(g_A^{a_i})$, and $g_A^{b_i}$.
3. A computes $a = \sum_{i=1}^{m} a_i$, $g_A^b = \prod_{i=1}^{m} g_A^{b_i}$, and $K_{AB} = g_A^{ba} \bmod p$. A then sends $\mathrm{sig}_{T_i}(g_A^{a_i})$ $(i = 1, \ldots, m)$ to B.
4. B verifies components $g_A^{a_i}$, and computes g_A^a and $K_{AB} = g^{ab} \bmod p$.

Thus users A and B can now communicate using session key K_{AB}. Further, $T_1, \ldots, T_m$ jointly (an (m, m)-threshold access structure), or $U_1, \ldots, U_n$ jointly (an (n, n)-threshold access structure), can provided interception authorities with the information to construct either a private send or a private receive key for A or B, or a copy of session key K_{AB} (see [13] for details).

Second Key Splitting Variant [4] (Mechanism 7). This time we assume that each set of home TTPs has agreed a common secret key (K_T and K_U respectively) and a key generation function f. This function takes as input the identity of two users and a secret TTP key, and outputs an integer. Let $f(A, B, K_T) = s_{TAB}$ and $f(A, B, K_U) = s_{UAB}$.

In this protocol any k out of $T_1, \ldots, T_m$ and any k out of $U_1, \ldots, U_n$ are collectively trusted, where $k \leq \min m, n$. We describe a simplified version of the protocol (omitting verification information and the explicit transfer of any public keys). From here on, all addition is modulo p, where p is a large prime. Further $p - 1$ has a large prime divisor q and g is an element of order q in $\mathbf{Z}_p$.

Protocol 2. With the setting as above, users A and B proceed as follows:

1. User A chooses s_A, publishes g^{s_A}, and generates shares $s_{A_1}, \ldots, s_{A_m}$ of a (k, m)-threshold scheme with secret s_A. For $i = 1, \ldots, m$, A sends s_{A_i} to T_i.
2. User B chooses s_B, publishes g^{s_B}, and generates shares $s_{B_1}, \ldots, s_{B_n}$ of a (k, n)-threshold scheme with secret s_B. For $j = 1, \ldots, n$, B sends s_{B_j} to U_j.
3. T_i publishes $g^{s_A s_{TAB}}$, and generates shares $s_{A_{i1}}, \ldots, s_{A_{in}}$ of a (k, n)-threshold scheme with secret s_{A_i}. For $j = 1, \ldots, n$, T_i sends $s_{A_{ij}} s_{TAB}$ to U_j.
4. U_j publishes $g^{s_B s_{UAB}}$, and generates shares $s_{B_{j1}}, \ldots, s_{B_{jm}}$ of a (k, m)-threshold scheme with secret s_{B_j}. For $i = 1, \ldots, m$, U_j sends $s_{B_{ji}} s_{UAB}$ to T_i.
5. Each TTP T_i sends $g^{s_B s_{UAB} s_{TAB}}$ to user A.
6. Each TTP U_j sends $g^{s_A s_{TAB} s_{UAB}}$ to user B.

Users A and B can now both compute the session key $K_{AB} = g^{s_A s_B s_{TAB} s_{UAB}}$. In either domain the key K_{AB} can be recovered by at least k home TTPs.

2.3 Efficiency of Split Escrow Variants

There are two different types of overhead encountered when implementing the protocols. Firstly, the advance agreement of information between TTPs on protocol initiation, and secondly, the performance of operations during protocol execution. In likely order of increasing communication cost, common protocol operations are generation of shares of a secret sharing scheme, and communication between users and their home TTPs, between TTPs from the same domain, and between TTPs from different domains.

Note that Protocol 1 requires every pair of TTPs from different domains to agree a common key on initiation, but during operation does not involve communication between any TTPs. In contrast, Protocol 2 makes extensive use of all but the third of the operations, but only requires each set of home TTPs (in the same domain) to agree a secret key in advance.

Another difference is that in Protocol 1 the type of key splitting used is (n, n)-threshold, whereas in Protocol 2 it is (k, n)-threshold. We proceed by generalising the type of key splitting used in both protocols, and in doing so pay particular attention to, and often improve, the resulting efficiency.

3 Generalising the First Key Splitting Variant

We now assume the existence of sets of home TTPs $T_1, \ldots, T_m$ and $U_1, \ldots, U_n$, as before, but allow the trusted subsets of TTPs to take a more general form, given by access structures Γ_T and Γ_U. Protocol 1 makes use of the fact that an (n, n)-threshold scheme can be set up without the assistance of a TTP (or *Mutually Trusted Authority*). To achieve this for more general access structures, we need to use MTA-free secret sharing [10, 11], which works as follows:

1. a subset $\{P_1, \ldots, P_k\} \in \Gamma$ of users establish a (k, k) threshold scheme among themselves (without needing a TTP), where user P_i has share s_i;
2. each user P_i generates shares of a secret sharing scheme with secret s_i and distributes these shares among a subset of the users.

The idea is to choose the various access structures, and on which subsets of users they are defined, in such a way that the resulting secret sharing scheme has the desired access structure. This is best illustrated by means of an example.

Example 1. In order to set up an MTA-free $(2, 3)$-threshold scheme among the users A,B and C, a possible construction is as follows:

1. A generates s_A, and B generates s_B. The secret is taken to be $s_A + s_B$;
2. A generates shares s_{AB}, s_{AC} of a $(2, 2)$-threshold scheme with secret s_A and distributes s_{AB} to B and s_{AC} to C. User B sends s_B to C.

In [11] some protocols are given for the design of suitable MTA-free secret sharing schemes. Several efficiency measures were proposed, one being the *linkage* of a protocol, which was defined to be the total number of communications between different users necessary to enable the scheme. In Example 1 the linkage is $2+1 = 3$, which can be shown to be optimal [11].

We now apply this technique to Protocol 1. Firstly, A and $T_1, \ldots, T_m$, and B and $U_1, \ldots, U_n$, both agree to publicly designate one set $\{T_1, \ldots, T_s\} \in \Gamma_T$ and $\{U_1, \ldots, U_t\} \in \Gamma_U$, respectively, to be *nominated sets*. The nominated sets of TTPs have a special role to play in the protocol and for reasons of efficiency (see Sect. 2.3) the best choice of nominated set is a minimal set of smallest size. The generalised protocol set-up is similar to Protocol 1 except that the private receive key b is computed as follows: each nominated U_j sends b_{ij} $(i = 1, \ldots, s)$ to B; user B computes $b_i = \sum_{j=1}^{t} b_{ij}$; user B computes $b = \sum_{i=1}^{s} b_i$. Also prior to communication, each nominated TTP U_j computes $b'_j = \sum_{i=1}^{s} b_{ij}$. The nominated TTPs $U_1, \ldots, U_t$ establish an MTA-free secret sharing scheme for Γ_U by generating shares that correspond to the secrets $b'_1, \ldots, b'_t$, and distributing these shares appropriately among $U_1, \ldots, U_n$.

Protocol 3. The protocol runs as follows:

1. A informs each nominated T_i that they want to communicate with B.
2. Each nominated T_i chooses a_i for A, computes b_{ij} $(j = 1, \ldots, t)$ and $g_A^{b_i} = \prod_{j=1}^{t} g_A^{b_{ij}}$. Each T_i sends a_i, $\text{sig}_{T_i}(g_A^{a_i})$, and $g_A^{b_i}$, to A.
3. $T_1, \ldots, T_s$ establish an MTA-free secret sharing scheme for Γ_T by generating shares that correspond to secrets $a_1, \ldots, a_s$ and distributing these shares appropriately among $T_1, \ldots, T_m$.
4. Each nominated T_i computes b_i. TTPs $T_1, \ldots, T_s$ establish an MTA-free secret sharing scheme for Γ_T by generating shares that correspond to secrets $b_1, \ldots, b_s$ and distributing these shares appropriately among $T_1, \ldots, T_m$.
5. A computes $a = \sum_{i=1}^{s} a_i$, $g_A^b = \prod_{i=1}^{s} g_A^{b_i}$, and $K_{AB} = g_A^{ba} \mod p$. User A then sends $\text{sig}_{T_i}(g_A^{a_i})$ $(i = 1, \ldots, s)$ to B.

6. B verifies $g_{\mathrm{A}}^{a_i}$, and computes g_{A}^{a} and $K_{\mathrm{AB}} = g^{ab} \bmod p$.

Users A and B can now communicate using key K_{AB}. Further, any set of TTPs belonging to Γ_{T} can jointly compute a and b, and any set of TTPs belonging to Γ_{U} can jointly compute b. The correctness of Protocol 3 follows from the correctness of Protocol 1. As with Protocol 1 there is no need for any communication between TTPs from different domains, however our generalisation of the splitting access structure comes at the expense of some communication between TTPs from the same domain. This communication can be minimised by using MTA-free secret sharing schemes with low linkages (see [11] and Sect. 7). For many applications this TTP communication could be off-line. Note that it is only strictly necessary that nominated TTP pairs $(\mathrm{T}_i, \mathrm{U}_j)$ agree on a secret key.

4 Generalising the Second Key Splitting Variant

We now consider generalising the key splitting in Protocol 2. In fact the generalisation is straightforward, as long as secret sharing schemes are used that offer *verifiability* (it should be possible for each participant to detect the submission of a wrong share by another participant), and *linearity* (if x is a share of secret s then $ax + b$ is a share of secret $as + b$).

Verification capabilities can be provided for any secret sharing scheme which has secret and shares $s, s_1, \ldots, s_n \in \mathbf{Z}_p$ (p prime) by the dealer broadcasting $(g^s, g^{s_1}, \ldots, g^{s_n})$, where g is publicly known. Thus, providing verification is not a problem, and as before we omit it from further protocol descriptions.

Secret sharing schemes that are linear are also easy to construct. In particular schemes that are *homomorphic* [1, 9] are useful, because schemes that are homomorphic to themselves (informally, two sets of shares, and secret, can be combined to obtain a third set of shares, and secret, in the same scheme) are linear [11]. Of particular interest are schemes that are homomorphic to themselves with respect to addition over $\mathbf{Z}_p$. Important and useful examples are *geometric schemes* (constructed using projective geometry). Geometric schemes can be constructed for all access structures [16] and are equivalently described in terms of vector spaces [3] and linear error-correcting codes [8].

Having observed that generalising Protocol 2 is fairly straightforward, we note that as stated the protocol is not very efficient. Perhaps the most serious drawback is the number of separate communications that needs to take place between TTPs from different domains, namely $2mn$. In the remaining sections we discuss several methods for improving efficiency.

5 Reducing Communication between Different Domains

5.1 A Simple Communication Reduction

Communication between TTPs from different domains can be reduced (compared to Protocol 2) by the following simple modification.

Protocol 4. Replace Steps 3 and 4 in Protocol 2 with:

3. Select the smallest minimal set $\{T_1, \ldots, T_k\} \in \Gamma_T$. For $i = 1, \ldots, k$, each T_i publishes $g^{s_A s_{TAB}}$, and generates shares $s_{A_{i1}}, \ldots, s_{A_{in}}$ of a secret sharing scheme for Γ_U with secret s_{A_i}. For $j = 1, \ldots, n$, T_i sends $s_{A_{ij}} s_{TAB}$ to U_j.
4. Select the smallest minimal set $\{U_1, \ldots, U_l\} \in \Gamma_U$. For $j = 1, \ldots, l$, each U_j publishes $g^{s_B s_{UAB}}$, and generates shares $s_{B_{j1}}, \ldots, s_{B_{jm}}$ of a secret sharing scheme for Γ_T with secret s_{B_j}. For $i = 1, \ldots, m$, U_j sends $s_{B_{ji}} s_{UAB}$ to T_i.

The only difference with respect to Protocol 2 is that not all the TTPs are involved in communication exchange during Steps 3 and 4. For the threshold splitting case considered in Protocol 2, the number of communications between TTPs is reduced from $2mn$ to $kn + lm$. The minor disadvantage with Protocol 4 is that some agreements have to be reached in advance regarding which minimal set is used in Steps 3 and 4, but if the communication saving is significant then this extra procedure will be worth the effort.

5.2 Using MTA-free Secret Sharing

Recall the idea of MTA-free secret sharing from Sect. 3. We propose a protocol which can be used if the following condition holds:

[C1] there exists an integer k such that Γ_T has a minimal set of size k and Γ_U contains a set of size k.

Protocol 5. Proceed as in Protocol 2 except replace Step 3 with:

3.1. Select the smallest minimal set $\{T_1, \ldots, T_k\} \in \Gamma_T$ such that there exists a set $\{U_1, \ldots, U_k\} \in \Gamma_U$. For $i = 1, \ldots, k$, each T_i publishes $g^{s_A s_{TAB}}$ and sends $s_{A_i} s_{TAB}$ to U_i;
3.2. $U_1, \ldots, U_k$ establish an MTA-free secret sharing scheme for Γ_U by generating shares that correspond to secrets $s_{A_1} s_{TAB}, \ldots, s_{A_k} s_{TAB}$ and distributing these shares appropriately among $U_1, \ldots, U_n$.

The number of communications from TTPs in A's domain to TTPs in B's domain has thus been reduced to k. If an equivalent condition applies in the reverse direction, namely that there exists an integer l such that Γ_U has a minimal set of size l and Γ_T contains a set of size l, then Step 4 can be similarly replaced. If, as in Protocol 2, Γ_T is (k, m)-threshold and Γ_U is (k, n)-threshold, then both conditions hold and the number of communications between TTPs in different domains is reduced from $2mn$ to $2k$.

The reduction of communications between TTPs from different domains in Protocol 5 comes at a cost of introducing communication between TTPs from the same domain. The number of such communications is given by the sum of the linkages of the MTA-free secret sharing schemes used in Steps 3 and 4. Thus to minimise this number we should again use MTA-free secret sharing schemes with low linkages (see [11] and Sect. 7).

We note that to achieve minimal linkages it is sometimes necessary for a user to broadcast his share to the other users [11]. This translates in Protocol 5 to one TTP U_i broadcasting $s_{A_i} s_{TAB}$ to the other TTPs U_j. As a broadcast is insecure we assume that all other parties can obtain $s_{A_i} s_{TAB}$. This information would seem of litle use to any other party except for user A who knows s_{A_i} and could thus determine s_{TAB}. It is not clear whether this is a problem because the main role of s_{TAB} is to provide a mask for the passing of s_{A_i} to the TTPs from the other domain. However, should it not be desirable that A can learn the value of s_{TAB} in this manner then the broadcasts, where appropriate, should be replaced by secure transmissions and hence the number of communications between TTPs from the same domain will be increased slightly.

In the event that [C1] only fails because $k > n$ then Protocol 5 can still be used. The extra TTPs T_i can either broadcast their shares in Step 3.1, or should send their information to some predetermined U_j, who either broadcasts the shares or securely distributes them among the other TTPs U_j. The latter method should be used if broadcasting raises the same previous concerns.

5.3 Factoring Access Structures

If [C1] is not satisfied then in certain cases it is still possible to make improvements. Firstly we recall a convenient representation of an access structure [16].

Example 2. Let Γ be $\{A, B, C\}, \{A, B, D\}, \{A, B, C, D\}$. The minimal sets are $\{A, B, C\}$ and $\{A, B, D\}$ and we represent this as $\Gamma = ABC + ABD$. This notation is convenient because we can replace the right hand side by any equivalent logical expression and use this as an equally valid representation of the access structure. So, for instance, we can also write $\Gamma = AB(C + D)$.

Let Γ be an access structure that can be represented in the form $\Gamma = \Gamma_1 \Gamma_2 \dots \Gamma_r$, where each Γ_i is an access structure (defined on some subset of participants). We refer to the Γ_i as *factors* of Γ. In Example 2, AB and $C + D$ are factors of Γ. We can use the factors of an access structure to obtain an efficient protocol if the following condition holds:

[C2] there exists a minimal set in Γ_T of size r and a factorisation of Γ_U of the form $\Gamma_U = \Gamma_{U_1} \dots \Gamma_{U_r}$.

Protocol 6. Proceed as in Protocol 2 except replace Step 3 with:

3. Select a minimal set $\{T_1, \dots, T_k\} \in \Gamma_T$ such that $\Gamma_U = \Gamma_{U_1} \dots \Gamma_{U_k}$. For $i = 1, \dots, k$, each TTP T_i publishes $g^{s_{A} s_{TAB}}$ and generates shares $s_{A_{i1}}, \dots, s_{A_{ir_i}}$ of a secret sharing scheme with access structure Γ_{U_i} and secret s_{A_i}. For $j = 1, \dots, r_i$, T_i sends $s_{A_{ij}} s_{TAB}$ to each U_{i_j}, where Γ_{U_i} is defined on $\{U_{i_1}, \dots, U_{i_{r_i}}\}$;

If the equivalent of [C2] holds from Γ_U to Γ_T then we can introduce a similar improvement to Step 4.

Example 3. Let Γ_T be $(2,3)$-threshold and Γ_U be $(3,3)$-threshold. Then we can write $\Gamma_U = \Gamma_{U_1}\Gamma_{U_2}$, where Γ_{U_1} is $(2,2)$-threshold on $\{U_1, U_2\}$ and Γ_{U_2} is $(1,1)$-threshold on U_3. Then Step 3 becomes:

3. Choose $\{T_1, T_2\}$. TTP T_1 generates shares $s_{A_{11}}, s_{A_{12}}$ of a $(2,2)$-threshold scheme with secret s_{A_1}, sends $s_{A_{11}}s_{TAB}$ to U_1 and sends $s_{A_{12}}s_{TAB}$ to U_2. TTP T_2 sends $s_{A_2}s_{TAB}$ to U_2.

For the access structures in Example 3, the number of communications from TTPs T_i to TTPs U_j using Protocols 2, 4, and 6, are 9, 6 and 3, respectively.

6 Reducing Other Protocol Operations

6.1 Reducing User to TTP Communication

In Protocol 2, communication between users and their home TTPs takes place in Steps 1, 2, 5 and 6. The only way of 'reducing' the communication in Steps 1 and 2 is to make sure that m and n are chosen to be as small as possible. Note that if Γ_T and Γ_U are threshold access structures, as in Protocol 2, then the size of m and n is not an indication of the security level (this is determined by the threshold value k), but is rather an indication of the system flexibility.

In Steps 5 and 6 of (the generalised) Protocol 2, and in all other variants discussed in this paper, a trivial saving in user to TTP communication overheads can be made by nominating just one of the TTPs T_i and U_j to pass on the final value to users A and B respectively. For Protocol 2 this reduces the number of communications between users and their TTPs from $2(m+n)$ to $m+n+2$.

6.2 Reducing the Generation of Secret Sharing Schemes

In this section we consider minimising the generation of secret sharing schemes in Protocol 2. Note that these protocols do not necessarily involve less communication between TTPs from different domains than the protocols of Sect. 5.

In the simple case that $\Gamma_T = \Gamma_U$ we can remove the need for generation of any secret sharing schemes in Steps 3 and 4 of Protocol 2 as follows:

Protocol 7. If $\Gamma_T = \Gamma_U$ it is sufficient to replace Steps 3, 4 of Protocol 2 with:

3. Each TTP T_i computes s_{TAB}, publishes $g^{s_A s_{TAB}}$ and sends $s_A s_{TAB}$ to U_i.
4. Each TTP U_j computes s_{UAB}, publishes $g^{s_B s_{UAB}}$ and sends $s_B s_{UAB}$ to T_j.

The difference between Protocol 7 and the use of Protocol 4 for transfer between two (k, n)-threshold schemes is that Protocol 7 involves n communications between TTPs from different domains (just k for Protocol 4), but no communication between TTPs from the same domain, or generation of secret sharing schemes $(kn - (1/2)k(k+1)$ and k respectively for Protocol 4).

Let Γ be an access structure defined on a set $\mathcal{P}$. Then a *cumulative map* for Γ is a set $\mathcal{S}$ and a mapping α from $\mathcal{P}$ to the collection of subsets of $\mathcal{S}$, such that for any $A \subseteq \mathcal{P}$, $\cup_{P \in A} \alpha(P) = \mathcal{S} \iff A \in \Gamma$. Let Γ^+ denote the collection of maximal sets that do not belong to Γ. Then in [16] it was shown that there exists a unique *minimal* cumulative map for Γ, where *minimal* is used in the sense that all other cumulative maps contain the minimal solution. The minimal cumulative map for Γ can be defined by letting $\mathcal{S} = \Gamma^+$ and then for any $P \in \mathcal{P}$, $\alpha(P) = \{B \in \Gamma^+ \mid P \notin B\}$.

Example 4. Let $\Gamma = AB + BCD$. Then Γ^+ consists of the sets $B_1 = \{A, C, D\}$, $B_2 = \{B, D\}$, $B_3 = \{B, C\}$. Thus let $\mathcal{S} = \{B_1, B_2, B_3\}$. The minimal cumulative map for Γ is $\alpha(A) = \{B_2, B_3\}$, $\alpha(B) = \{B_1\}$, $\alpha(C) = \{B_2\}$, $\alpha(D) = \{B_3\}$.

Note that we can easily construct a cumulative map for Γ which has $|\mathcal{S}| > |\Gamma^+|$ from the minimal cumulative map for Γ. The following protocol does not need the generation of any secret sharing schemes in Step 3. It can be used if:

[C3] there exists a minimal set in Γ_T that is at least the size of Γ_U^+.

Protocol 8. Proceed as in Protocol 2 except replace Step 3 with:

3. Select the smallest minimal set $\{T_1, \ldots, T_k\} \in \Gamma_T$ such that $k \geq |\Gamma_U^+|$. Let $\mathcal{S}$ (of size k) and α be a cumulative map for Γ_U. For $i = 1, \ldots, k$, each T_i publishes $g^{s_A s_{TAB}}$ and sends $s_{A_i} s_{TAB}$ to each U_i that belongs to $\alpha(U_i)$;

If the equivalent of [C3] applies from Γ_U to Γ_T^+ then we can introduce a similar change to Step 4. As the number of TTPs U_i increases, the number of sets in Γ_U^+ generally increases at an exponential rate, so Protocol 8 is likely to be useful only if m is large, relative to n. For *connected* access structures Γ (every participant belongs to at least one minimal set of Γ), the number of sets in Γ^+ is at least the number of participants. Thus, generally [C3] only holds if [C1] holds. It follows that Protocol 8 is an alternative to Protocol 5 for use when minimising the generation of secret sharing schemes, rather than as an option for cases when [C1] does not hold.

7 Threshold Transferral

In this section Γ_T and Γ_U are threshold access structures. We aim to first minimise the number of communications between TTPs from different domains, and describe solutions that achieves this, while also keeping communication between TTPs from the same domain, and generation of secret sharing schemes, to a minimum. We recall a result from [11]:

Result 9. The minimum linkage for an MTA-free (k, n)-threshold scheme is $nk - (1/2)k(k+1)$, and can be realised by the *Contraction Construction* [11].

Example 1 has the minimum possible linkage of 3 for a $(2, 3)$-threshold scheme and is an example of application of the Contraction Construction. There are three cases of threshold transferral to consider:

Case 1. (k, m)-*threshold to* (n, n)-*threshold.* Use Protocol 6. This needs a total of max k, n communications between TTPs from different domains, involves no communications between TTPs from the same domain, and needs generation of one secret sharing scheme (by one of the sending TTPs) if $k < n$.

Case 2. (k, m)-*threshold to* (l, n)-*threshold,* $(k \geq l, l < n)$. Use Protocol 5. This needs k communications between TTPs from different domains. Using the Contraction Construction it needs $nk - (1/2)k(k-1)$ communications between (receiving) TTPs from the same domain, and needs generation of k secret sharing schemes (by receiving TTPs).

Case 3. (k, m)-*threshold to* (l, n)-*threshold,* $(k < l < n)$. Use Protocol 4. This needs kn communications between TTPs from different domains, involves no communication between TTPs from the same domain, and needs generation of k secret sharing schemes (by sending TTPs).

Given that the complete version of any of the described protocols involves a transfer from the TTPs T_i to the TTPs U_j and then from the TTPs U_j to the TTPs T_i, we identify in Table 1 five cases for consideration over a complete protocol (where in each direction one of Cases 1, 2 and 3 applies). For each Type

Table 1. Types of Threshold Transferral

Type	Transfer between thresholds			Conditions
1	(m, m)	and	(n, n)	$m \neq n$
2	(m, m)	and	(l, n)	$l < m, l < n$
3	(m, m)	and	(l, n)	$m < l < n$
4	(k, m)	and	(l, n)	$k < m, l < k, l < n$
5	(k, m)	and	(k, n)	

shown in Table 1 we compare the use of the Protocols suggested in Cases 1, 2 and 3 with Protocol 2. The necessary number of communications are given in Table 2. As can be seen from Table 2, for each type the number of communications

Table 2. Efficiencies of Threshold Transferral

	Using Protocol 2			Using Cases 1,2,3		
Type	Diff Dom	Same Dom	SSS Gen	Diff Dom	Same Dom	SSS Gen
1	$2mn$	0	$m + n$	$2 \max m, n$	0	1
2	$2mn$	0	$m + n$	$m + \max l, m$	$ln - (1/2)l(l+1)$	$l + 1$
3	$2mn$	0	$m + n$	$mn + \max l, m$	0	m
4	$2mn$	0	$m + n$	$k + lm$	$ln - (1/2)l(l+1)$	$2l$
5	$2mn$	0	$m + n$	$2k$	$k(m + n - k - 1)$	$2k$

between TTPs from different domains when using Cases 1, 2 and 3 is no more, and usually significantly less, than if Protocol 2 had been used. Further the total number of communications between TTPs (from the same and different domains) is generally less than for Protocol 2. The number of secret sharing schemes needing generated is always less than for Protocol 2.

Note that the use of Protocol 8 is not recommended for transfer between threshold access structures (when [C3] applies). The reason is that the sizes of set S and the sizes of the sets $\alpha(P_i)$ are relatively large for threshold access structures (compared to other access structures on the same number of participants). In all cases the simple change suggested in Sect. 4.1 can be used to reduce communication between users and their home TTPs.

References

1. J. Benaloh. Secret sharing homomorphisms: keeping shares of a secret secret. *Adv. in Cryptology - CRYPTO'86, Lecture Notes in Comput. Sci*, **263** (1987) 251–260.
2. G. R. Blakley. Safeguarding cryptographic keys. *Proceedings of AFIPS 1979 National Computer Conference*, **48** (1979) 313–317.
3. E. F. Brickell. Some ideal secret sharing schemes. *J. Combin. Math. Combin. Comput.*, **9** (1989) 105–113.
4. L. Chen, D. Gollmann and C.J. Mitchell. Key escrow in mutually mistrusting domains. *Proceedings of 1996 Cambridge Workshop on Security Protocols, Lecture Notes in Comput. Sci.*, **1189** (1997) 139–153.
5. D.E. Denning and D.K. Branstad. A taxonomy for key escrow encryption systems. *Communications of the ACM*, **39**(3) (1996) 34–40.
6. Y. G. Desmedt and Y. Frankel. Homomorphic zero-knowledge threshold schemes over any finite abelian group. *SIAM J. Disc. Math.*, **4** (1994) 667–679.
7. W. Diffie and M. Hellman. New directions in cryptography. *IEEE Transactions on Information Theory*, **22** (1976) 644–654.
8. M. van Dijk, On the information rate of perfect secret sharing schemes, *Des. Codes Cryptogr.*, **6** (1995) 143–169.
9. Y. Frankel and Y. Desmedt. Classification of ideal homomorphic threshold schemes over finite Abelian groups. *Adv. in Cryptology – EUROCRYPT'92, Lecture Notes in Comput. Sci.*, **658** (1992) 25–34.
10. I. Ingemarsson and G. J. Simmons. A protocol to set up shared secret schemes without the assistance of a mutually trusted party. *Adv. in Cryptology – EUROCRYPT'90, Lecture Notes in Comput. Sci.*, **473** (1991) 266–282.
11. W.-A. Jackson, K.M. Martin and C.M. O'Keefe. Mutually Trusted Authority free Secret Sharing Schemes. *J. Cryptology*, to appear.
12. N. Jefferies, C. Mitchell and M. Walker. A proposed architecture for trusted third party services. *Cryptography: Policy and Algorithms, Lecture Notes in Comput. Sci.*, **1029** (1996) 98–104.
13. N. Jefferies, C. Mitchell and M. Walker. Practical solutions to key escrow and regulatory aspects. *Proceedings of Public Key Solutions '96*, Zurich, September 1996.
14. S. Micali. Fair cryptosystems. *MIT Technical Report*, MIT/LCS/TR-579.c, (1994).
15. A. Shamir. How to share a secret. *Comm. ACM*, **22**(11) (1979) 612–613.
16. G. J. Simmons, W.-A. Jackson, and K. Martin. The geometry of shared secret schemes. *Bull. Inst. Combin. Appl.*, **1** (1991) 71–88.

Multi Dimensional Compartment Schemes

Johannes Maucher
Dept. of Information Technology
University of Ulm
Germany

Abstract

We define a multi dimensional compartment scheme, which is a secret sharing scheme where each participant acts not only as a member of one party, but as a representative of one party on each dimension. The secret can be reconstructed whenever on each dimension at least one representative of a predetermined number of distinct parties contribute its private share. It is also shown how a non-complete multi dimensional compartment scheme can be realized.

1 Introduction

A secret sharing scheme is a procedure, where the key distributor (dealer) splitts a secret s into a set of b private shares $C = \{c_0, \ldots, c_{b-1}\}$. Any participant $P_k \in \mathcal{P}, |\mathcal{P}| \leq b$, receives a private share c_k and whenever an authorized subset of participants put together their private shares the secret s can be reconstructed. The set of all authorized subsets of participants is denoted by Γ and the set of all subsets of $\mathcal{P}$, which cannot obtain any information on the secret is denoted by Δ. The access structure is then defined by (Γ, Δ) and it is called complete, if $\Delta = \Gamma^c = \{Z \mid Z \notin \Gamma\}$. A secret sharing scheme where the access structure consists of all subsets, containing at least t participants is called a (t, b)−threshold scheme, introduced independently by Shamir [3] and Blakely [1]. However the vast variety of possible applications demand for secret sharing schemes, realizing different access structures. For example in a compartment (or multiparty) scheme the participants are divided into N disjoint parties $\mathcal{B}_1, \ldots, \mathcal{B}_N$. To each party $\mathcal{B}_i$ is assigned a threshold t_i, and the secret can be reconstructed whenever in at least K parties $\mathcal{B}_{i_j}$ the number of participants, which contribute their private shares is at least t_{i_j}.

In this paper we consider a secret sharing scheme where there exists b_l disjoint parties $\mathcal{B}_0^{(l)}, \ldots, \mathcal{B}_{b_l-1}^{(l)}$ on each dimension l. Any participant P belongs to one party $\varphi^l(P)$ on each dimension $l \in \{1, \ldots, m\}$.

The map of the participants to the parties on the different dimensions can be visualized in the 2 - dimensional case by the following array :

$$
\begin{array}{ccccc}
P_0 & P_1 & \cdots & P_{b_1-1} & \} \ \ \mathcal{B}_0^{(2)} \\
P_{b_1} & P_{b_1+1} & \cdots & P_{2b_1-1} & \} \ \ \mathcal{B}_1^{(2)} \\
\vdots & \vdots & \ddots & \vdots & \vdots \ \ \vdots \\
\underbrace{P_{(b_2-1)b_1}} & \underbrace{P_{(b_2-1)b_1+1}} & \cdots & \underbrace{P_{b_2b_1-1}} & \} \ \ \mathcal{B}_{b_2-1}^{(2)} \\[2mm]
\mathcal{B}_0^{(1)} & \mathcal{B}_1^{(1)} & \cdots & \mathcal{B}_{b_1-1}^{(1)} &
\end{array}
$$

The participant P_k in column i and row j belongs to the party $\varphi^1(P_k) = \mathcal{B}_i^{(1)}$ on the first dimension and to the party $\varphi^2(P_k) = \mathcal{B}_j^{(2)}$ on the second dimension.

An m-dimensional compartment scheme, which we formally define in the next section is a secret sharing scheme where there exists a threshold $t_l \leq b_l$ on each dimension l and the secret is reconstructable whenever the participants which contributed their shares belong to at least t_l distinct parties on each dimension l.

For a possible application of m-dimensional compartment schemes we consider the following example.

Let the participants be representatives of b_2 different states. The representatives of each state are ministers of b_1 distinct departments. Therefore on the second dimension ministers of the same state belong to the same party $\mathcal{B}_j^{(2)}$ and on the first dimension ministers of the same department, but different state, belong to the same party $\mathcal{B}_i^{(1)}$. Imposing on this constellation an m-dimensional compartment scheme provides that the secret can be reconstructed whenever representatives of at least t_2 different states and at least t_1 different departments contribute their private shares.

2 Definition of the m-Dimensional Compartment Scheme

Any participant $P_k \in \mathcal{P}$ belongs to one party $\varphi^l(P_k)$ on each dimension $l \in \{1,\ldots,m\}$. The components t_l of the m-dimensional vector $\bar{t} = (t_1,\ldots,t_m)$ determine the thresholds and the components $b_l \geq t_l$ of the vector $\bar{b} = (b_1,\ldots,b_m)$ determine the number of different parties on each dimension l. We replace the index k of participant P_k by an m-dimensional vector $\bar{u}(k) = (u_1(k),\ldots,u_m(k))$, such that $P_k = P_{\bar{u}(k)}$ and $P_{\bar{u}(k)}$ and $P_{\bar{u}(z)}$ belong to the same party on dimension l if the $l.th$ component of their index vectors coincide, i.e. $\varphi^l(P_{\bar{u}(k)}) = \varphi^l(P_{\bar{u}(z)}) = \mathcal{B}_\xi^{(l)}$, if $u_l(k) = u_l(z) = \xi$. The private share which is given to participant $P_{\bar{u}(k)}$ is denoted by c_k.

Definition 1 : (**m-dimensional compartment scheme**) *Let* $\gamma = \{P_{\bar{u}(1)}, \ldots, P_{\bar{u}(r)}\} \subseteq \mathcal{P}$ *be a set of participants which contributed their private shares. The set of parties on dimension l, which have at least one representative in γ is denoted by*

$$\Phi^l(\gamma) = \bigcup_{P_{\bar{u}(i)} \in \gamma} \varphi^l(P_{\bar{u}(i)}) = \bigcup_{P_{\bar{u}(i)} \in \gamma} \mathcal{B}^l_{u_l(i)}$$

and the number of distinct parties in $\Phi^l(\gamma)$ is $n^{(l)}(\gamma) = |\Phi^l(\gamma)|$. Then a $(\bar{t}, \bar{b})$-m-dimensional compartment scheme $((\bar{t}, \bar{b})$-mdcs) is defined by

$$\Gamma = \left\{\gamma \mid n^{(l)}(\gamma) \geq t_l , \ \forall \ l \in \{1, \ldots, m\}\right\}.$$

A constellation γ which is authorized to reconstruct the secret s in a $(\bar{t}, \bar{b})$-mdcs is therefore a set of participants, such that these participants belong to more than t_l parties on each dimension l. Note that the definition of the $(\bar{t}, \bar{b})$-mdcs is general in the sense that it does not specify the set $\Delta \subseteq \Gamma^c$.

Example 1 : In a 2-dimensional compartment scheme we can arrange the participants in a $b_2 \times b_1$ array such that participants in the same row belong to the same party on dimension 2 and participants in the same column belong to the same party on dimension 1. For a $(\bar{t}, \bar{b})$-mdcs with $\bar{t} = (3, 2)$ and $\bar{b} = (4, 3)$ the participants are arranged as follows :

$$
\begin{array}{llll}
P_{(0,0)} & P_{(1,0)} & P_{(2,0)} & P_{(3,0)} \quad \} \ \mathcal{B}^{(2)}_0 \\
P_{(0,1)} & P_{(1,1)} & P_{(2,1)} & P_{(3,1)} \quad \} \ \mathcal{B}^{(2)}_1 \\
P_{(0,2)} & P_{(1,2)} & P_{(2,2)} & P_{(3,2)} \quad \} \ \mathcal{B}^{(2)}_2 \\
\underbrace{\phantom{P_{(0,0)}}} & \underbrace{\phantom{P_{(1,0)}}} & \underbrace{\phantom{P_{(2,0)}}} & \underbrace{\phantom{P_{(3,0)}}} \\
\mathcal{B}^{(1)}_0 & \mathcal{B}^{(1)}_1 & \mathcal{B}^{(1)}_2 & \mathcal{B}^{(1)}_3
\end{array}
$$

For the subset $\gamma_1 = \{P_{(0,0)}, P_{(1,0)}, P_{(2,0)}, P_{(3,0)}\}$ we have $\Phi^1(\gamma_1) = \{\mathcal{B}^{(1)}_0, \mathcal{B}^{(1)}_1, \mathcal{B}^{(1)}_2, \mathcal{B}^{(1)}_3\}$, $n^{(1)}(\gamma_1) = 4 \geq t_1$, but as $\Phi^2(\gamma_1) = \{\mathcal{B}^{(2)}_0\}$, $n^{(2)}(\gamma_1) = 1 < t_2$ the constellation γ_1 does not belong to Γ.

3 Construction of an m-Dimensional Compartment Scheme

In this section we represent an algorithm for generating and distributing the private shares c_k in a non-complete $(\bar{t}, \bar{b})$-mdcs. As the proposed mdcs is closely related to the original (t, b)-threshold scheme, introduced by Shamir in [3], we start with a short description of Shamir's threshold scheme : The private shares $c_k \in GF(q)$, $k \in \{0, \ldots, b-1\}$ are derived from the secret $s \in GF(q)$ by evaluating a polynomial $a(x) = a_0 + a_1 x + \cdots + a_{t-1} x^{t-1}$ of degree t with coefficients in $GF(q)$ at the $b < q$ places $\alpha^0, \alpha^1, \ldots, \alpha^{b-1}$, where α is a primitive element in $GF(q)$. The coefficient a_0 is the secret s and all the other coefficients of $a(x)$ are chosen randomly by the dealer. The private share for participant P_k is then $c_k = a(x_k = \alpha^k)$. The secret s can be reconsructed by interpolation of the polynomial $a(x)$ whenever at least t pairs (x_k, c_k) are known. Usually only c_k is the private share and x_k which corresponds to participant P_k is publicly known. Based on this threshold scheme the basic idea for the construction of an m-dimensional compartment scheme can be described as follows.

Let the secret $\bar{s} = (s_1, \ldots, s_m)$ and the share $\bar{c}_k = (c_{k,1}, \ldots, c_{k,m})$ for participant P_k be m-dimensional vectors. The share components $c_{k,l}$ are derived from the secret component s_l by setting $a_0^{(l)} := s_l$ and evaluating a polynomial $a^{(l)}(x) = a_0^{(l)} + a_1^{(l)} x + \cdots + a_{t_l-1}^{(l)} x^{t_l-1}$ at the places $x_{k,l} = \alpha^i$ for all dimensions $l \in \{1, \ldots, m\}$. However a share component $c_{k,l}$ is not only given to one participant but to all participants which belong to a common party on dimension l, i.e. to all participants P_k with index vector $\bar{u}(k)$ and $u_l(k) = i$. Algorithm 1 describes this key distribution for an $(\bar{t}, \bar{b})$-mdcs more formally. Note that for all l the condition $t_l \leq b_l < q$ must be satisfied.

Algorithm 1 :

Input : $\bar{s} = (s_1, \ldots, s_m)$, $\bar{t} = (t_1, \ldots, t_m)$, $\bar{b} = (b_1, \ldots, b_m)$
Output : $\bar{c}_k = (c_{k,1}, \ldots, c_{k,m})$ for all $P_k \in \mathcal{P}$

For l from 1 to m **do**
- Set $a_0^{(l)} := s_l$

- Choose $a_1^{(l)}, \ldots, a_{t_l-1}^{(l)} \in GF(q)$ by random

- **For** i from 0 to $b_l - 1$ **do**
  - $c_l(i) = a_0^{(l)} + a_1^{(l)}(\alpha^i)^1 + \ldots + a_{t_l-1}^{(l)}(\alpha^i)^{t_l-1}$
  - Give the share component $c_l(i)$ to all participants $P_{\bar{u}(k)}$ with $u_l(k) = i$

- **end**
end

Example 2 :
For $q = 5$ and α a primitive element in $GF(5)$ the private shares for the participants in the $(\bar{t}, \bar{b})$-2-dimensional compartment scheme of example 1 are derived from the secret $\bar{s} = (\alpha, \alpha^3)$ as follows: Set $a_0^{(1)} := s_1 = \alpha$, $a_0^{(2)} := s_2 = \alpha^3$ and choose for example $a_1^{(1)} := \alpha$, $a_2^{(1)} := \alpha^2$, $a_1^{(2)} := 1$. For all $i \in \{0, \ldots, 3\}$ compute $c_1(i) := a^{(1)}(x = \alpha^i)$ and give it to all participants $P_{(i,u_2)}$ and for all

$j \in \{0,\ldots,2\}$ compute $c_2(j) := a^{(2)}(x = \alpha^j)$ and give it to all participants $P_{(u_1,j)}$. The private shares of the participants are then

$$\begin{array}{cccccccccccc}
P_{(0,0)} & P_{(1,0)} & P_{(2,0)} & P_{(3,0)} & P_{(0,1)} & P_{(1,1)} & P_{(2,1)} & P_{(3,1)} & P_{(0,2)} & P_{(1,2)} & P_{(2,2)} & P_{(3,2)} \\
\binom{\alpha^3}{\alpha^2} & \binom{\alpha}{\alpha^2} & \binom{\alpha^2}{\alpha^2} & \binom{\alpha^3}{\alpha^2} & \binom{\alpha^3}{0} & \binom{\alpha}{0} & \binom{\alpha^2}{0} & \binom{\alpha^3}{0} & \binom{\alpha^3}{\alpha} & \binom{\alpha}{\alpha} & \binom{\alpha^2}{\alpha} & \binom{\alpha^3}{\alpha}
\end{array}$$

Theorem 1 : : *The access structure (Γ, Δ) of a secret sharing scheme, where the private shares are distributed as in Algorithm 1 is the access structure of a $(\bar{t}, \bar{b})$-multi dimensional compartment scheme, with Γ as defined in Definition 1 and $\Delta = \gamma \mid n^{(l)}(\gamma) < t_l, \ \forall l \in \{1,\ldots,m\}$.*

Proof : From the proof of Shamir's (t, b) - threshold scheme we know that for a given dimension l the secret component s_l can be reconstructed whenever at least t_l distinct points $(x_{k,l}, c_{k,l})$, where $c_{k,l} = a^{(l)}(x_{k,l})$ and $x_{k,l} = \alpha^i$, are known. All participants which belong to the same party on dimension l carry the same information on this dimension, because they all correspond to the same point $(x_{k,l}, c_{k,l})$ on this dimension. Distinct points for a given dimension correspond to participants of distinct parties on this dimension. Therefore the participants which contributed their shares must belong to at least t_l distinct parties in order to reconstruct the secret component s_l. The secret componentes can be reconstructed on each dimension indepentendly and the secret $\bar{s}$ is available, if the reconstruction of all components s_l has been successfull. $\qquad\square$

An important measure of the security of a secret sharing scheme is the probability Q of a successfull estimation of the secret by a non-authorized constellation of participants. A non-authorized constellation is any subset of participants which does not belong to the set Γ. A secret sharing scheme is called *complete* if no non-authorized constellation can get any information about the secret s. The proposed m-dimensional compartment scheme is obviously non-complete. A non-authorized constellation $\gamma \notin \Delta$, which is able to reconstruct $r < l$ components of the secret vector $\bar{s}$ can estimate the secret with the success probability

$$Q = (q^{l-r})^{-1}$$

and therefore has more information about the secret than an outstanding person or a non-authorized constellation which belongs to the set Δ.

4 Conclusions

We defined a multi dimensional compartment scheme which is well structured and easy to describe. Also many applications for such a scheme are considerable. We represented a very straightforward construction of such a multi dimensional compartment scheme. However the constructed scheme is non-complete and the construction of complete schemes should be considered in further investigations.

238

References

[1] G.R. Blakely : Safeguarding Kryptographic Keys, *Proc. AFIPS 1979 Natl. Computer Conf., New York, vol. 48, pp.313-317, June 1979.*

[2] E.F. Brickell : Some Ideal Secret Sharing Schemes, *J. Combin. Math. and Combin. Comput., 9, pp. 105-113 , 1989.*

[3] A. Shamir : How to Share a Secret, *Massachusetts Institute of Technology, Technical Report MIT/LCS/TM-134, May 1979.*

[4] G.J. Simmons : An Introduction to Shared Secret and/or Shared Control Schemes and their Application, *Contemporary Cryptology, IEEE Press, pp. 441-497, 1991.*

Evaluation of Standard Approximation to Log-Likelihood Ratio Addition in the MAP Algorithm, and Its Application in Block Code ('Turbo') Iterative Decoding Algorithms

S.McManus & P.G.Farrell
Communications Research Group
University of Manchester
UK

Abstract

This paper examines the log-likelihood addition aspects of the Turbo Decoding of product codes. Simple Parity Check (SPC) codes are used as the Constituent Codes (CCs) of the Turbo encoder in an array code format. The general approximation used in the log-likelihood addition of the soft values of the code bits is evaluated and some results are presented. Two iterative ('Turbo') decoding algorithms for block codes are compared, only one of which uses the previously mentioned approximation. The first is set out in Hagenauer (1994), which uses the dual code method to implement the MAP decoder, and the second was developed in Pyndiah (1994), which uses an algorithm presented in Chase (1972). The results presented show the affect of the approximation on decoder performance. The outcome differs somewhat from that which was expected.

1. Introduction to Turbo Codes

It can be said that turbo codes are in fact an amalgamation of several different advanced coding techniques which when brought together form particularly powerful codes. These coding techniques include parallel or serial concatenation of the CCs, soft-decision demodulation, data interleaving and soft iterative decoding (i.e. the use of soft decision information in the iteration process).

It should be noted that generally a notational convention in this paper is that upper case denotes a vector, and lower case an element of a vector. Also, U is *transmitted* systematic bits, P^h and P^v are *transmitted* parity bits. Y, Y^h and Y^v are *received* systematic bits, Encoder1 parity and Encoder2 parity respectively.

The notation (n,k) refers to the number of input bits (message bits), k, and the number of output bits (code bits), n, which characterise a code.

1.2 A Turbo Coding Framework

A turbo code framework consists of the parallel (or serial) concatenation of two (or more) systematic codes, be they convolutional or block codes, or a mixture of both, and was introduced, in its parallel, convolutional form, in Berrou (1993). The serial configuration was suggested, before the 'discovery ' of turbo codes, in Forney (1966), and has been used in a hard decision format for many years. Soft demodulation is used at the receiver, to provide confidence information about the received code bits. All values used in the decoding process are represented as soft decisions. The function of the interleaver, P, is to ensure that input sequences that give rise to low-weight codewords from Encoder1 are rearranged to enable Encoder2 to produce high-weight codewords (see Fig.1.). Information in low-weight codewords is more susceptible to error. If that same information is also combined in high-weight codewords then cross referencing of the two received sets of parity bits will enable correction of more errors. The two decoders share information about the received data and by iteration the turbo decoder output asymptotically approaches the originally transmitted data.

A simplified version of a parallel concatenated turbo codec can be seen in Fig.1. The information bits are input into Encoder1 directly, which produces a parity output (P^h). After permutation the Information bits are input into Encoder2, which produces a parity output (P^v). It could be said that the two parities together form a 'turbo parity.' The information bits, P^h and P^v are combined together to form the 'turbo codeword' which is transmitted across the channel. It is easy to see that the turbo code is systematic.

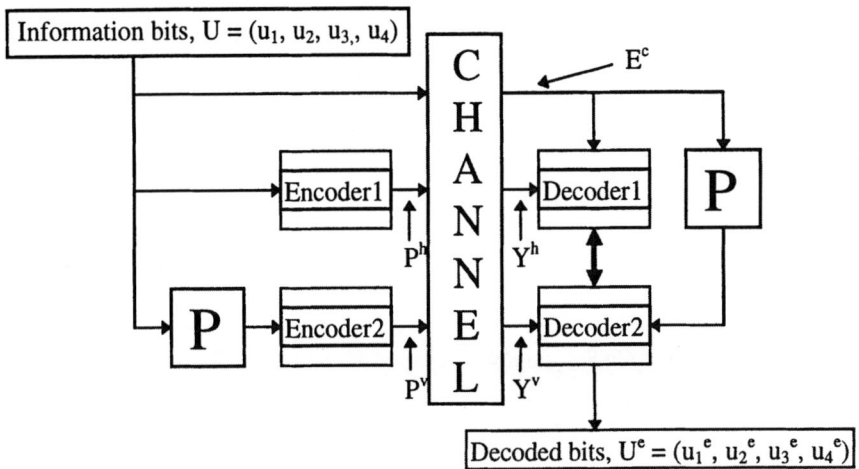

Fig.1. a Parallel concatenated Turbo codec.

In order to understand the decoding process it is necessary to appreciate the method used to pass error information between the two decoders. The decoders receive the relevant parity information and the information bits from the channel. Note that the information bits are interleaved (permuted) before they are input to Decoder2. The parity contains information about the systematic bits used in its generation. This information is called Extrinsic Information. To extract the Extrinsic Information about a particular systematic bit, we must first remove the information pertaining to the other systematic bits from that parity. This process can be viewed as a reversal of the parity check mechanism. The Extrinsic Information generated by Decoder1 - from the received, systematic bits E^c, and Encoder1 parity, Y^h - E^h, can be interleaved and passed to Decoder2 and used as a priori information in the generation of the Extrinsic Information - from the systematic bits, E^c, and the received parity, Y^v, - E^v. E^v can then be used in the next iteration as a priori information for Decoder1, and so on.... This process is represented on Fig.1. by the bi-directional arrow and can be best understood with the aid of a diagram. The different parts of the decoding process, for a simple (8,4) Turbo code with SPC CCs in an array code format, are illustrated in Fig.2. The shaded areas indicate the information used to generate $L(Ue)_j$. The notation and Fig.2. are from Hagenauer (1996).

A summary of notation follows.

1.3 Log-likelihood Algebra

A Log-likelihood Ratio (LLR) is defined as :

$$L(X) = \log_e \frac{P_X(x = +1)}{P_X(x = -1)} \qquad (1.0)$$

where P_X indicates the probability that X equals the value of x. $L(X)$ is the 'soft value' referred to earlier. The soft value consists of a hard decision and a confidence level. The sign of the soft value is the hard decision, and the magnitude of the soft value is how much confidence there would be in that hard decision, i.e. the confidence level.

To represent the addition of two log-likelihood ratios we will use the symbol '$+$', which is defined in (1.1) :

$$L(x_1) + L(x_2) = L(x_1 \oplus x_2) \qquad (1.1)$$

and for statistically independent random variables X_1 and X_2:

$$L(x_1 \oplus x_2) \equiv \log_e \frac{1 + e^{L(x1)}.e^{L(x2)}}{e^{L(x1)} + e^{L(x2)}} \qquad (1.2)$$

$$\equiv \text{sign} (L(x_1)) \cdot \text{sign} (L(x_2))$$
$$\cdot \min(|L(x_1)|, |L(x_2)|) \qquad (1.3)$$

Additional rules include :

$$L(x) + \infty = L(x),\ L(x) + -\infty = -L(x)\ \text{and}\ L(x) + 0 = 0 \qquad (1.4)$$

We can use the approximation given in (1.3) to calculate the addition of log-likelihood ratios. This summary was first presented in Hagenauer (1996).

1.4 Decoding Using Iterative Decoding and the MAP Algorithm

The received information that was transmitted over the channel is the primary estimate of the original information. This is represented by $E^c = L_c \cdot y$, where L_c is the reliability of the channel ($L_c = 4.E_S/N_0$ for the Gaussian channel) and Y is the received systematic bits, (y_1, y_2, y_3, y_4). $Y^h = (y_1{}^h, y_2{}^h)$ and $Y^v = (y_1{}^v, y_2{}^v)$ are the

E^c
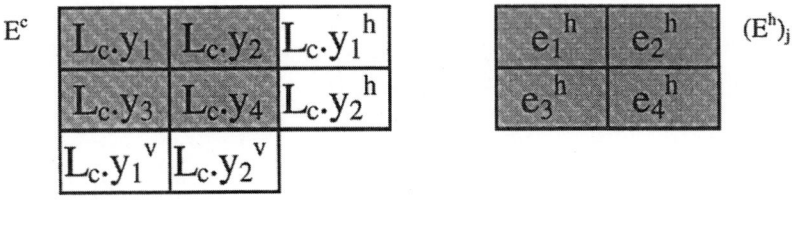
$(E^h)_j$

$(E^v)_j$

$L(U^e)_j$

Fig.2. Structure of Decoding.

received parity estimates. Using log-likelihood algebra we can generate secondary, $(E^h)_j$, and tertiary, $(E^v)_j$, estimates of the original information. E^c, $(E^h)_j$ and $(E^v)_j$ are all multiple 'soft values.' The MAP algorithm enables us to combine these separate estimates into one:

$$L(U^e)_j = E^c + (E^h)_j + (E^v)_j \qquad (1.5)$$

where $(U^e)_j$ is $(u_1^e, u_2^e, u_3^e, u_4^e)_j$, the overall estimate of U at the end of iteration 'j' of the algorithm; $(E^h)_j$ is $(e_1^h, e_2^h, e_3^h, e_4^h)_j$, the horizontal Extrinsic Information for iteration 'j'; and $(E^v)_j$ is $(e_1^v, e_2^v, e_3^v, e_4^v)_j$, the vertical Extrinsic Information for iteration 'j'.

The 'h' and the 'v' represent horizontal and vertical. This implies a transpose interleave of the information bits, which is the function of interleaver P.

The key to this estimation process is the generation of Extrinsic Information from the information and parity bits of the received words. The parity bits contain information about U, and in reversing the parity check mechanism, we extract this information and label it the Extrinsic Information.

In the horizontal parity case:

Remember: $\qquad u_1 \oplus u_2 = p_1^h.$ $\qquad\qquad (1.6)$
it is also true to say: $\qquad u_2 \oplus p_1^h = u_1.$

Or, with received information: $\qquad y_2 \oplus y_1^h = u_1.$ $\qquad\qquad (1.7)$

We can repeat the process for the vertical case where:

$$u_1 \oplus u_3 = p_1^v. \quad (1.8)$$

and so: $$u_3 \oplus p_1^v = u_1.$$

Again with received information: $$y_3 \oplus y_1^v = u_1 \quad (1.9)$$

Remember, it is valid to say that using the received information and parity bits (Y, Y^h and Y^v) we can estimate the original information (U). We now convert these estimates into log-likelihood ratios in order to generate the extrinsic information, in this case part of $(E^h)_1$:

$$L(y_2 \oplus y_1^h) = L(y_2) + L(y_1^h) = (L_c.y_2) + (L_c.y_1^h) = (e_1^h)_1. \quad (1.10)$$

It is only for the first half-iteration of iteration '1' that there is no a priori information available to modify the received information bit soft value, in this case $L(y_2)$. For the second half of iteration '1', the generation of $(E^v)_1$:

$$L(y_3 \oplus y_1^v) = [L(y_3) + (e_3^h)_1] + L(y_1^v) = [(L_c.y_3) + (e_3^h)_1] + (L_c.y_1^v)$$
$$= (e_1^v)_1. \quad (1.11)$$

Inside the square brackets of equation (1.11), the value of $L(y_3)$, the received soft value, is modified by the addition of the extrinsic information, $(e_3^h)_1$, as a priori information.

For ensuing iterations we use the extrinsic information generated in the previous half-iteration as the a priori probability for the current half-iteration. This enables us to reconstruct secondary and tertiary estimates of all the information bits from their neighbouring information bits and the parity check bits. As iterations progress the values of these estimates modify the values contained in the overall estimate $L(U^e)_j$, and as the number of iterations performed increases (as $j \to \infty$) the overall estimate $(U^e)_j$ converges to the original information U. In reality the number of iterations is quite small.

2. The Log-likelihood Addition Approximation

It becomes clear that equation (1.2) used to calculate log likelihood ratio (LLR) addition, (the 'equation result' as it shall be known) is a relatively complex formula to calculate, especially in a DSP hardware implementation. In a software solution the calculation is easier but by no means less problematic. Consider a 'p'-dimensional SPC turbo code with k input bits and n output bits. If we performed iterative 'turbo' decoding for 'j' iterations, say, then we would have to calculate [(k × p) × j] lots of extrinsic information. That is, we would use the LLR addition equation [(k × p) × j] times. If we have a higher rate code, then as the number of log-likelihood additions is proportional to k, we would have to perform a substantially higher number of log-likelihood additions. Also the more iterations required, again a higher number of log-likelihood additions are needed. All these extra calculations take more and more time.

To make the MAP as flexible as possible and as applicable as possible, a faster approximation was necessary, one that removed the need for logarithmic or exponential calculations, but at the same time enabled the MAP algorithm to work with a minimum loss of performance. To this end the approximation of equation (1.3) was developed by Hagenauer (1994). From now on this approximation will be referred to as the 'approx-result.' Nothing has been mentioned in the literature as to the accuracy of the approx-result, except that it is a 'good' approximation. To verify its validity, a program, 'log_like_approx.cpp', has been written that calculates the approx-result and the equation result for each LLR addition, and compares the two sets of values for a range of inputs, in this case the range of both inputs was from 0.1 to 3.0. A small selection of the outputs of both equations (1.2) and (1.3) for a range of $L(u1)$ input values of 0.1, 1.0, 2.0, and 3.0 are given in the appendix, Fig.s A.1 to A.4.. It can be seen that the difference between the two sets of outputs is quite large when the two input values are the same or similar. As one of the two inputs increases in value (say $L(u1)$), the difference in the output of the two equations decreases across the range of the second input $(0.1 <= L(u2) <= 3.0)$, but the worst case remains that of similar inputs. The best case for minimising error of the approx-result output is when one input value is small, and one input value is large.

It should be noted that the sign of the inputs is not taken into account here (all inputs and results are shown for the positive signs case), as if the same input magnitudes are used and the input signs are varied through all four cases (plus, plus; plus, minus; minus, plus; minus, minus) then, the result is always the same magnitude, irrespective of the signs of the input LLR values. For reference, when both inputs are of the same sign, the output is always positive, when the inputs are of opposite signs, the output is always negative.

If we consider a received turbo codeword, of a rate ½ (8, 4) turbo code with SPC CCs, then there are four information bits and four parity bits included in the codeword. On transmission, the LLR values of the code bits are all at their optimum confidence levels (say +3.0, -3.0). After corruption by noise in passing over the channel, the confidence levels of some of the LLR values of the code bits will vary. If we consider an error in an information bit, i.e. there is just enough noise present on the channel to change the bit's level at that instant of transmission (say from +3.0 to -0.1), then the confidence level of the received systematic bit will be very low. This combined with a high confidence level, parity bit will produce an extrinsic output bit whose value is in error by approximately 9%. This is a reasonable margin of error. In the case where both parity and information bits have been corrupted by noise to low confidence values, the approximation yields its least accurate results, with an error of almost 2000%. This high figure should be viewed in the light of the fact that the error is measured as a percentage of the equation result. Also if the BER is 10^{-3}, say, the probability of having two bits in error in the same codeword would be 10^{-6}, and the probability of those two bits being a related information and parity bit pair, is even smaller.

2.1 Effect of LLR Addition Approximation on Decoder Performance

Simulations were carried out using the approximation to LLR addition and the mathematically derived equation for LLR addition incorporated into the Iterative decoding stages. The previously mentioned rate ½, (8,4) Turbo code with SPC CCs in an array code format was used as it would become increasingly difficult to calculate the derived equation-result for much more complicated CCs. Results are given in the appendix in Fig.s A.5 to A.7. In Fig. A.7 it can be seen that the approximation appears to marginally improve the performance of the decoder compared to the derived equation.

A possible explanation for this is the fact that when the approximation is in error, the result produced is of the correct sign, but of a higher magnitude. This can in some error situations actually produce corrected errors in instances where the derived equation output would not be of sufficient magnitude to enable the correction of the same error situation. An interesting question to ask is: are there other functions that can approximate to the ML equation-result performance, with positive error ? i.e. an error that improves the magnitude of a correct decision with respect to the equation-result. Another point to consider is, why does the opposite not happen and cancel out the effect of the approximation ? i.e., Surely there should be, on average, an equal number of situations where the equation-result is of small enough magnitude to not cause an erroneous result, whereas the approx-result is larger and therefore should cause an error that wasn't originally there.

3. A Comparison Between two Block Turbo Decoding Algorithms

Before comparing the two algorithms it is necessary to briefly describe the Pyndiah algorithm (as it shall be known in this paper) of Pyndiah (1994). The Algorithm developed in Pyndiah (1994) was based on the Chase Algorithm, presented in Chase (1972). Pyndiah et al have applied Chase's ideas to the iterative decoding of block codes. Consider the vector R, which characterises a received codeword. It can be represented as the combination of the transmitted codeword vector, E, and a noise vector, N, generated on an AWGN channel.

Simply put,

$$R = E + N, \tag{1.12}$$

where $R = (r_1, r_2, \ldots r_f, \ldots, r_n)$, $E = (e_1, e_2, \ldots e_f, \ldots e_n)$, and $N = (n_1, n_2, \ldots n_f, \ldots n_n)$.

The LLR of the decison d_j, LLR_j, is defined as :

$$LLR_j = \log_e \frac{\Pr\{e_j = +1/R\}}{\Pr\{e_j = -1/R\}} \tag{1.13}$$

by much manipulation LLR$_j$ can be expressed as :

$$LLR_j \cong 1/2\sigma^2.[r_j + \sum_{f=1,f\neq j}^{n} r_f.c_f^{min(+1)}.p_f] \qquad (1.14)$$

where

$$p_f = 0 \text{ if } c_f^{min(+1)} = c_f^{min(-1)}$$

OR $\qquad\qquad\qquad\qquad\qquad\qquad\qquad\qquad (1.15)$

$$p_f = 1 \text{ if } c_f^{min(+1)} \neq c_f^{min(-1)}$$

see (Pyndiah 1994) for a good exposition of the derivation. LLR$_j$ normalised with respect to $2/\sigma^2$ is equal to :

$$r'_j = r_j + w_j \qquad \{1.16\}$$

where :

$$w_j = \sum_{f=1,f\neq j}^{n} r_f.c_f^{min(+1)}.p_f \qquad (1.17)$$

w_j can be viewed as performing the same function as the renowned 'extrinsic' information in turbo codes, i.e. that of a corrective factor applied to the soft input of the constituent decoder. r_j is the soft input to the decoder, and r'_j is the soft output. W(m) is the matrix of values w_j at time m, R(m) is the matrix of values r_j at time m, and R'(m) is the matrix of values r'_j at time m. It is possible to say :

$$R'(m) = R(m) + \alpha(m).W(m) + W(m+1) \qquad (1.18)$$

where $\alpha(m)$ is a damping coefficient used to minimise the effect of W(m) during the first iterations when W(m) is not so reliable. Equation (1.18) bears a remarkable resemblance to equation (1.5) of the Hagenauer algorithm. W(m) is the extrinsic information from decoder m, and W(m+1) is the extrinsic information from decoder (m+1). After the calculation of R'(m), W(m+1) would be output to the succeeding decoder for the next iteration of the decoding process. R'(m) can be used to evaluate the progress of the decoding somewhat like L(U^e) in the Hagenauer algorithm.

The Hagenauer Algorithm uses the MAP algorithm with a Dual code implementation (Hagenauer 1996). For the dual code it is possible to reduce complexity and suboptimal approximations are realisable, with the inclusion of only a limited number of the codewords of the dual code, for example. The Pyndiah Algorithm is suboptimal, and reasonably complex compared to a simple block decoder. Pyndiah et al have outlined simplifications that can be made to the block turbo decoder (Pyndiah 1996). These simplifications amount to a reduction in complexity by a factor of ten. The Hagenauer decoding algorithm (without approximation) is superior in performance to the Pyndiah unsimplified algorithm by 0.4 dB at a Signal to Noise Ratio (SNR) of 10^{-5}. If we compare the same implementation of the Hagenauer algorithm to the reduced complexity Pyndiah algorithm, the Hagenauer algorithm is approximately 1.1 dB better at the same SNR. Remember that the Pyndiah algorithm is offering a rather substantial reduction in complexity compared to the dual code implemented MAP.

It is interesting to note that both algorithms use the same interleaver, i.e. a row/column or transpose interleaver, where columns become rows and rows become columns. The Pyndiah algorithm uses a complete product code structure including the check on checks bit at the end of the right hand column of the product code matrix, unlike the Hagenauer algorithm. It would be interesting to see if a full product code implementation of Hagenauer's algorithm would provide any improvement in performance. The use of W(m) at the interface of the decoders, as opposed to the full LLR, necessitates the use of the empirically derived modifying factors $\alpha(m)$ and β. Although, with low complexity, suboptimal decoders, these factors are necessary to increase the performance (Hoeher 1997).

It is possible to directly compare the Hagenauer and Pyndiah algorithms. If we examine Fig. A.8, we can see that the Hagenauer algorithm performs slightly better. The Hagenauer curve is for Hamming (63,57,3) CCs and the Pyndiah curve is for BCH (63,57,3) CCs, which for the purposes of this comparison can be considered to be equivalent.

4. Conclusions

In this paper we have provided an introduction to block turbo codes, specifically those developed by Hagenauer. A study has been made of the mechanics of the LLR addition approximation and its performance. The algorithms of Hagenauer and Pyndiah have been compared.

When the idea for this paper was formulated, the assumption was made that the Hagenauer approximation would have some drawbacks, and improvements would be required. This was supported by the fact that the approximation was arbitrarily chosen. When a comparison of the outputs of the approx-result and equation-result were made (see Fig.s A.1 to A.4) the assumption appeared to be validated. The next logical step was to observe the affect of the approximation on decoder performance. The suprising result was that the approximation outperformed the equation it approximates to. The approximation is superior to the equation result by approximately 0.2 dB at a BER of 2×10^{-5}. At higher noise levels the improvement in performance is less (see Fig. A.7). We have provided a preliminary explanation for these results and aim to study this phenomena further. More work needs to be carried out on the effect of the approximation over many iterations and for more complicated CCs, which will require substantial computing power.

The approximation itself is a reasonable one, with approximately nine percent error in the best case. The worst case is a very large error but the probability of it happening is very small, approximately 1×10^{-7}, and even in this case the error only serves to reinforce the confidence level of the decision made by the full equation.

248

Appendix-A

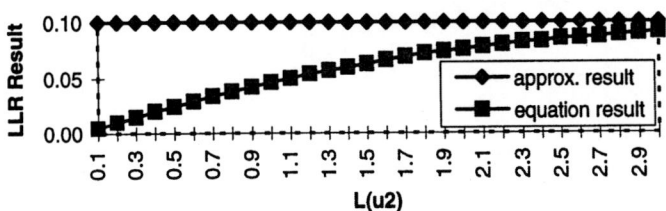

Fig. A.1. $L(u1) = 0.1$, $L(u2) = 0.1 - 3.0$

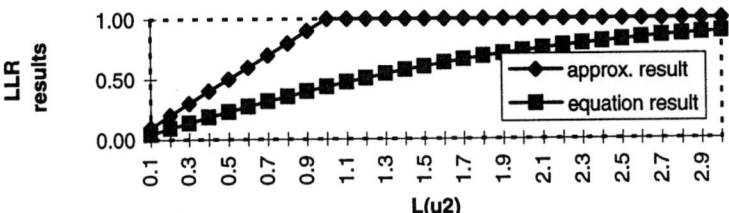

Fig.A.2. $L(u1) = 1.0$, $L(u2) = 0.1 - 3.0$

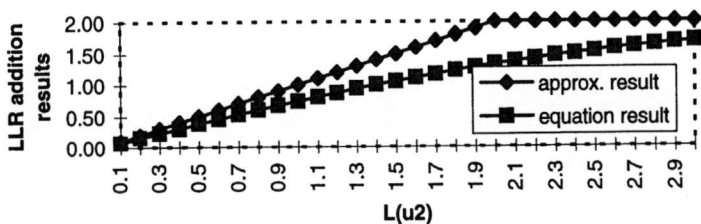

Fig.A.3. $L(u1) = 2.0$, $L(u2) = 0.1 - 3.0$

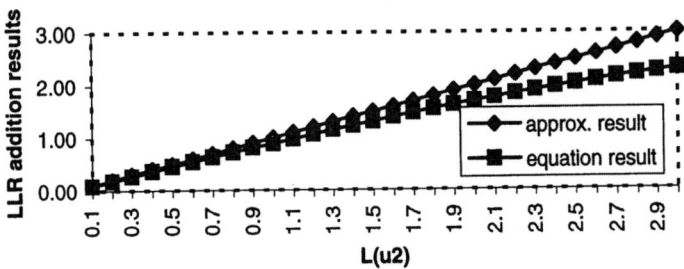

Fig.A.4. $L(u1) = 3.0$, $L(u2) = 0.1 - 3.0$

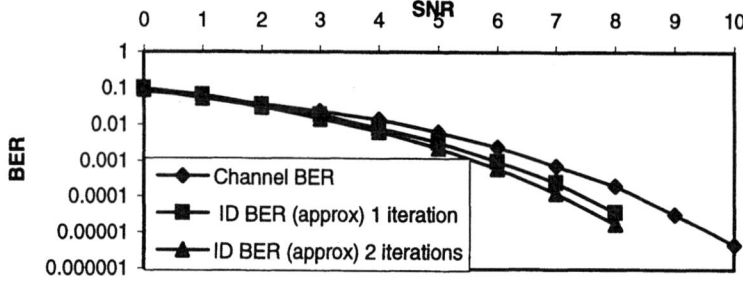

Fig.A.5. *BER of the Turbo decoder using the approximated method of LLR addition*

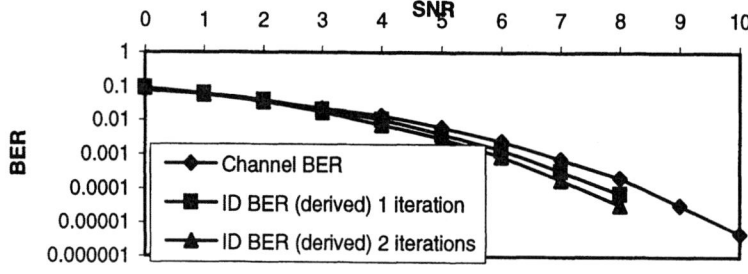

Fig.A.6. *BER of the Turbo decoder using the equation (derived) method of LLR addition*

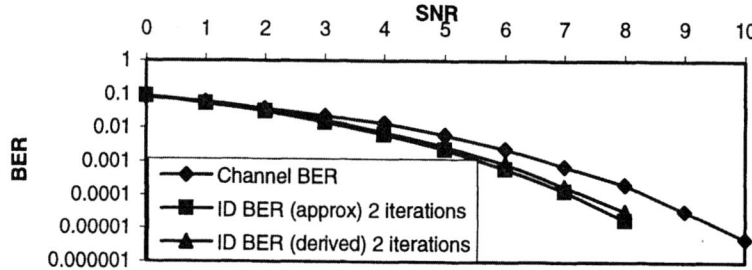

Fig.A.7. *BER of the Turbo decoder showing performance comparison between decoders using the approximated and derived methods of LLR addition.*

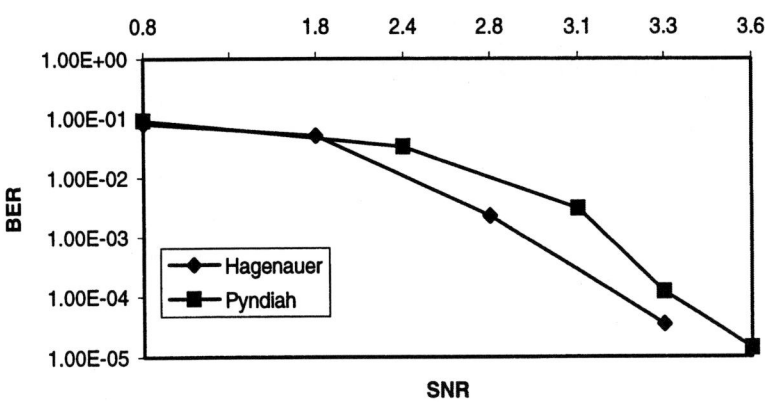

Fig.A.8. *Comparison of the performance of the Hagenauer and Pyndiah Algorithms.*

References :

Berrou, C., Glavieux, A., & Thitimajshima, P., "Near Shannon limit error correcting coding and decoding: Turbo Codes", IEEE Int. Conf. on Comm.s, 1993, (ICC'93), Geneva, pp.1064 - 1070, May 1993.

Forney Jr., G.D., "Concatenated Codes", M.I.T., Cambridge, MA, USA, 1966.

Hagenauer, J., Robertson, P., & Papke, L., "Iterative ('turbo') decoding of systematic convolutional codes with the MAP and SOVA algorithms", ITG Conference, Munich, Germany, pp.1-9, 26-28 Oct.1994.

Hagenauer, J., "Iterative Decoding of Binary Block and Convolutional Codes", IEEE Trans. on Information Theory, Vol.42(2), pp.429-445, March 1996.

Hoeher, P., "New Iterative ('Turbo') Decoding Algorithms", Proceedings of International Symposium on Turbo Codes & Related Topics., Brest, France, pp.63-70, 3rd - 5th September 1997.

Mohammadi, A.H.S., & Khandani, A.K., "Unequal Error Protection on Turbo-encoder Output bits", Electronic Letters, Vol.33(4), pp.273-274, 13th Feb. 1997.

Multiuser Coding Based on Detecting Matrices for Synchronous-CDMA Systems

Wai Ho Mow

Division of Communication Engineering, School of EEE, Block S1,
Nanyang Technological University, Nanyang Avenue, Singapore 639798

Abstract. A multiuser coding approach to the design of spreading sequences for synchronous CDMA systems in which each receiver makes use of the knowledge of the spreading sequences and idle status of all users was recently introduced. For interuser-interference-free communications using sequences of length L, the system capacity (i.e. maximum number of simultaneous users) of the resultant system is about $L \log_2(L)/2$ while that of conventional system is only L. In this work, it is shown that in this approach the synchronous CDMA system is equivalent to the classical multiuser binary adder channel. Consequently, the multiuser codes based on ternary detecting matrices can be used as the spreading sequences. A new recursive construction of ternary detecting matrices which generalizes and/or improves the best previously known constructions is presented. Comparing with previous works, a remarkable improvement on the system capacity is obtained.

1 Introduction

For each user in a conventional synchronous code-division multiple-access (S-CDMA) system, any data bit to be transmitted is used to modulate the ± 1-valued spreading sequence of that particular user before transmission. Assuming symbol and sequence synchronization, the channel essentially outputs the sum of these data-modulated sequences from all users. At the receiver, a matched filter for the spreading sequence of the target user is used to demodulate the data transmitted by that particular user. Interuser interference can be totally avoided only if the set of all spreading sequences are mutually orthogonal. This implies that the number of simultaneous users in the system is at most equal to the sequence length. In existing S-CDMA systems such as those following the IS-95 standard, Walsh sequences are typically used as the spreading sequences.

While in the conventional S-CDMA the knowledge of the spreading sequences of other users and of the users' idle status are not required, the use of such knowledge at the receiver can significantly improve the system capacity, i.e. the maximum number of simultaneous users supported. Recently, Khachatrian and Martirossian [1] presented a multiuser coding and joint detection approach to designing such a S-CDMA system and showed that for a sequence length of 2^n, it is possible to accommodate $n2^{n-1} - 1$ simultaneous users without interuser interference. For large sequence length L, an impressive improvement factor of $\log_2(L)/2$ is achieved. In this work, we shall show that further improvement is possible if two different spreading sequences are assigned to each user to represent bit 0 and bit 1 respectively, instead of using sequence inversion keying.

We mention in passing that Wu and Chang [2] recently showed that for the S-CDMA system in [1], if the users' idle status is not available at the receivers, interuser-interference-free S-CDMA is still achievable with an asymptotic improvement factor of $\log_3(L)/2$.

2   S-CDMA and Multiuser Binary Adder Channel

The S-CDMA system described is very similar to the classical multiuser binary adder channel (MBAC) considered in [3]. A MBAC outputs the (integer-valued) sum of the zero-one codewords from all users. Each user is assigned two codewords representing data bits 0 and 1 respectively. The problem is to design a *uniquely decodable* multiuser code such that the transmitted codewords of all users can be identified from their sum output by the MBAC.

Let w_{i0} and w_{i1} be the codewords of the ith user representing data bits 0 and 1 respectively. Interpret a L-bit codeword as a L-dimensional zero-one vector, and define the two $L \times M$ zero-one matrices W_0 and W_1 by writing the ith column as w_{i0} and w_{i1} respectively. Here M represents the number of users. Further define the *difference matrix* D as $W_1 - W_0$ so that D is a $\{-1, 0, 1\}$-valued (or ternary) matrix. Let x be a M-dimensional zero-one column vector, whose ith element is the data bit transmitted by the ith user. Then the output of MBAC can be conveniently represented by $Dx + W_0\mathbf{1}_M$, where $\mathbf{1}_n$ is the n-dimensional all-one column vector. For the code to be uniquely decodable, we require for any zero-one vectors x and y, $Dx = Dy$ only if $x = y$. We shall call any matrix satifying this condition a *detecting matrix*.

For the S-CDMA system of interest, the set of spreading sequences is ± 1-valued instead of zero-one. It can be represented by $2(Dx + W_0\mathbf{1}_M) - M\mathbf{1}_L$ and is also uniquely decodable as long as D is a detecting matrix. Therefore, both the sequence design problem of S-CDMA systems and the coding problem of multiuser binary adder channel can be reduced to the problem of constructing detecting matrices. Namely, for any given L, we want to construct a $L \times M$ ternary detecting matrix with M as large as possible. (In the terminology of S-CDMA, we want to maximize the number M of simultaneous users supported for any given sequene length L.)

So far, the discussion has not considered the effect of idle users. When a user is idle, no signal is transmitted. However, the decoder always assumes the presence of a valid spreading sequence from every user. To ensure correct decoding, the receiver generates for each idle user a valid spreading sequence and add it to the channel output. This guarantees that the processed channel output can be decoded in the way as if there is no idle user. This complementary approach requires the knowledge of all users' idle status and was adopted in [1].

3   Ternary Detecting Matrices

As a class of combinatorial objects, detecting matrices were first investigated by mathematicians in the early 1960's. The interest in detecting matrices was originated from the coin-weighing problem of Söderberg and Shapiro [4].[1] In the same paper, they constructed a class of zero-one detecting matrices using Kronecker product. Subsequently, there were many constructions of zero-one detecting matrices by others.

The first construction of $2^k \times 2^{k-1}(k+2)$ ternary detecting matrices was apparently published by Cantor and Mills [5] in 1966. In connection with the study of coding for MBAC, Chang and Weldon [3] presented a construction in 1979. Ferguson [7] gave a generalization three years later. However, the sizes of matrices in these three constructions are the same. Later, Chang [8] introduced a shortening technique to devise new detecting matrices from the construction in [3]. This allows the number of rows of the matrix to be any integer.

It is not difficult to see that the construction in [3] is in fact equivalent to that of Cantor and Mills [5]. Further, we note that the construction of Söderberg and Shapiro [4] can easily be adapted from the zero-one matrices to the ternary ones. The proof is almost verbatim. The following theorem summarizes this.

Theorem 1. *Let G be an $k \times k$ invertible ternary matrix such that there exists an $k \times k$ ternary matrix F with $FG \equiv O_k$, where $\equiv$ means "is congruent modulo-2", and O_k denotes the $k \times k$ all-zero matrix. Also, let H be an $km \times sm$ ternary matrix such that $\mathrm{Rank}_2((F \otimes I_m)H) = sm$, where $\mathrm{Rank}_2(X)$ denotes the rank of X over $\mathrm{GF}(2)$, $s = \mathrm{Rank}_2(F)$, $\otimes$ is the Kronecker product operator, and I_m denotes the $m \times m$ identity matrix. If the $m \times n$ matrix D is a ternary detecting matrix, so is the $km \times (kn + sm)$ matrix $[G \otimes D | H]$.*

Now put

$$F = G = \begin{bmatrix} 1 & 1 \\ 1 & -1 \end{bmatrix}$$

so that $k = 2$, $s = 1$, and

$$FG = \begin{bmatrix} 2 & 0 \\ 0 & 2 \end{bmatrix} \equiv O_2. \tag{1}$$

Also, let

$$H = \begin{bmatrix} H_1 \\ H_0 \end{bmatrix}$$

with H_0 and H_1 both $m \times m$ matrices so that

$$(F \otimes I_m)H = \begin{bmatrix} I_m & I_m \\ I_m & -I_m \end{bmatrix} \begin{bmatrix} H_1 \\ H_0 \end{bmatrix} = \begin{bmatrix} H_1 + H_0 \\ H_1 - H_0 \end{bmatrix}$$

and hence $\mathrm{Rank}_2((F \otimes I_m)H) = \mathrm{Rank}_2(H_1 - H_0)$. Then the following corollary is resulted, which coincides with the construction of Ferguson [7].

---

[1] The problem can be equivalently stated in a set-theoretic terminology. Similarly, any detecting matrix can be expressed as a detecting set and vice versa. See for example, [6] and [5].

Corollary 2. *Let H_0 and H_1 be $m \times m$ ternary matrices such that $\mathrm{Rank}_2(H_1 - H_0) = m$. If the $m \times n$ matrix D is a ternary detecting matrix, so is the $2m \times (2n + m)$ matrix*

$$\begin{bmatrix} D & D\,H_1 \\ D & -D\,H_0 \end{bmatrix}.$$

Note that the construction of Cantor and Mills [5] simply corresponds to the case when $H_1 = I_m$ and $H_0 = O_m$.

We now present a new recursive construction of ternary detecting matrices, which is more general than Corollary 2. The construction also allows the number of rows to be any integer like Chang's shortening technique, but is generally more efficient.

Theorem 3. *If the $m \times n$ matrix D and the $m_2 \times n_2$ matrix D_2 are both ternary detecting matrices with $m \leq m_2$, so is the $(m + m_2) \times (n + n_2 + m)$ matrix*

$$\begin{bmatrix} D_2^{[m]} & D_1\ H_1 \\ D_2^{[m]} & D_0\ H_0 \\ D_2^{(m_2-m]} & Y_0\ Y \end{bmatrix}$$

where D_0 and D_1 are two ternary matrices such that $D_1 - D_0 = 2D$, H_0 and H_1 are both $m \times m$ ternary matrices such that $\mathrm{Rank}_2(H_1 - H_0) = m$, and Y_0 and Y are any $(m_2 - m) \times n$ and $(m_2 - m) \times m$ ternary matrices respectively. Here, $X^{[k]}$ denotes the submatrix consisting of the top k rows of X, $X^{(k]}$ denotes the submatrix consisting of the bottom k rows, and $\mathrm{Rank}_2(X)$ denotes the rank of matrix X over $\mathrm{GF}(2)$.

Proof: First of all, note that given any ternary detecting matrix D, it is always possible to set

$$\begin{aligned} D_1(i,j) &= -D_0(i,j) = 1, & &\text{if } D(i,j) = 1, \\ D_1(i,j) &= -D_0(i,j) = -1, & &\text{if } D(i,j) = -1, \\ D_1(i,j) &= D_0(i,j) = -1, 0 \text{ or } 1, & &\text{if } D(i,j) = 0, \end{aligned}$$

so that $D_1 - D_0 = 2D$ with both D_0 and D_1 being ternary. Here, $X(i,j)$ denotes the element in the ith row and the jth column of matrix X.

By definition, a $m \times n$ matrix D is detecting if for any two zero-one n-dimensional vectors x and y, $Dx = Dy$ implies $x = y$. An alternative definition is that for any ternary n-dimensional vector e, $De = \mathbf{0}_m$ implies $e = \mathbf{0}_n$. This can be easily seen by setting $e = x - y$. For the sake of convenience, the latter definition will be used in this proof.

Let e_0, e_1 and e_2 be arbitrary n_2-, n- and m-dimensional ternary column vectors respectively. Also, let r_0, r_1 and r_2 be m-, m- and $(m_2 - m)$-dimensional column vectors respectively so that

$$\begin{bmatrix} r_0 \\ r_1 \\ r_2 \end{bmatrix} = \begin{bmatrix} D_2^{[m]} & D_1\ H_1 \\ D_2^{[m]} & D_0\ H_0 \\ D_2^{(m_2-m]} & Y_0\ Y \end{bmatrix} \begin{bmatrix} e_0 \\ e_1 \\ e_2 \end{bmatrix}. \tag{2}$$

Suppose that r_0, r_1 and r_2 are all-zero vectors. It suffices to prove that e_0, e_1 and e_2 are all all-zero vectors. From (2),

$$r_0 - r_1 = 2De_1 + (H_1 - H_0)e_2 \equiv (H_1 - H_0)e_2 \pmod 2. \tag{3}$$

The square matrix $(H_1 - H_0)$ is invertible over GF(2) because $\mathrm{Rank}_2(H_1 - H_0) = m$ by hypothesis. Therefore, $e_2 \equiv \mathbf{0}_m \pmod 2$, which implies that $e_2 = \mathbf{0}_m$ since e_2 is ternary. Now, by (3),

$$De_1 = \frac{1}{2}(r_0 - r_1 - (H_1 - H_0)e_2) = \mathbf{0}_m,$$

which implies $e_1 = \mathbf{0}_n$ because D is detecting. Finally, it follows from (2) and the fact that r_1, r_2, e_1 and e_2 are all all-zero vectors that

$$D_2 e_0 = \begin{bmatrix} r_1 \\ r_2 \end{bmatrix} - \begin{bmatrix} D_0 & H_0 \\ Y_0 & Y \end{bmatrix} \begin{bmatrix} e_1 \\ e_2 \end{bmatrix} = \mathbf{0}_m.$$

As D_2 is detecting, this implies that $e_0 = \mathbf{0}_{n_2}$. The proof is thus complete.

□

Corollary 2 corresponds to the special case when $m = m_2$ and $D_2 = D_1 = -D_0 = D$. Setting $m_2 = 2^s$ and $m = 1, 2, \cdots, 2^s$, the theorem gives $k \times (B(k) + k)$ ternary detecting matrices for any k. Here, $B(k)$ denotes the number of ones in the binary expansions of all positive integers less than k, and is defined by $B(1) = 0$ and

$$B(2^s + i) = B(2^s) + B(i) + i, \quad i = 1, 2, \cdots, 2^s, \; s = 0, 1, \cdots, \tag{4}$$

Note that $B(2^k) = 2^{k-1}k$.

The number $T(k)$ of columns of the matrices obtained by Chang's shortening technique [8] is given by

$$T(k) = (i + 1)2^{i-1} - j - \sum_{n=0}^{i-2} j_n(n + 2)2^{n-1},$$

where i, j and j_n are defined by

$$k = 2^i - j, \qquad 1 \le j \le 2^{i-1}$$

and

$$j - 1 = \sum_{n=0}^{i-2} j_n 2^n, \qquad j_n \in \{0, 1\}.$$

The values of $T(k)$ and $B(k) + k$ for k the number of rows up to 16 are shown in Table 1.

A comparison shows that the new construction is generally an improvement over Chang's technique, in particular for $k=7$, 11, 13, 14 and 15 in the table. Besides, unlike Chang's shortening technique, the new construction is recursive in nature and hence has an efficient recursive decoding algorithm.

Table 1. Numbers of columns of ternary detecting matrices with k rows obtained by Chang's shortening technique [8], $T(k)$, and by Theorem 3, $B(k) + k$.

k	1	2	3	4	5	6	7	8	9	10	11	12	13	14	15	16
$T(k)$	1	3	5	8	10	13	15	20	22	25	27	32	34	37	39	48
$B(k) + k$	1	3	5	8	10	13	16	20	22	25	28	32	35	39	43	48

4 Concluding Remarks

A S-CDMA system in which each receiver makes use of the knowledge of the spreading sequences and the idle status of all users to increase the system capacity was investigated. Though the spreading sequences are ± 1-valued instead of zero-one, it was recognized that the system is equivalent to the classical multiuser binary adder channel. As a consequence, good construction of ternary detecting matrices can be used to design the spreading sequences. New construction of ternary detecting matrices that generalizes and/or improves the best previously known results was presented.

As a comparison, consider the case of a sequence length of 8. For interuser-interference-free S-CDMA, the maximum number of simultaneous users supported for the conventional system is 8 and that for the system of Khachatrian and Martirossian [1] is 13, while for the S-CDMA system considered here, the system capacity is increased to 20.

Acknowledgements

Part of the work was done when the author was visiting the Department of Electrical and Computer Engineering, University of Waterloo, Waterloo, Ontario, Canada in 1995 under the support of Croucher Fellowship. He is indebted to Professor Ian F. Blake for some inspiring discussions on the topic during this period.

References

1. G. H. Khachatrian and S. S. Martirossian, "A New Approach to the Design of Codes for Synchronous-CDMA Systems," *IEEE Trans. Inform. Theory*, vol. IT-41, pp. 1503–1506, 1995.
2. Y. W. Wu and S. C. Chang, "Coding Scheme for Synchronous-CDMA Systems," *IEEE Trans. Inform. Theory*, vol. IT-43, pp. 1064–1067, 1997.
3. S.-C. Chang and E. J. Weldon, Jr., "Coding for T-User Multiple-Access Channels," *IEEE Trans. Inform. Theory*, vol. IT-25, pp. 684–691, 1979.
4. S. Söderberg and H. S. Shapiro, "A combinatory detection problem," *Amer. Math. Monthly*, vol. 70, pp. 1066–1070, 1963.

5. D. G. Cantor and W. H. Mills, "Determining a Subset from Certain Combinatorial Properties," *Canad. J. Math*, vol. 18, pp. 42–48, 1966.
6. B. Lindström, "On a combinatory detection problem," *Publ. Hung. Acad. Sci.*, vol. 9, pp. 195–207, 1964.
7. T. J. Ferguson, "Generalized T-User Codes for Multiple-Access Channels," *IEEE Trans. Inform. Theory*, vol. IT-28, pp. 775–778, 1982.
8. S.-C. Chang, "Further Results on Coding for T-User Multiple-Access Channels," *IEEE Trans. Inform. Theory*, vol. IT-30, pp. 411–415, 1984.

Enumeration of Convolutional Codes and Minimal Encoders

Conor P. O'Donoghue and Cyril J. Burkley
Dept. of Electronic Engineering
University of Limerick
Limerick
Ireland
Email : odonoghuec@ul.ie

1 Introduction

In a landmark paper [1] on the algebraic structure of convolutional codes, Forney introduced two canonical forms for convolutional encoders. The *minimal encoder* is a feedback-free encoder with a feedback-free inverse and is consequently guaranteed to be non-catastrophic. Its main advantage is that, in the obvious realisation (controller canonical form), the dimension of its state space is equal to the overall constraint length and is no greater than that of any equivalent encoder realisation. Furthermore, it has the property that it encodes short information sequences into short codewords and hence the number of decoding errors associated with short channel error sequences is minimised. The *systematic encoder*, on the other hand, in general, contains feedback but has a (trivial) feedback-free inverse and hence is also non-catastrophic. In addition, it can always be realised with complexity equal to that of any equivalent minimal encoder. A significant advantage of the systematic encoder is that, with some convention as to which columns will contain the identity matrix, there is a unique systematic encoder associated with each convolutional code.

Minimal encoders have been enumerated by Rosenberg [2], Forney [3] and Shusta [4] for rate $R = 1/n$. The different enumerators obtained are due to the fact that each author considers a slightly different ensemble of encoders. Shusta extends his results to rate k/n encoders but only succeeds in obtaining bounds on the enumerators which turn out to be rather loose especially for high rate codes. Furthermore, the bounds are not in closed form for the general rate k/n case. Each rate of interest requires a cumbersome z-transform and inverse transform to obtain the bound in closed form.

In this paper we present a detailed analysis of equivalent encoders and determine the exact structure of the $k \times k$ matrix relating two equivalent minimal encoders. Using this result, we enumerate the minimal encoders generating a given convolutional code. Then, beginning with a very general ensemble of encoders, we derive enumerators for all rate k/n minimal encoders, canonical systematic encoders and convolutional codes with arbitrary constraint lengths. These enumerators are of practical interest since, in contrast to block codes, good convolutional codes are not, in general, found by algebraic construction, but rather by exhaustive computer search [5, 6, 7]. Hence, for example, the number of distinct codes generated by a given ensemble of encoders is a useful reference in defining the efficiency of rejection rules in a code search. In addition, these enumerators are of theoretical interest in deriving random coding bounds for fixed convolutional codes [8] and determining the relative size of subclasses of convolutional codes with desirable properties.

2 Equivalent encoders

The fact that each code, $\mathcal{C}$, has a unique canonical systematic encoder suggests the systematic encoder as the best canonical form to use in the enumeration of convolutional codes. However, except for rate $1/n$ and $(n-1)/n$ codes, the minimal order of $\mathcal{C}$ is not readily found from the systematic encoding matrix making the choice of a suitable ensemble difficult. For this reason our preference is for the minimal encoder. But, because it is not unique, our first task is the enumeration of the minimal encoders generating $\mathcal{C}$. We give several new lemmas on equivalent encoders which are required to prove some of the results in the sequel.

Lemma 1 *Let G be an arbitrary $k \times n$ polynomial encoder with ordered constraint lengths $v = (v_1, v_2, \ldots, v_k)$ such that $v_1 \geq v_2 \geq \cdots \geq v_k$. Let $\mathbf{G}$ be some equivalent minimal encoder with ordered minimal indices $\mu = (\mu_1, \mu_2, \ldots, \mu_k)$ such that $\mu_1 \geq \mu_2 \geq \cdots \geq \mu_k$. Then*

$$\mu_i \leq v_i \qquad 1 \leq i \leq k \tag{1}$$

Remark 1 *Lemma 1 is a generalisation of a well known result [1] which states that the overall constraint length of any encoder is lower bounded by the overall constraint length of an equivalent minimal encoder.*

The following lemma relates the high order coefficient matrix of an encoder to that of an equivalent minimal encoder.

Lemma 2 *Let $\mathbf{G}$ be some $k \times n$ minimal matrix with constraint lengths $\mu = (\mu_1, \ldots, \mu_k)$ and high order coefficient matrix $[\mathbf{G}]_h$ and let A be some $k \times k$ nonsingular $F_q[D]$-matrix. Define $G := A\mathbf{G}$ and let $v = (v_1, \ldots, v_k)$ denote the constraint lengths of G. Then*

$$[G]_h = [A]_{hh}[\mathbf{G}]_h \tag{2}$$

where $[A]_{hh} := \|a_{ij}[v_i - \mu_j]\|$, the $k \times k$ matrix of the coefficients of $D^{v_i - \mu_j}$ in $a_{ij}(D)$.

We proceed to determine the structural properties of the $k \times k$ $F_q[D]$-matrix, A, relating two equivalent minimal encoders and then enumerate the minimal encoders generating a given rate k/n convolutional code with the following theorem.

Theorem 3 *Let C be an arbitrary rate k/n convolutional code with minimal indices $\mu = (\mu_1, \ldots, \mu_k)$, Without loss of generality we may assume that the elements of μ are ordered such that $\mu_1 \geq \mu_2 \geq \cdots \geq \mu_k$. Then the number of equivalent minimal encoders with ordered constraint lengths generating C is*

$$N_{eq}(\mu) = \left[\prod_{i=1}^{L} \prod_{i=0}^{k_i-1} (q^{k_i} - q^i) \right] q^{\sum_{i=1}^{L-1} \sum_{j=i+1}^{L} k_i k_j (U_i - U_j + 1)} \tag{3}$$

where $(U_1, \ldots, U_L)$ are the distinct constraint lengths in μ s.t. $U_1 > \cdots > U_L$, and k_i is the frequency of occurrence of U_i in μ.

3 Enumeration of minimal encoders

The enumeration of minimal encoders requires the selection of an appropriate ensemble of encoders which turns out to be of critical importance. Other authors have chosen restrictive ensembles requiring encoders to satisfy most of the conditions necessary for minimality. We adopt a novel approach in choosing a very general ensemble and contend that this is the most appropriate one since the enumeration procedure is not direct.

Definition 1 *Define $\mathcal{E}_v$ to be the ensemble of all $k \times n$ $F_q[D]$-matrices with ordered constraint lengths $v = (v_1, v_2, \ldots, v_k)$, $v_1 \geq v_2 \geq \cdots \geq v_k$, and full rank high order coefficient matrix, $[G]_h$.*

We show that the cardinality of $\mathcal{E}_v$ is

$$|\mathcal{E}_v| = \left[\prod_{j=0}^{k-1} (q^n - q^j) \right] q^{nv} \tag{4}$$

where $v := \sum_{i=1}^{k} v_i$ is the overall constraint length. The encoder equivalence relation induces an equivalence partition in $\mathcal{E}_v$. The cardinality of each equivalent class is given by the following theorem

Theorem 4 *Let G be some $k \times n$ minimal encoder with ordered constraint lengths, μ, such that $\mu_1 \geq \mu_2 \geq \cdots \geq \mu_k$ and let v be any set of ordered*

constraint lengths such that $v_1 \geq v_2 \geq \cdots \geq v_k$. Then the number of encoders in $\mathcal{E}_v$ equivalent to $\mathcal{G}$ is given by:-

$$N_A(\boldsymbol{\mu}, \boldsymbol{v}) = \left[\prod_{i=1}^{k} q^{k-i}(q^{i-n_i} - 1) \right] q^{\sum_{i=1}^{k} \sum_{n_i+1}^{k}(v_i - \mu_j)} \tag{5}$$

where

$$n_i := \max\{j | \mu_j > v_i, 1 \leq j \leq k\} \tag{6}$$

Collecting the results so far, we obtain the following recursive relation for the number of minimal encoders

$$\sum_{\mu_k=0}^{v_k} \sum_{\mu_{k-1}=\mu_k}^{v_{k-1}} \cdots \sum_{\mu_1=\mu_2}^{v_1} N_A(\boldsymbol{\mu}, \boldsymbol{v}) N_m(\boldsymbol{\mu}) / N_{eq}(\boldsymbol{\mu}) = |\mathcal{E}_v| \tag{7}$$

where $N_m(\boldsymbol{\mu})$ is the number of $k \times n$ minimal encoders with ordered constraint lengths, $\boldsymbol{\mu} = (\mu_1, \ldots, \mu_k)$. Solving for $N_m(\boldsymbol{\mu})$ gives our main theorem.

Theorem 5 (Main Theorem) *Let $N_m(\boldsymbol{v})$ denote the total number of rate $k \times n$ minimal encoders with constraint lengths $\boldsymbol{v}$. Without loss of generality, the elements of $\boldsymbol{v}$ may be assumed to be ordered such that $v_1 \geq v_2 \geq \cdots \geq v_k$. Then*

$$N_m(\boldsymbol{v}) = \left[\prod_{i=0}^{k-1} (q^n - q^i) \right] \left[\prod_{j=t+1}^{k} (q^n - q^j) \right] q^{n(v-k+t)} \tag{8}$$

where

$$t = \begin{cases} k_L & \text{if } V_L = 0 \\ 0 & \text{otherwise} \end{cases} \tag{9}$$

A proof is given only in the special case where $v_1 > v_2 > \cdots > v_k$. The proof of the general case is considerably more involved but will be available on request.

4 Enumeration of convolutional codes

Finally, we enumerate convolutional codes

Theorem 6 *The total number of rate $k \times n$ convolutional codes with minimal indices[1] $(v_1, \ldots, v_k)$ is*

$$N_c(\boldsymbol{v}) = \frac{\left[\prod_{i=0}^{k-1}(q^n - q^i) \right] \left[\prod_{i=t+1}^{k}(q^n - q^i) \right] q^{n(v-k+t)}}{\prod_{i=1}^{L} \left[\prod_{j=0}^{k_i-1}(q^{k_i} - q^j) \right] q^{\sum_{i=1}^{L-1} \sum_{j=i+1}^{L} k_i k_j (V_i - V_j + 1)}} \tag{10}$$

where $(V_1, \ldots, V_L)$ are the distinct constraint lengths in $\boldsymbol{v}$ s.t. $V_1 > \cdots > V_L$, and k_i is the frequency of occurrence of V_i in $\boldsymbol{v}$.

---

[1]Note that the order of the minimal indices of a convolutional code has no physical significance as equivalent minimal encoders need not have constraint lengths in the same order.

Corollary 7 *Assuming some convention as to which columns will contain the identity matrix, there is a unique systematic encoder associated with each convolutional code and hence $N_c(v)$ also enumerates the class of canonical systematic encoders.*

5 Numerical results

Selected results for rate $R = 2/5$ and $R = 4/5$ codes are shown in Table 1. $\Delta_{k,n}(v)$ denotes the fraction of zero-delay row proper encoders with constraint lengths v that are non-catastrophic and hence minimal. It can be shown that, for all rates, at least half the encoders in this ensemble are minimal.

The figures for $N_{eq}(v)$ indicate that the number of equivalent minimal encoders increases exponentially with k. In general, for fixed k, $N_{eq}(v)$ is greater the more unevenly distributed the constraint lengths. This would seem to suggest that minimal encoders are unsuitable for code searches unless a canonical minimal encoder such as the echelon form derived by Forney [9] is used. On the other hand, the minimal encoder allows for the easy exclusion of codes with poor distance properties. For example, for $R = 2/5$ and $v = 5$, it can be shown [6] that $d_\infty \leq 11$. Assuming that the best codes meet the bound with equality, then only encoders with $v = (3, 2)$ need be included in the search since $v_i < 2$ implies the existence of codewords of weight less than 11. Therefore, for small k, the minimal encoder may be the best canonical form on which to base a code search.

6 Conclusions

A full analysis of equivalent encoders is presented which is subsequently used to enumerate the minimal encoders generating a convolutional code. We extend the known results on rate $1/n$ minimal encoders to the rate k/n case and derive enumerators for minimal encoders, canonical systematic encoders and convolutional codes. Numerical results show that the majority of codes have minimal encoders whose overall constraint length is more of less evenly distributed among the k rows and these codes are more likely to have good distance properties.

References

[1] G.D. Forney, Jr., "Convolutional codes I: Algebraic structure," *IEEE Trans. Inform. Theory*, vol. IT-16, pp. 720-738, Nov. 1970. Correction in *IEEE Trans. Inform. Theory*, vol. IT-17, No. 3, p. 360, May 1971.

[2] W.J. Rosenberg, *Structural Properties of Convolutional Codes*, Ph.D. dissertation, University of California, Los Angeles, 1971.

Rate 2/5				
v	$N_m(v)$	$\Delta_{k,n}(v)$	$N_c(v)$	$N_{eq}(v)$
$(1,0)$	26,040	0.93	6,510	4
$(2,0)$	833,280	0.93	104,160	8
$(1,1)$	781,200	0.90	130,200	6
$(3,0)$	26,664,960	0.93	1,666,560	16
$(2,1)$	24,998,400	0.90	6,249,600	4
$(4,0)$	853,278,720	0.93	26,664,960	32
$(3,1)$	799,948,800	0.90	99,993,600	8
$(2,2)$	799,948,800	0.90	133,324,800	6
$(5,0)$	27,304,919,040	0.93	426,639,360	64
$(4,1)$	25,598,361,600	0.90	1,599,897,600	16
$(3,2)$	25,598,361,600	0.90	6,399,590,400	4

Rate 4/5				
v	$N_m(v)$	$\Delta_{k,n}(v)$	$N_c(v)$	$N_{eq}(v)$
$(1,0,0,0)$	9,999,360	0.67	930	10,752
$(2,0,0,0)$	319,979,520	0.67	3,720	86,016
$(1,1,0,0)$	239,984,640	0.57	26,040	9,216
$(3,0,0,0)$	10,239,344,640	0.67	14,880	688,128
$(2,1,0,0)$	7,679,508,480	0.57	312,480	24,576
$(1,1,1,0)$	6,719,569,920	0.53	624,960	10,752
$(4,0,0,0)$	327,659,028,480	0.67	59,520	5,505,024
$(3,1,0,0)$	245,744,271,360	0.57	1,249,920	196,608
$(2,2,0,0)$	245,744,271,360	0.57	1,666,560	147,456
$(2,1,1,0)$	215,026,237,440	0.53	17,498,880	12,288
$(1,1,1,1)$	201,587,097,600	0.52	9,999,360	20,160

Table 1: Enumerators for binary codes for $R = 2/5$ and $R = 4/5$.

[3] G.D. Forney, Jr., Unpublished material.

[4] T.J. Shusta, "Enumeration of minimal convolutional encoders," *IEEE Trans. Inform. Theory*, vol. IT-23, No. 1, pp. 127-132, Jan. 1977.

[5] E. Paaske, "Short binary convolutional codes with maximal free distance for rates 2/3 and 3/4," *IEEE Trans. Inform. Theory*, vol. IT-20, No. 5, pp. 683-9, Sept. 1974.

[6] D.G. Daut, J.W. Modestino and L.D. Wismer, "New short constraint length convolutional code constructions for selected rational rates," *IEEE Trans. Inform. Theory*, vol. IT-28, No. 5, pp.794-800, Sept. 1982.

[7] M. Cedervall and R. Johannesson, "A fast algorithm for computing distance spectrum of convolutional codes," *IEEE Trans. Inform. Theory*, vol. IT-35, No. 6, pp. 1146-1159, Nov. 1989.

[8] G. Seguin, "A random coding bound for fixed convolutional codes of rate $1/n$," *IEEE Trans. Inform. Theory*, vol. IT-40, No. 4, pp. 1668-1670, Sept. 1994.

[9] G.D. Forney, Jr., "Minimal bases of rational vector spaces, with applications to multivariable linear systems," *SIAM J. Control*, vol. 13, pp.493-520, May 1975.

On Using Carmichael Numbers for Public Key Encryption Systems

R.G.E. Pinch

Queens' College, Silver Street
Cambridge CB3 9ET
U.K.

Abstract. We show that the inadvertent use of a Carmichael number instead of a prime factor in the modulus of an RSA cryptosystem is likely to make the system fatally vulnerable, but that such numbers may be detected.

1   Introduction

Huthnance and Warndof [12] comment that if one of the factors of the modulus in an RSA cryptosystem [23] is a Carmichael number rather than a prime, the cryptosystem will still work as expected. In this note we show that for the currently popular choice of 512 bit moduli, this would bring the modulus within the range of the elliptic curve factoring method [15], and that the most probable Carmichael numbers which might arise from currently popular primality tests can be factored very quickly by the $p+1$ method [26].

2   Primality tests

Many primality tests in current use [18],[19] are based on some form of the Fermat criterion: P is composite unless $b^{P-1} \equiv 1 \bmod P$. A *Carmichael number* is a composite number which satisfies the Fermat condition for all b coprime to P: hence the Fermat test is no more likely to reveal compositeness in this case than trial division.

The Fermat condition can be strengthened at little extra computational cost to the "strong" or Miller–Rabin criterion: putting $P - 1 = 2^r s$, with s odd, the sequence

$$b^s, b^{2s}, b^{2^2 s}, \ldots, b^{2^r s} = b^{P-1} \quad \bmod P$$

should end in 1 (the Fermat condition) and the last term which is not 1 (if any) should be $-1 \bmod P$. There are no analogues of Carmichael numbers for this test; indeed, for composite P the proportion of $b \bmod P$ for which the criterion holds is less than $1/4$. In each case we refer to a composite number which passes the test as a *pseudoprime* for that test.

Version 2 of the popular PGP implementation [27],[28] of the RSA cryptosystem uses four rounds of the Fermat test, with $b = 3, 5, 7$ and 11, and declares a number prime if it passes all four tests. The RSA Inc. B/SAFE toolkit uses four rounds of the Miller–Rabin test [13].

The composite numbers most likely to satisfy these criteria, and hence be falsely accepted as primes were classified by Damgård, Landrock and Pomerance [7],[8]. They showed that the most likely class of composites to escape detection are Carmichael numbers with three prime factors.

Huthnance and Warndof observed that RSA encryption and decryption [12] can be regarded as an elaborate form of the Fermat test, and hence that if an RSA modulus is formed inadvertently using a Carmichael number instead of a prime factor, then the cryptosystem will still function correctly. They argue that it is therefore unnecessary to pursue any more elaborate primality tests.

We maintain in this paper that this is dangerous advice.

3 Carmichael numbers

We record from [17] that a Carmichael number N has the following properties: N is square-free; N has at least three prime factors; $p \mid N$ implies $p - 1 \mid N - 1$ and $p < \sqrt{N}$.

A simple method of generating Carmichael numbers is due to Chernick [6]. If the three factors in $N = (6k + 1)(12k + 1)(18k + 1)$ are simultaneously prime, then N is a Carmichael number. It is not yet known whether there are infinitely many Carmichael numbers of Chernick form, although this would follow from the more general conjecture of Dickson [11].

We are interested in Carmichael numbers with three prime factors. Their distribution has been studied by Balasubramanian and Nagaraj [4], who show that the number of such Carmichael numbers up to x is at most $O(x^{5/14+\epsilon})$. If $N = pqr$ is a Carmichael number then we have $p - 1 = ha$, $q - 1 = hb$ and $r - 1 = hc$ where a, b, c are coprime and $habc \mid N - 1$. So h must satisfy the congruence $h(ab + bc + ca) \equiv -(a + b + c) \bmod abc$. The Chernick form is the case $a = 1$, $b = 2$, $c = 3$, leading to $h \equiv 0 \bmod 6$. We see that most values of (a, b, c) will lead to a possible congruence for $h \bmod abc$, whose smallest solution may be expected to be of the same order as abc.

Using the methods of [17] we have been able to compute the Carmichael numbers up to 10^{18} with three prime factors[1]. There are 35585 of them (compared with 24 739 954 287 740 860 primes: Deléglise and Rivat [9],[10]).

Of these 35585 Carmichael numbers, 783 correspond to the Chernick form with parameters $(a, b, c) = (1, 2, 3)$ and a total of 4091 have $(a, b) = (1, 2)$. There were no other cases with $b/a = 2$. We may expect small values of abc to predominate since in general we may expect h to about the same order as abc and thus N to be at least $(abc)^2$.

---

[1] The tables are available at `ftp://ftp.dpmms.cam.ac.uk/pub/Carmichael`

4 Detecting Carmichael numbers with three prime factors

Let us consider how to guard against the possiblity that a number P generated during the construction of an RSA modulus is composite: we consider especially the case of detecting Carmichael numbers.

We should point out that this event is somewhat unlikely. The exact computations for 18-digits numbers shows that Carmichael numbers are extremely rare compared to primes.

We can apply a primality proving method such as ECPP [3] or APR-CL [2] to P. Such methods are applicable in general and quite practical for numbers of hundreds of decimal digits.

We note that one prime factor of a Carmichael number P has to be less that $P^{1/3}$. For an RSA modulus of 512 bits, we can assume that P is about 2^{256} and so has a factor at most 2^{86}, about 10^{26}. Such factors may be routinely extracted by the elliptic curve factoring method (ECM) of Lenstra [15]. For a modulus of 1024 bits, the smallest factor of P could be over 50 digits and this is just beyond the present extreme of the ECM (47 digits in a 135-digit factor of $5^{256} + 1$ by Montgomery in December 1995).

We note that Carmichael numbers of any fixed form, for example, with given values of (a, b, c) but unknown h can be detected quickly by solving a single equation in h as a real variable. We demonstrate in the next section that a Carmichael number with parameters satisfying $b/a = 2$ is also easy to detect, even as a factor.

5 Detecting numbers divisible by a Carmichael number

Let us suppose that N is an RSA modulus divisible by a pseudoprime P. It is likely that P is a Carmichael number with three prime factors, say $P = pqr$; and again likely that the parameters (a, b, c) are all small.

Let us concentrate on the case $b = 2a$: for example, $a = 1$, $b = 2$. We saw that this occurred in over 10% of such numbers up to 10^{18}. We have $p = 2g + 1$, $q = 4g + 1$ for some g, so that $q + 1 = 2p$. So $2N$ is a multiple of $q + 1$, and hence q can be extracted immediately by the $p + 1$ factoring method of Williams [26]. Knowledge of q immediately yields g and hence p. We finally need to extract r. If N is 512 bits long then N/pq will be at most 100 digits and so can be factored by general purpose methods. Since r is congruent to 1 modulo a factor of g which is probably g itself, one could also use the Brent–Pollard modification [5] of Pollard's rho method [20],[21] for finding factors in given congruence classes.

6 Rare events and standards

Rivest [22] described numerical experiments using probabilistic primality tests for finding primes for potential use in an RSA cryptosystem and concluded that in practice the probability of obtaining a composite number was within acceptable limits. The estimates of Damgård, Landrock and Pomerance [7],[8] show that for the size of primes likely to be used in practice the probability of inadvertently finding a composite number is negligible (less than 2^{-100}, say). Some authors, such as Landrock [14] and Silverman [25] have concluded that there is no point in including requirements for primality proofs in standards such as ANSI X9.31.

We suggest that such requirements are not pointless. If an RSA cryptosystem is being used to sign digital documents, it may be to the advantage of a user to deliberately weaken his published modulus in order to subsequently repudiate a document, which he has in fact signed, by asserting that the weakness arose by chance and that an intruder has discovered the weakness. (Such a situation might arise, for example, in the context of electronic share trading.) Imposing checks against such weaknesses in the standard immediately eliminates this possibility, and avoids the necessity for contesting the plausibility of such claims in court.

7 Conclusion

Reliance on a probabilistic primality test for construction of an RSA modulus is likely to lead to a choice of factors which if not prime leave the system fatally weak. Proving the primality of the factors is reasonably fast, guards against an admittedly unlikely event and can prevent some forms of cheating.

References

1. L.M. Adleman and M.-D.A. Huang (eds.), *Algorithmic number theory*, Lecture notes in Computer Science, vol. 877, Berlin, Springer–Verlag, 1994, Proceedings, first international symposium, Ithaca, NY, May 1994.
2. L.M. Adleman, C. Pomerance, and R. Rumely, *On distinguishing prime numbers from composite numbers*, Ann. Math. 17 (1983), 173–206.
3. A.O.L. Atkin and F. Morain, *Elliptic curves and primality proving*, Math. Comp. 61 (1993), 29–68, Lehmer memorial issue.
4. R. Balasubramanian and S.V. Nagaraj, *Density of Carmichael numbers with three prime factors*, Math. Comp. 66 (1997), no. 220, 1705–1708.
5. R.P. Brent and J.M. Pollard, *Factorization of the eighth Fermat number*, Math. Comp. 36 (1981), no. 154, 627–630.
6. J. Chernick, *On Fermat's simple theorem*, Bull. Amer. Math. Soc. 45 (1939), 269–274.
7. I. Damgård and P. Landrock, *Improved bounds for the Rabin primality test*, Cryptography and coding III (M. Ganley, ed.), IMA conference series (n.s.), vol. 45, Institute of Mathematics and its Applications, Oxford University Press, 1993, Proceedings, 3rd IMA conference on cryptography and coding, Cirencester, December 1991., pp. 117–128.
8. I. Damgård, P. Landrock, and C. Pomerance, *Average case error estimates for the strong probable prime test*, Math. Comp. 61 (1993), 177–194, Lehmer memorial issue.

9. M. Deléglise and J. Rivat, *Computing $\pi(x)$, $M(x)$ and $\Psi(x)$*, In Adleman and Huang [1], Proceedings, first international symposium, Ithaca, NY, May 1994, p. 264.

10. ———, *Computing $\pi(x)$: the Meissel, Lehmer, Lagarias, Miller, Odlyzko method*, Math. Comp. **65** (1996), no. 213, 235–245.

11. L.E. Dickson, *A new extension of dirichlet's theorem on prime numbers*, Messenger of Mathematics **33** (1904), 155–161.

12. E.D. Huthnance and J. Warndof, *On using primes for public key encryption systems*, Appl. Math. Lett. **1** (1988), no. 3, 225–227.

13. B.S. Kaliski jr, *How RSA's toolkits generate primes*, Tech. Report 003-903028-100-000-000, RSA Laboratories, Redwood City, CA, 18 Feb 1994.

14. P. Landrock, *Proper key generation*, Cryptomathic Bull. **1** (1996), at URL `http://www.cryptomathic.dk/ matt/news.html`.

15. H.W. Lenstra jr, *Factoring integers with elliptic curves*, Annals of Math. **126** (1987), 649–673.

16. A.J. Menezes and S.A. Vanstone (eds.), *Advances in cryptology — CRYPTO '90*, Lecture notes in Computer Science, vol. 537, Berlin, Springer–Verlag, 1991.

17. R.G.E. Pinch, *The Carmichael numbers up to 10^{15}*, Math. Comp. **61** (1993), 381–391, Lehmer memorial issue.

18. ———, *Some primality testing algorithms*, Notices Amer. Math. Soc. **40** (1993), no. 9, 1203–1210.

19. ———, *Some primality testing algorithms*, The Rhine workshop on computer algebra, Karlsruhe, March, 1994 proceedings (J. Calmet, ed.), February 1994, pp. 2–13.

20. J.M. Pollard, *Theorems on factorization and primality testing*, Proc. Cambridge Philos. Soc. **76** (1974), 521–528.

21. ———, *A Monte Carlo method for factorization*, BIT **15** (1975), 331–334.

22. R.L. Rivest, *Finding four million large random primes*, In Menezes and Vanstone [16], pp. 625–626.

23. R.L. Rivest, A. Shamir, and L. Adleman, *A method for obtaining digital signatures and public-key cryptosystems*, Communications of the ACM **21** (1978), no. 2, 120–126, Reprinted as [24].

24. ———, *A method for obtaining digital signatures and public-key cryptosystems*, Communications of the ACM **26** (1983), no. 1, 96–99, Reprint of [23].

25. R.D. Silverman, *Fast generation of random, strong RSA primes*, Cryptobytes **3** (1997), no. 1, 9–13.

26. H.C. Williams, *A $p + 1$ method of factoring*, Math. Comp. **39** (1982), 225–234.

27. P.R. Zimmerman, *The official PGP user's guide*, MIT Press, 1995, 0-262-74017-6.

28. ———, *PGP source code and internals*, MIT Press, 1995.

Hash Functions and MAC Algorithms
Based on Block Ciphers

B. Preneel*

Katholieke Universiteit Leuven
Department Electrical Engineering-ESAT
Kardinaal Mercierlaan 94
B–3001 Heverlee, Belgium.
Email : bart.preneel@esat.kuleuven.ac.be
www.esat.kuleuven.ac.be/~preneel

Abstract. This paper reviews constructions of hash functions and MAC algorithms based on block ciphers. It discusses the main requirements for these cryptographic primitives, motivates these constructions, and presents the state of the art of both attacks and security proofs.

1 Introduction

Hash functions and MAC algorithms are important tools to protect information integrity. Together with digital signature schemes, they play an important role for securing our electronic interactions, in applications such as electronic cash and electronic commerce. This paper discusses how to reduce the construction of these cryptographic primitives to that of another primitive, namely a block cipher. The focus is on the relation between the security and performance of the primitives. Section 2 contains (informal) definitions of hash functions and MAC algorithms. Next a general model for these functions is presented in Sect. 3. Section 4 lists brute force attacks on hash functions, and overviews constructions for hash functions based on block ciphers. This is followed by a similar treatment of MAC algorithms in Sect. 5. Finally some concluding remarks are made.

2 Definitions

2.1 Hash functions

Hash functions were introduced in cryptography by W. Diffie and M. Hellman together with the concept of digital signatures. The basic idea is to sign a short 'digest' or 'fingerprint' of the message rather than the message itself.

Hash functions are used in a more general context to replace the protection of the integrity of a large amount of data (in principle of arbitrary size) by a

---

* F.W.O. postdoctoral researcher, sponsored by the National Fund for Scientific Research – Flanders (Belgium).

short string of fixed length (m bits), the hash result. Applications require that the evaluation of the hash function is 'efficient'. Moreover, the hash function should be publicly known, and should not require any secret parameter (these hash functions have also been called MDCs or Manipulation Detection Codes). For cryptographic applications, one imposes three security requirements, that can be stated informally as follows:

preimage resistance: for essentially all outputs, it is 'computationally infeasible' to find any input hashing to that output;

2nd-preimage resistance: it is 'computationally infeasible' to find a second (distinct) input hashing to the same output as any given input;

collision resistance: it is 'computationally infeasible' to find two colliding inputs, i.e., x and $x' \neq x$ with $h(x) = h(x')$.

A hash function that is preimage resistant and 2nd-preimage resistant is called *one-way*; a hash function that satisfies the three security properties is called *collision resistant*. Collision resistance may be required for digital signatures to preclude repudiation by the sender: if she can find a collision pair (x, x'), she can later claim that she has signed x' rather than x. Not all applications need collision resistance; hash functions that are only one-way are more efficient in terms of computation and storage (cf. Sect. 4.1).

Hash functions have also been used to construct new digital signature schemes (where the hash function is an integral part of the design). Other applications include the commitment to information without revealing it, the protection of pass-phrases, tick payments, key derivation, and key agreement. It should be pointed out that hash functions have often been used in applications that require new security properties, for which they have not been evaluated.

2.2 MAC algorithms

The banking world used MACs already in the seventies. A MAC is a hash function that takes as input a second parameter, the secret key. The sender of a message appends the MAC to the information; the receiver recomputes the MAC and compares it to the transmitted version. He accepts the information as authentic only if both values are equal.

The goal is that an eavesdropper, who can modify the message, does not know how the update the MAC accordingly. Hence the main security property for a MAC is that for someone who does not know the key, it should be computationally infeasible to predict the MAC value for a given message. Unlike digital signatures, MAC algorithms can only be used between mutually trusting parties, but they are faster to compute and require less storage than digital signatures.

The attack model is as follows: an opponent can choose a number of inputs x_i, and obtain the corresponding MAC value $h_K(x_i)$ (his choice of x_i might depend on the outcome of previous queries, i.e., the attack may be adaptive). Next he has to come up with an input x ($\neq x_i$, $\forall i$) and the value $h_K(x)$, which has to be correct with probability significantly larger than $1/2^m$, with m the number of bits in the MAC result. If the opponent succeeds in finding such a value, he is said to be capable of an *existential forgery*. If the opponent can choose the value of x, he is said to be capable of a *selective forgery*. If the success probability of the attack is close to 1, the forgery is called *verifiable*. An alternative strategy is to try to recover the secret key K from a number of message/MAC pairs. A *key recovery* is more devastating than a forgery, since it allows for arbitrary selective forgeries.

A MAC is said to be secure if it is 'computationally infeasible' to perform an existential forgery under an adaptive chosen text attack. Note that in many applications only known text attacks are feasible. For example, in a wholesale banking application one could gain a substantial profit if one could choose a single message and obtain its MAC. Nevertheless, it seems prudent to work with a strong definition.

3   General model

Almost all hash functions are iterative processes, which repeatedly apply a simple compression function f. Therefore they are called *iterated* hash functions. The input x is padded to a multiple of the block size, and is divided into t blocks denoted x_1 through x_t. It is then processed block by block.

The intermediate result is stored in an n-bit ($n \geq m$) *chaining variable* denoted with H_i:

$$H_0 = IV \qquad\qquad H_i = f(H_{i-1}, x_i), \quad 1 \leq i \leq t \qquad\qquad h(x) = g(H_t).$$

Here g denotes the *output transformation*.

The goal of the designer is to derive the security properties of the hash function h from those of the compression function f. In order to exclude trivial attacks, it is important to fix the value of IV and to precode the message. The simplest way of coding is to append an extra block at the end with the length of the input in bits. R. Merkle and I. Damgård showed that under these conditions, collision resistance of f is sufficient for collision resistant of h [6, 23]. Note that it is not a necessary condition, i.e., collisions for the compression function do not necessarily lead to collisions for the hash function. X. Lai and J. Massey proved a similar property for (2nd) preimage resistance [20]; in this case the converse holds (in a weaker form).

Similar to hash functions, MACs are often built using a compression function. The output transformation g is in this case often more important. The secret key may be introduced in the IV, in the compression function f, and/or in the output transformation g.

4   Hash functions

This section first discusses generic attacks on hash functions, i.e., attacks that depend only on the size of the hash result, and not on the internal structure of the hash function. Then constructions of hash functions based on block ciphers are reviewed.

4.1 Generic attacks

Two brute force attacks on hash functions can be identified.

(2nd) preimage attack. In order to find (2nd) preimages, one can pick an arbitrary input and evaluate h. The success probability of a single trial is $1/2^m$, with m the number of bits in the hash result, which implies that on average 2^{m-1} attempts are required. This attack can be applied off-line and in parallel. If it is sufficient to find a (2nd) preimage for one out of t given values, the success probability is increased with a factor of t. This can be avoided by parameterising the hash function for each input. This attack can be precluded by choosing values of m between
64 bits (marginally secure) to 128 bits (security for 20 years or more even against a determined opponent).

collision attack. Finding collisions is much easier than finding preimages: one needs only about $\sqrt{2^m} = 2^{m/2}$ evaluations of the hash function (as this results in about 2^m pairs of hash values) [36]. This phenomenon is related to the 'birthday paradox,' which states that in a group of 23 people, the odds are 1:2 that there are two people with the same birthday. Note that it is easy to impose that the colliding message are restricted to a small set with a prescribed format, and that one can implement this attack in parallel with minimal storage requirements [34]. Collision resistance requires that m is between 128 bits (marginally secure) and 160 bits (15 years or more).

4.2 Constructions

Hash functions based on block ciphers have been popular in part for historical reasons, as designers tried to use the Data Encryption Standard (DES, [9]) also for hashing. This reduces the design and evaluation effort, and results in compact implementations. It also allows to transfer the trust in DES (or in any other block cipher) to a hash function. This is quite important since many hash functions have been broken. However, this approach has some disadvantages. First, the use of a block cipher may create additional *export problems*.

Moreover, the use of a block cipher in this application requires *stronger properties* from the block cipher. Indeed, it might be that the block cipher has certain properties that do not affect its security level for encryption, but create serious problems in hashing modes. Examples are the (semi-)weak keys of DES [7, 25] and the quasi weak keys and weak hash keys [15]. Another problem is that differential cryptanalysis can be adopted to this setting; for DES this has been explored in [32].

This construction is only *efficient* if the key scheduling of the block cipher is not too slow, as every iteration typically requires a key change. Note that current dedicated hash functions (such as RIPEMD-160 [8] and SHA-1 [10]) run at about 50 Mbit/s on a 90 MHz Pentium. DES achieves 17 Mbit/s on the same machine; other block ciphers are up to twice as fast.

However, the performance including key schedule is about 2 to 10 times slower, and some constructions need more than one encryption to process a single block. As a consequence, dedicated hash functions are for the time being 5...10 times faster than hash functions based on block ciphers with a comparable security level.

In the rest of this section (except for Sect. 4.6) it will be assumed that the block cipher has block length and key length of m bits. The *hash rate* of a hash function based on an m-bit block cipher is defined as the number of m-bit message blocks hashed per encryption.

4.3 Single block length hash functions

For these hash functions the size of the hash result is equal to the block size of the block cipher. All these schemes have rate 1. The first cryptographic hash function was probably the construction proposed by M.O. Rabin [31]: $H_i = E_{x_i}(H_{i-1})$. Here $E_K(x)$ denotes the encryption of plaintext x using the key K. The fact that this hash function is not collision resistant when used with DES ($m = 64$) was pointed out by G. Yuval in [36]. R. Merkle showed that finding a (2nd) preimage requires only $2^{m/2}$ operations (using a meet-in-the-middle attack).

The first secure construction was the hash function of Matyas, Meyer, and Oseas: $H_i = E_{H_{i-1}}(x_i) \oplus x_i$.

This scheme has been included in ISO/IEC 10118–2 [14], with an additional mapping from the ciphertext space to the key space (cf. Sect. 4.4 for an example). Its dual is known as the Davies-Meyer scheme, although it probably should be attributed to S.M. Matyas and C.H. Meyer:

$H_i = E_{x_i}(H_{i-1}) \oplus H_{i-1}$.

A variant of the two schemes was proposed by the author (in '89) and by Miyaguchi et al. [24]: $H_i = E_{H_{i-1}}(x_i) \oplus x_i \oplus H_{i-1}$.

In [27], the author has analyzed all 64 schemes of this general form, and came to the conclusion that 12 of these are secure. They can be constructed from the Matyas-Meyer-Oseas scheme and the Preneel-Miyaguchi scheme by applying an affine transformation to the inputs. It is conjectured that for these schemes no shortcut attacks exist, which implies that a collision attack requires about $2^{m/2}$ operations and a (2nd) preimage attack about 2^m operations. However, since most block ciphers have a block length of $m = 64$ bits, collisions can be found in only 2^{32} operations. Therefore hash functions with a larger hash result are needed.

4.4 Double block length hash functions

The aim of *double block length* hash functions is to achieve a higher security level against collision attacks. Ideally a collision attack on such a hash function should require 2^m operations, and a (2nd) preimage attack 2^{2m} operations. An important class of proposals is of the following form:

$$H_i^1 = E_{A_i^1}(B_i^1) \oplus C_i^1$$
$$H_i^2 = E_{A_i^2}(B_i^2) \oplus C_i^2,$$

where A_i^1, B_i^1, and C_i^1 are binary linear combinations of H_{i-1}^1, H_{i-1}^2, x_i^1, and x_i^2 and where A_i^2, B_i^2, and C_i^2 are binary linear combinations of H_{i-1}^1, H_{i-1}^2, x_i^1, x_i^2, and H_i^1. The hash result is equal to the concatenation of H_t^1 and H_t^2. Several hash functions in this class have been published as individual proposals between '89 and '93. First it was shown by Hohl et al. that the security level of the *compression function* of these hash functions is at most that of a single block length hash function [11]. Next L. Knudsen, X. Lai, and B. Preneel showed that for all hash functions in this class, a (2nd) preimage attack requires at most 2^m operations, and a collision attack requires at most $2^{3m/4}$ operations (for most schemes this can be reduced to $2^{m/2}$) [17].

Several schemes of rate less than 1 have been proposed. From the few that have survived, the most important ones are MDC-2 and MDC-4 with hash rate 1/2 and 1/4 respectively. MDC-2 can be described as follows:

$$T_i^1 = E_{H_{i-1}^1}(x_i) \oplus x_i = LT_i^1 \parallel RT_i^1 \qquad H_i^1 = LT_i^1 \parallel RT_i^2$$
$$T_i^2 = E_{H_{i-1}^2}(x_i) \oplus x_i = LT_i^2 \parallel RT_i^2 \qquad H_i^2 = LT_i^2 \parallel RT_i^1 .$$

MDC-2 has been included in Part 2 of ISO/IEC 10118-2 [14]. The standard does not specify the block cipher; it also requires the specification of two mappings u, v from the ciphertext space to the key space such that $u(IV^1) \neq v(IV^2)$. For DES, these mappings from 64 to 56 bits drop the parity bits in every byte and fix the second and third key bits to 01 and 10 respectively (to preclude attacks based on (semi-)weak keys).

One iteration of MDC-4 consists of two MDC-2 steps, where the plaintexts in the second instance are equal to respectively H_{i-1}^2 and H_{i-1}^1; the keys are the same for both instances. The security level of the hash functions MDC-2 and MDC-4 (with fixed IV's) and of their compression functions is listed in Table 1. These attacks are described in [19, 20]. Note that the compression function is not very strong and that the protection of the hash function against collision attacks is not very high if DES is used.

	hash function		compression function	
	collision	(2nd) preimage	collision	(2nd) preimage
MDC-2	2^m	$2^{3m/2}$	$2^{m/2}$	2^m
MDC-4	2^m	$2^{7m/4}$	$2^{3m/4}$	$2^{3m/2}$
MDC-2 (DES)	2^{55}	2^{83}	2^{28}	2^{54}
MDC-4 (DES)	2^{56}	2^{109}	2^{41}	2^{90}

Table 1. Security level for MDC-2 and MDC-4 based on a block cipher with equal block and key length and based on DES.

4.5 Schemes with a 'security proof'

Two constructions have a 'proof of security' for the collision resistance of the compression function. They both start from the assumption that the Matyas-Meyer-Oseas single block length hash function (or one of its variants) is secure. R. Merkle describes several constructions that achieve a security level of 2^m (2^{55} for DES) against collision attacks [23]. The fastest version has rate 0.27 for DES. The simplest scheme (with rate 1/18.3 for DES) can be described as follows:

$$H_i = \text{chop}_{16}\left[E_{0\|H^1_{i-1}}(H^2_{i-1}\|x_i) \oplus (H^2_{i-1}\|x_i) \,\|\, E_{1\|H^1_{i-1}}(H^2_{i-1}\|x_i) \oplus (H^2_{i-1}\|x_i)\right].$$

Here H_{i-1} is a string consisting of 112 bits, the leftmost 55 bits of which are denoted H^1_{i-1}, and the remaining 57 are denoted H^2_{i-1}; x_i consists of 7 bits only. This hash function is similar to MDC-2, but has a secure compression function at the cost of a low speed. Note that in fact additional measures have to be taken to preclude attacks based on weak keys.

The constructions proposed by L.R. Knudsen and B. Preneel [18, 19] offer better trade-offs between rate and security level; moreover the security level can be higher than 2^m. They have the disadvantage that they require a larger internal memory, and an output transformation to preclude partial collisions i.e., collisions where only part of the outputs collide. The security proof assumes that the best attack on a number of 'coupled' single block length hash functions (i.e., with dependent inputs) is a brute force attack. This coupling is achieved by encoding the input bits to the compression function with a code over $GF(2^2)$ (or, in general $GF(2^t)$, $t \geq 2$) to obtain the plaintext and key input of the individual single block length hash functions. For an $[n, k, d]$ code, the number of parallel instances of a single block length hash function equals n and a collision requires at least $2^{(d-1)/2}$ encryptions. Table 2 gives the properties of the constructions using codes over $GF(2^4)$.

code	rate	collision
$[6, 4, 3]$	1/4	$\geq 2^m$
$[8, 6, 3]$	1/2	$\geq 2^m$
$[12, 10, 3]$	2/3	$\geq 2^m$
$[9, 6, 4]$	1/3	$\geq 2^{3m/2}$
$[16, 13, 4]$	5/8	$\geq 2^{3m/2}$

Table 2. Parameters of constructions for hash functions based on codes over $GF(2^4)$.

4.6 Block ciphers with longer key lengths

R. Merkle observed already in [22] that if the key length of a block cipher is larger than the block size, it can be used as the compression function of a single block length hash function by just fixing the plaintext, and considering the mapping from key to ciphertext: $H_i = E_{H_{i-1} \parallel x_i}(C)$, with C a constant. X. Lai and J. Massey have extended these constructions in [20] to double block length hash functions. L. Knudsen and the author provide improved constructions with security proof for this case as well [19].

5 MAC algorithms

As in the previous section, first generic attacks are discussed, and then an overview of constructions for MAC algorithms is given.

5.1 Generic attacks

The attacks discussed in this section depend only on the size of the parameters of the MAC, and not on its internal structure: brute force key search, guessing of the MAC, and a birthday forgery attack.

An **exhaustive key search** consists of running through the key space and checking whether a key corresponds to the known message-MAC pairs. About k/m values are sufficient to determine the key uniquely; for most applications this value lies between 1 and 4. The expected computational effort is 2^{k-1} MAC evaluations, where k is the number of (effective) key bits.

Brute force key search can be precluded by choosing a sufficiently large key. A value of 56 bits offers only marginal security, while 75 to 90 bits is sufficient for 15 years or more. Currently finding a 56-bit key in a period of 1 year requires an investment of 50 000\$, and it can be done in a few months by using idle cycles on the Internet, as was demonstrated in the Spring of 1997; with a 1 million US\$ investment, the search time can be reduced to a few hours [5, 35]. Moreover, one also has to take into account the empirical observation that the computing power for a given cost is multiplied by four every 3 years (Moore's 'Law').

It is important to recover a MAC key within its active lifetime, which can be very short in communications applications. After that period, the key is completely useless. However, one can mount the attack during the complete lifetime of the system; it is sufficient to feed to the search machine the current text/MAC pairs. In order to assess the feasibility of this attack, one has to compare the cost of an attack with the profit which can be made by recovering one or more keys during the lifetime of the system.

Another attack strategy is to pick an arbitrary input and **guess the MAC value** by choosing an m-bit string uniformly at random. For a good MAC algorithm, one expects that the success probability equals $1/2^m$ (here the probability is taken over all keys). A related strategy consists in guessing the value of the key, and computing the MAC value. Its success probability is $1/2^k$. The success probability of the combined attacks is equal to $1/2^{\min(k,m)}$. The value of k is typically larger than that of m to preclude a brute force key search. This forgery attack is clearly not verifiable. However, one should take it into account when selecting a MAC algorithm. The value of m depends on the expected profit of a successful attempt, as well as on the number of trials that are allowed by the system, i.e., the way the system reacts to wrong MAC values. For most applications $m = 32 \ldots 64$ is sufficient to render this attack uneconomical.

Recently a **birthday forgery attack** has been developed that applies to all iterated MAC algorithms (B. Preneel and P.C. van Oorschot, [29]). The basic idea behind the attack is the birthday paradox. Its feasibility depends on the bitsizes n of the chaining variable and m of the MAC result, and the nature of the output transformation g. If $n = m$, and g is a permutation, the attack requires one chosen text and about $2^{n/2}$ known message-MAC pairs. If $n > m$, an additional number of about 2^{n-m} chosen message-MAC pairs is required. The attack requires at least one chosen text, hence in applications where any access to the MAC algorithm with the correct key is precluded, it should not be considered a problem. Also, the forged message is of a special form: if the MAC is known for three messages of the form x, x', and $x\|y$, one can forge the MAC for $x'\|y$. Note that further optimisations of the attack are possible, that reduce the number of known texts. A more serious concern is that for certain MAC algorithms, the forgery attack can be extended to a key recovery attack.

This attack motivates the use of a MAC result that is smaller than the size of the internal memory n. It can be precluded by appending a sequence number at the *end* of every message. Such a sequence number if useful to prevent replay attacks as well [7]. An alternative is to randomise the output transformation.

5.2 Constructions for MACs

MAC algorithms based on block ciphers have been proposed already in the late seventies. Together with DES, **CBC-MAC** based on DES became very popular, especially in the banking world. It has been standardized in several documents comprising ANSI X9.9 [1], ANSI X9.19 [2], ISO 8731 [12], and ISO/IEC 9797 [13]. Only during the last years, a thorough security analysis has been performed. The compression function of CBC-MAC has the following form:

$$H_i = E_K(H_{i-1} \oplus x_i), \quad 1 \le i \le t,$$

with $H_0 = 0$. An output transformation is needed to preclude the following simple forgery: given $h_K(x)$, $h_K(x\|y)$, and $h_K(x')$, one knows that $h_K(x'\|y') = h_K(x\|y)$ if $y' = y \oplus h_K(x) \oplus h_K(x')$.

One approach is for g to select the leftmost m bits. However, L.R. Knudsen has shown that the simple attack can be extended to this case [16]. It requires then approximately $2^{(n-m)/2}$ chosen texts and 2 known texts.

A stronger and widely used alternative is to replace the processing of the last block by a two-key triple encryption (with keys $K_1 = K$ and K_2); this is commonly known as the **ANSI retail MAC**, since it first appeared in ANSI X9.19:

$$g(H_t) = E_{K_1}(D_{K_2}(H_t)).$$

Here D denotes decryption. This mapping requires little overhead, and has the additional advantage that it precludes an exhaustive search against the 56-bit DES key.

All these variants are vulnerable to the birthday forgery attack, that requires a single chosen message and about 2^{32} known messages (if DES is used with $m = 64$). If $m = 32$, an additional 2^{32} chosen messages are required. Note that with a fast DES implementation on a PC, this number of texts can be collected in a single day. Bellare et al. provide a proof of security for CBC-MAC [3], i.e., they establish a lower bound to break the system under certain assumptions on the block cipher. It almost matches the upper bound provided by the birthday forgery attack.

For the ANSI retail MAC, 2^{32} known texts allow for a key recovery requiring $3 \cdot 2^k$ encryptions, compared to 2^{2k} encryptions for exhaustive search [30]. If DES is used, this implies that key recovery may become feasible. Another key recovery attack needs only a single known text, but requires about 2^k MAC verifications. Moreover, it reduces the effective MAC size from $\min(m, 2k)$ to $\min(m, k)$. It seems possible to preclude these key recovery attacks by introducing also a triple-DES encryption in the first iteration, but further research is necessary to confirm this.

An alternative to CBC-MAC is **RIPE-MAC**, that adds a feedforward [33]:

$$H_i = E_K(H_{i-1} \oplus x_i) \oplus x_i, \ \ 1 \le i \le t.$$

It has the advantage that the round function is harder to invert (even for someone who knows the secret key). An output transformation is needed as well.

XOR-MAC is another scheme based on a block cipher [4]. It is a randomized algorithm and its security can again be reduced to that of the block cipher. It has the advantage that it is parallellisable and that it is incremental, i.e., small modifications to the message (and to the MAC) can be made at very low cost. The use of random bits clearly helps to improve security, but it has a cost in practical implementations. Also, the performance is typically 30% to 50% slower than CBC-MAC.

Note that cryptanalysis of the underlying block cipher can often be extended to an attack on the MAC, in spite of the fact that an attacker obtains less information than in conventional cryptanalysis on the ECB mode. Examples can be found in the literature for linear and differential cryptanalysis of CBC-MAC based on DES [26, 28].

6 Concluding remarks

The construction of a new cryptographic primitive (hash function, MAC algorithm) based on an existing primitive (a block cipher) is a common paradigm in cryptography. This requires a very good understanding of the security of both primitives, and of additional assumptions that have to be made. For MAC algorithm, significant progress has been made, both from the side of the attacks (upper bound on the security level) and from the side of the constructions (lower bound on the security level based on certain assumptions). For hash functions, the cryptanalytic side has achieved some successes, and has resulted in new constructions with some provable properties. However, important open problems remain such as the formalization of the security of single block length hash functions.

It is likely that DES will be replaced in a few years by the AES (with a 128-bit block size). It will then be possible to achieve a higher security level using the existing constructions. However, it is an interesting open problem to improve the security and performance of constructions that use the DES or other 64-bit block ciphers.

References

1. ANSI X9.9 (revised), *"Financial Institution Message Authentication (Wholesale),"* American Bankers Association, April 7, 1986.
2. ANSI X9.19 *"Financial Institution Retail Message Authentication,"* American Bankers Association, August 13, 1986.
3. M. Bellare, J. Kilian, P. Rogaway, "The security of cipher block chaining," *Advances in Cryptology, Proc. Crypto'94, LNCS 839,* Y. Desmedt, Ed., Springer-Verlag, 1994, pp. 341–358.
4. M. Bellare, R. Guérin, P. Rogaway, "XOR MACs: new methods for message authentication using block ciphers," *Advances in Cryptology, Proc. Crypto'95, LNCS 963,* D. Coppersmith, Ed., Springer-Verlag, 1995, pp. 15–28.
5. M. Blaze, W. Diffie, R.L. Rivest, B. Schneier, T. Shimomura, E. Thompson, M. Wiener, "Minimal key lengths for symmetric ciphers to provide adequate commercial security. A Report by an Ad Hoc Group of Cryptographers and Computer Scientists," January 1996.
6. I.B. Damgård, "A design principle for hash functions," *Advances in Cryptology, Proc. Crypto'89, LNCS 435,* G. Brassard, Ed., Springer-Verlag, 1990, pp. 416–427.
7. D. Davies, W. Price, *Security for Computer Networks,* 2nd ed., Wiley, 1989.
8. H. Dobbertin, A. Bosselaers, B. Preneel, "RIPEMD-160: a strengthened version of RIPEMD," *Fast Software Encryption, LNCS 1039,* D. Gollmann, Ed., Springer-Verlag, 1996, pp. 71–82.
9. FIPS 46, *Data encryption standard,* NBS, U.S. Department of Commerce, Washington D.C., Jan. 1977.
10. FIPS 180-1, *Secure hash standard,* NIST, US Department of Commerce, Washington D.C., April 1995.
11. W. Hohl, X. Lai, T. Meier, C. Waldvogel, "Security of iterated hash functions based on block ciphers," *Advances in Cryptology, Proc. Crypto'93, LNCS 773,* D. Stinson, Ed., Springer-Verlag, 1994, pp. 379–390.
12. ISO 8731:1987, *Banking – approved algorithms for message authentication, Part 1, DEA, Part 2, Message Authentication Algorithm (MAA).*
13. ISO/IEC 9797:1993, *Information technology - Data cryptographic techniques - Data integrity mechanisms using a cryptographic check function employing a block cipher algorithm.*

14. ISO/IEC 10118:1994, *"Information technology – Security techniques – Hash-functions, Part 1: General and Part 2: Hash-functions using an n-bit block cipher algorithm,"*.

15. L.R. Knudsen, "New potentially 'weak' keys for DES and LOKI," *Advances in Cryptology, Proc. Eurocrypt'94, LNCS 950*, A. De Santis, Ed., Springer-Verlag, 1995, pp. 419–424.

16. L.R. Knudsen, "Chosen-text attack on CBC-MAC," *Electronics Letters*, Vol. 33, No. 1, 1997, pp. 48–49.

17. L.R. Knudsen, X. Lai, B. Preneel, "Attacks on fast double block length hash functions," *Journal of Cryptology*, in print.

18. L.R. Knudsen, B. Preneel, "Hash functions based on block ciphers and quaternary codes,"
Advances in Cryptology, Proc. Asiacrypt'96, LNCS 1163,
K. Kim and T. Matsumoto, Eds., Springer-Verlag, 1996, pp. 77–90.

19. L.R. Knudsen, B. Preneel, "Fast and secure hashing based on codes,"
Advances in Cryptology, Proc. Crypto'97, LNCS 1294,
B. Kaliski, Ed., Springer-Verlag, 1997, pp. 485–498.

20. X. Lai, J.L. Massey, "Hash functions based on block ciphers," *Advances in Cryptology, Proc. Eurocrypt'92, LNCS 658*, R.A. Rueppel, Ed., Springer-Verlag, 1993, pp. 55–70.

21. S.M. Matyas, C.H. Meyer, J. Oseas, "Generating strong one-way functions with cryptographic algorithm," *IBM Techn. Disclosure Bull.*, Vol. 27, No. 10A, 1985, pp. 5658–5659.

22. R. Merkle, *"Secrecy, Authentication, and Public Key Systems,"* UMI Research Press, 1979.

23. R. Merkle, "One way hash functions and DES," *Advances in Cryptology, Proc. Crypto'89, LNCS 435*, G. Brassard, Ed., Springer-Verlag, 1990, pp. 428–446.

24. S. Miyaguchi, M. Iwata, K. Ohta, "New 128-bit hash function," *Proc. 4th International Joint Workshop on Computer Communications*, Tokyo, Japan, July 13–15, 1989, pp. 279–288.

25. J.H. Moore, G.J. Simmons, "Cycle structure of the DES for keys having palindromic (or antipalindromic) sequences of round keys,"
IEEE Trans. on Software Engineering, Vol. SE–13, No. 2, 1987, pp. 262–273.

26. K. Ohta, M. Matsui, "Differential attack on message authentication codes,"
Advances in Cryptology, Proc. Crypto'93, LNCS 773,
D. Stinson, Ed., Springer-Verlag, 1994, pp. 200–211.

27. B. Preneel, R. Govaerts, J. Vandewalle, "Hash functions based on block ciphers: a synthetic approach,"
Advances in Cryptology, Proc. Crypto'93, LNCS 773,
D. Stinson, Ed., Springer-Verlag, 1994, pp. 369–379.

28. B. Preneel, M. Nuttin, V. Rijmen, J. Buelens, "Cryptanalysis of the CFB mode of the DES with a reduced number of rounds,"
Advances in Cryptology, Proc. Crypto'93, LNCS 773,
D. Stinson, Ed., Springer-Verlag, 1994, pp. 212–223.

29. B. Preneel, P.C. van Oorschot, "MDx-MAC and building fast MACs from hash functions,"
Advances in Cryptology, Proc. Crypto'95, LNCS 963,
D. Coppersmith, Ed., Springer-Verlag, 1995, pp. 1–14.

30. B. Preneel, P.C. van Oorschot, "A key recovery attack on the ANSI X9.19 retail MAC," *Electronics Letters*, Vol. 32, No. 17, 1996, pp. 1568–1569.

31. M.O. Rabin, "Digitalized signatures," in *"Foundations of Secure Computation,"* R. Lipton, R. DeMillo, Eds., Academic Press, New York, 1978, pp. 155–166.

32. V. Rijmen, B. Preneel, "Improved characteristics for differential cryptanalysis of hash functions based on block ciphers,"
Fast Software Encryption, LNCS 1008,
B. Preneel, Ed., Springer-Verlag, 1995, pp. 242–248.

33. RIPE, *"Integrity Primitives for Secure Information Systems. Final Report of RACE Integrity Primitives Evaluation (RIPE-RACE 1040),"* *LNCS 1007,* A. Bosselaers and B. Preneel, Eds., Springer-Verlag, 1995.

34. P.C. van Oorschot, M.J. Wiener, "Parallel collision search with application to hash functions and discrete logarithms," *Proc. 2nd ACM Conference on Computer and Communications Security,* ACM, 1994, pp. 210–218. Final version to appear in Journal of Cryptology.

35. M.J. Wiener, "Efficient DES key search," *Technical Report TR-244,* School of Computer Science, Carleton University, Ottawa, Canada, May 1994. Presented at the rump session of Crypto'93.

36. G. Yuval, "How to swindle Rabin," *Cryptologia,* Vol. 3, No. 3, 1979, pp. 187–189.

Witness Hiding Restrictive Blind Signature Scheme

Cristian Radu, René Govaerts and Joos Vandewalle

Katholieke Universiteit Leuven, Laboratorium ESAT
Kardinaal Mercierlaan 94, B-3001 Heverlee, Belgium
Cristian.Radu@esat.kuleuven.ac.be

Abstract. In this paper, we propose a restrictive blind signature scheme with enhanced security for the signer. It is suitable to be used as a basic cryptographic primitive in the design of a privacy protecting off-line electronic payment system. The solution is derived from the transformation of a witness hiding proof of knowledge. In order to blindly sign a message, preserving a certain invariant structure of this message, a proof of knowledge of a representation is diverted. The choice of a witness hiding proof system improves the security and provability of the restrictive blind signature scheme. This is the main contribution of our solution. The cost paid for enhanced security is reasonable with regard to the overall efficiency of the scheme.

1 Introduction

A blind signature scheme [3] is the cryptographic primitive that is used by a certification authority (CA) in order to issue validated pseudonyms $(\pi, V_{CA}(\pi))$ to Alice. The first requirement of this scheme is that the CA should not be able to link a pair pseudonym/validator to a particular signing protocol that has led to this pair. However, if Alice is restricted to using the secret key μ, corresponding to the pseudonym π, in only one signing protocol, then the CA requires a stronger condition. In order to get a validator $V_{CA}(\pi)$ on a pseudonym, Alice is compelled to apply a limited set of blinding operations with parameters of her own choice on a pair $(\pi_0, V_{CA}(\pi_0))$, which is also known to the CA. Furthermore, if $I(\pi_0) = i_{Alice}$ then $I(\pi) = i_{Alice}$, where π is the blind version of π_0, I is a blinding-invariant function, and i_{Alice} encodes the real identity of Alice. The attribute *restrictive* refers to this kind of blind signature scheme. When the (one-time) pseudonym π represents an electronic coin, the above condition allows the multiple-spending detector to derive the real identity of Alice if she tries to sign in multiple payment transactions using the same secret key μ of a coin [2].

The starting point in the design of our scheme is the *basic proof system* proposed by Chaum and Pedersen in their interactive signature scheme and its blind version [4]. This three-step protocol is a proof of knowledge of a discrete logarithm, which simultaneously satisfies a two-dimensional predicate. It is adapted from the Schnorr identification scheme [14]. The basic proof system is complete and sound, but is just *conjectured* to be witness hiding. Therefore, the motivation of our work is to investigate the possibility of designing an interactive signature scheme and its restrictive blind version, starting from a *provable* witness hiding proof system. This is derived from the basic proof system, using a "security amplification" similar to that applied by Okamoto to obtain the Identification Scheme 1 [10] (provably as secure as the discrete logarithm problem in groups of prime order) from the Schnorr identification protocol [14]. The choice of a witness hiding proof system improves the security and provability of the proposed restrictive blind signature scheme. The blinding-invariant function is of a similar type as the function used by Brands in his restrictive blind signature scheme [2]. The cost paid for enhanced security is reasonable with respect to the overall efficiency of the scheme.

The remainder of the paper is organized as follows. In Section 2, the notation, definitions, and the main cryptographic assumptions are introduced. Section 3 describes the witness hiding proof system and the corresponding interactive signature scheme derived from it. The security of these primitives is evaluated. The interactive signature scheme is the starting point from which, in Section 4, the proposed restrictive blind signature scheme is obtained. Its security is also considered. Finally, some concluding remarks are presented.

2 Preliminaries

For any prime p the set of positive integers smaller than p is denoted $\mathbb{Z}_p^*$. Let q be a prime dividing $p-1$. For reasons relating to the intractability of computing discrete logarithms in groups of prime order, the length of p should be at least 512 bits, and the length of q should be at least 160 bits. G_q denotes the unique subgroup of $\mathbb{Z}_p^*$ of order q, for which polynomial-time algorithms are known to determine equality of elements, to test membership, to compute inverses, to multiply, and to randomly select elements. In expressions involving elements in G_q, the reduction modulo p is not explicitly mentioned.

We denote by $X = (x_1, x_2, x_3)$ an element in G_q^3, by $\alpha = (\alpha_1, \alpha_2, \alpha_3)$, $\beta = (\beta_1, \beta_2, \beta_3)$ elements in $\mathbb{Z}_q^3$, and by c a constant in $\mathbb{Z}_q$. We also use the notation listed below:

$$\mathbf{0} = (0, 0, 0) \in \mathbb{Z}_q^3,$$
$$c\alpha = (c\alpha_1 \bmod q, c\alpha_2 \bmod q, c\alpha_3 \bmod q) \in \mathbb{Z}_q^3,$$
$$\alpha \cdot \beta = (\alpha_1\beta_1 + \alpha_2\beta_2 + \alpha_3\beta_3) \bmod q$$
$$\hat{\alpha} = \alpha \cdot (1, 1, 0) = (\alpha_1 + \alpha_2) \bmod q \in \mathbb{Z}_q,$$
$$\alpha + \beta = ((\alpha_1 + \beta_1) \bmod q, (\alpha_2 + \beta_2) \bmod q, (\alpha_3 + \beta_3) \bmod q) \in \mathbb{Z}_q^3,$$
$$X^\alpha = \prod_{i=1}^{3} x_i^{\alpha_i} = x_1^{\alpha_1} x_2^{\alpha_2} x_3^{\alpha_3} \bmod p \in G_q.$$

For any $X \in G_q^3$, $\alpha, \beta \in \mathbb{Z}_q^3$, and $c \in \mathbb{Z}_q$ it holds that :

P1 $X^{\alpha+\beta} = X^\alpha X^\beta$,

P2 $(X^\alpha)^c = X^{c\alpha}$,

P3 $\hat{\alpha} + c\hat{\beta} = (\alpha + c\beta) \cdot (1, 1, 0)$.

285

Definition 1. Let $G = (g_1, g_2, g_3), g_i \in G_q \setminus \{1\}, g_i \neq g_j$ for all $i \neq j \in \{1, 2, 3\}$, denote a generator-tuple in G_q^3. A representation of $h \in G_q$ with respect to G is denoted by $rep_G(h) = \mathbf{s} = (s_1, s_2, s_3) \in \mathbb{Z}_q^3$, satisfying $h = G^\mathbf{s} = \prod_{i=1}^{3} g_i^{s_i}$.

The security of the interactive signature scheme, which is introduced in the next section, relies on the following two cryptographic assumptions:

Assumption 2.1 *Finding the unique index* $\log_\gamma h \in \mathbb{Z}_q$ *of* $h \in G_q$ *with respect to* $\gamma \in G_q \setminus \{1\}$ *is the* discrete logarithm problem (DLP). *An algorithm is said to solve the DLP if, for inputs* $\gamma \neq 1, h$ *generated uniformly at random, it outputs* $\log_\gamma h$ *with at least non-negligible probability of success. The* discrete logarithm assumption (DLA) *states that there is no probabilistic polynomial-time algorithm that solves the DLP.*

Assumption 2.2 *Finding a representation* $rep_G(h) \in \mathbb{Z}_q^3$ *of* $h \in G_q$ *with respect to the generator-tuple* $G \in G_q^3$ *is the* representation problem for groups of prime order. *An algorithm is said to solve this problem if:*

- *for inputs* (G, h) *generated uniformly at random, it outputs* $rep_G(h)$, *or*
- *for input* G *generated uniformly at random it outputs a non-trivial representation of the unit in* G_q, $rep_G(1) \neq \mathbf{0}$,

with at least non-negligible probability of success. The representation assumption *states that there is no probabilistic polynomial-time algorithm that solves the representation problem.*

It was proved by Brands in [2] that the computational difficulty of the representation problem is equivalent to the DLP. We resume this result by the following proposition:

Proposition 2. *The representation problem for groups of prime order has the same computational difficulty as the DLP.*

3   Interactive Signature Scheme

An interactive signature scheme can be derived from a proof of knowledge of a representation, using the procedure introduced by Chaum and Pedersen in [4]. The interactive signature scheme allows a signer, represented by the public key $(p, q, G = (g_1, g_2, g_3), h)$ and the corresponding secret key $rep_G(h) = \mathbf{s} = (s_1, s_2, s_3) \in \mathbb{Z}_q^3$ so that $G^\mathbf{s} = h \in G_q \setminus \{1\}$, to interactively sign a message $m \in G_q$. The signature on m consists of $z = m^{\hat{s}}$ and a proof of knowledge of a representation of h with respect to G that verifies z for a given m. This proof of knowledge is provided by a witness hiding proof system where the signer plays the role of the prover and the recipient of the signature plays the role of the verifier.

Given (p, q, G, h, m, z) as a common input, the following three-step protocol is used to prove knowledge of a witness $\mathbf{s} = (s_1, s_2, s_3) \in \mathbb{Z}_q^3$ with respect to a two-dimensional predicate $G^\mathbf{s} = h, m^{\hat{s}} = z$:

1. The prover $\mathcal{P}$ randomly generates $\mathbf{w} = (w_1, w_2, w_3) \in \mathbf{Z}_q^3$ and sends the commitment $a \leftarrow G^{\mathbf{w}}$, $b \leftarrow m^{\hat{w}}$ to the verifier $\mathcal{V}$.
2. $\mathcal{V}$ randomly generates a challenge $c \in \mathbf{Z}_q$ and sends it to $\mathcal{P}$.
3. $\mathcal{P}$ computes the response $\mathbf{r} = (r_1, r_2, r_3) \leftarrow \mathbf{w} - c\mathbf{s}$ and sends it to $\mathcal{V}$.
4. $\mathcal{V}$ accepts the proof if $a = G^{\mathbf{r}}h^c$ and $b = m^{\hat{r}}z^c$.

The security of this three-step protocol is analyzed, considering the possibilities of cheating for $\mathcal{P}$ and $\mathcal{V}$. Let $\tilde{\mathcal{P}}$ (respectively $\tilde{\mathcal{V}}$) denote a fraudulent $\mathcal{P}$ (respectively $\mathcal{V}$). $\tilde{\mathcal{P}}$ ($\tilde{\mathcal{V}}$) may deviate from the protocol in computing $a, b, z, \mathbf{r}$ (respectively c). $\tilde{\mathcal{P}}$ knows neither the secret $\mathbf{s}$ nor the sum of its components. $\tilde{\mathcal{V}}$ can "learn" from $\mathcal{P}$'s proofs.

Proposition 3. *The three-step protocol is an interactive proof system.*

Proof. A three-step protocol is an interactive proof system if it is complete and sound [6]. If $\mathcal{P}$ is honest, in the sense that she knows a representation of h with respect to G, verifying also z for a given m, then $\mathcal{V}$ always accepts $\mathcal{P}$'s proof of knowledge. The following equalities hold:

$$G^{\mathbf{r}}h^c = G^{\mathbf{r}}(G^{\mathbf{s}})^c \overset{\mathbf{P2}}{=} G^{\mathbf{w}-c\mathbf{s}}G^{c\mathbf{s}} \overset{\mathbf{P1}}{=} G^{\mathbf{w}} = a,$$

$$m^{\hat{r}}z^c = m^{\hat{r}}(m^{\hat{s}})^c = m^{\hat{r}+c\hat{s}} \overset{\mathbf{P3}}{=} m^{(\mathbf{r}+c\mathbf{s})\cdot(1,1,0)}$$

$$= m^{(\mathbf{w}-c\mathbf{s}+c\mathbf{s})\cdot(1,1,0)} \overset{\mathbf{P3}}{=} m^{\hat{w}} = b.$$

Hence the protocol is complete.

The fraudulent $\tilde{\mathcal{P}}$ can cheat by guessing the correct $c \in \mathbf{Z}_q$ and by sending, with an arbitrary $\mathbf{w} \in \mathbf{Z}_q^3$, the following items to $\mathcal{V}$: $a \leftarrow G^{\mathbf{w}}h^c$, $b \leftarrow m^{\hat{w}}z^c$, and $\mathbf{r} \leftarrow \mathbf{w}$. The probability of success for this attack is $1/q$.

One still has to prove that this success probability cannot be increased, unless computing a valid representation $rep_G(h)$ is easy. Suppose $\tilde{\mathcal{P}}$ is able to convince $\mathcal{V}$ in time T with a probability at least $2^t/q$, for an integer $t \geq 1$. Then, $\tilde{\mathcal{P}}$ chooses in a suitable way a and b. With regard to these items, $\tilde{\mathcal{P}}$ must be able to correctly answer at least 2^t different challenges. Let c and c' be two such challenges, which can be found in expected time $2^{1-t}qT$. Let $\mathbf{r}$ and $\mathbf{r}'$ be the corresponding responses that verify the following relationships: $\mathbf{r} = \mathbf{w} - c\mathbf{s}$ and $\mathbf{r}' = \mathbf{w} - c'\mathbf{s}$. Because the probability that $c' - c \equiv 0 \bmod q$ is less than 2^{-t}, then with a probability at least $1 - 2^{-t}$ the fraudulent $\tilde{\mathcal{P}}$ is able to compute a valid representation $\mathbf{s} = (c' - c)^{-1}(\mathbf{r} - \mathbf{r}')$. This is in contradiction to Assumption 2.2 and, by Proposition 2, it is also in contradiction to Assumption 2.1. This proves the soundness of the three-step protocol.

The following proposition shows that the prover that uses an incorrect z in the interactive proof system is successful with negligible probability.

Proposition 4. *If $z \neq m^{\hat{s}}$ in the interactive proof system, then the verifier accepts with a probability at most $1/q$.*

Proof. Let us assume that $\tilde{\mathcal{P}}$ chooses $\boldsymbol{\xi} = (\xi_1, \xi_2, \xi_3) \in \mathbb{Z}_q^3$ and computes z as $z = m^{\hat{\xi}} \neq m^{\hat{s}}$. The proof works by contradiction. Suppose $\tilde{\mathcal{P}}$ is able to convince $\mathcal{V}$ in time T with a probability of at least $2^t/q$, for an integer $t \geq 1$. Then, $\tilde{\mathcal{P}}$ chooses in a suitable way a and b. With these items, $\tilde{\mathcal{P}}$ must be able to correctly answer at least 2^t different challenges. Let c and c' be two such challenges, which can be found in expected time $2^{1-t}qT$. Let $\mathbf{r}$ and $\mathbf{r}'$ be the corresponding responses that verify the following relationships: $b = m^{\hat{r}}z^c = m^{\hat{r}'}z^{c'}$ and $a = G^{\mathbf{r}}h^c = G^{\mathbf{r}'}h^{c'}$. From these relationships one can derive that $\mathbf{s} = (c' - c)^{-1}(\mathbf{r} - \mathbf{r}')$, and so, $\hat{s} = (c' - c)^{-1}(\hat{r} - \hat{r}')$. This implies that $z = m^{\hat{s}}$, which contradicts the assumption.

It is important to prove that the interactive proof system is witness hiding [7]. Informally, this property states that participating in the protocol does not help $\mathcal{V}$ to compute new witnesses to the input which he did not know at the beginning of the protocol. In order to prove this feature for the proposed proof system, we first analyze whether it is witness indistinguishable or not. Informally, this property is fulfilled if an infinitely powerful $\mathcal{V}$, who knows all the witnesses to the statements to be proved, cannot tell which witness the prover $\mathcal{P}$ is actually using (for a formal definition see [7]).

Proposition 5. *The interactive proof system is witness indistinguishable.*

Proof. In order to prove this fact one may assume that an infinitely powerful $\mathcal{V}$ is able to derive $\mu, \eta, \alpha_2, \alpha_3 \in \mathbb{Z}_q$ such that $h = g_1^{\mu}$, $z = m^{\eta}$, $g_2 = g_1^{\alpha_2}$, and $g_3 = g_1^{\alpha_3}$. Then, the statements being proved:

$$h = G^{\mathbf{s}} \to g_1^{\mu} = g_1^{s_1}g_2^{s_2}g_3^{s_3} = g_1^{s_1+\alpha_2 s_2 + \alpha_3 s_3},$$
$$z = m^{\hat{s}} \to m^{\eta} = m^{s_1+s_2},$$

can be written as:

$$\begin{pmatrix} \mu \\ \eta \end{pmatrix} = \begin{pmatrix} 1 & \alpha_2 & \alpha_3 \\ 1 & 1 & 0 \end{pmatrix} \cdot \begin{pmatrix} s_1 \\ s_2 \\ s_3 \end{pmatrix}.$$

Since G is a generator-tuple in G_q^3, $g_2 \neq g_1$ and $\alpha_2 \neq 1$. The rank of the matrix is 2 and there are q different witnesses $\mathbf{s} = (s_1, s_2, s_3) \in \mathbb{Z}_q^3$ that satisfy the predicate $G^{\mathbf{s}} = h, m^{\hat{s}} = z$. Consequently, even an infinitely powerful $\tilde{\mathcal{V}}$ cannot determine from a tape of conversations $tape = \{(a, b, \mathbf{r}, c)^{(1)}, (a, b, \mathbf{r}, c)^{(2)}, \ldots\}$ between $\mathcal{P}$ and $\tilde{\mathcal{V}}$ which representation $\mathbf{s}$ is actually used by $\mathcal{P}$.

To prove this fact, for two different witnesses $\mathbf{s}, \mathbf{s}^*$, satisfying $h = G^{\mathbf{s}} = G^{\mathbf{s}^*}$ and $z = m^{\hat{s}} = m^{\hat{s}^*}$, one shows that $\tilde{\mathcal{V}}$ cannot determine from $tape$ which witness was used by $\mathcal{P}$. Indeed, for any conversation $(a, b, \mathbf{r}, c) \in tape$, when $\mathbf{w}^* = \mathbf{w} + c(\mathbf{s} - \mathbf{s}^*)$, the following equalities hold: $a = G^{\mathbf{w}} = G^{\mathbf{w}^*}$, $b = m^{\hat{w}} = m^{\hat{w}^*}$, and $\mathbf{r} = \mathbf{w} - c\mathbf{s} = \mathbf{w}^* - c\mathbf{s}^*$. Since $\mathbf{w}$ and $\mathbf{w}^*$ follow the same uniform distribution, the above equalities prove that even an infinitely powerful $\mathcal{V}$ cannot distinguish between the two possible witnesses $\mathbf{s}$ and $\mathbf{s}^*$.

The following theorem links the strength of the proposed proof system to the discrete logarithm problem (see Assumption 2.1).

Proposition 6. *The interactive proof system is witness hiding under the discrete logarithm assumption.*

Proof. Let $\mathcal{A}$ be a probabilistic polynomial-time algorithm that, given $p, q, G = (g_1, g_2, g_3)$ and $h = G^{\mathbf{s}} = \prod_{i=1}^{3} g_i^{s_i}$, outputs a, b, m and z, then takes a challenge $c \in \mathbb{Z}_q$ as input and finally outputs $\mathbf{r} \in \mathbb{Z}_q^3$.

Assuming that $\mathcal{A}$ can convince $\mathcal{V}$ with non-negligible probability of success ε, then a non-trivial representation of the unit $1 \in G_q$ can be derived as follows. Choose $\mathbf{s}$ at random in $\mathbb{Z}_q^3$ and define $h = G^{\mathbf{s}}$. Run $\mathcal{A}$, with p, q, G and h as input. By the proof of Proposition 3, if $\mathcal{A}$ has non-negligible probability of success ε, then one can find in $2/\varepsilon$ (expected) tries two different challenges c, c' that can be answered correctly by $\mathcal{A}$. (Each time $\mathcal{A}$ is reset to the state immediately following the sending of a, b, m and z.) Denoting the responses by $\mathbf{r}, \mathbf{r}'$ it holds that a valid witness $\mathbf{s}^*$ can be computed as $(c' - c)^{-1}(\mathbf{r} - \mathbf{r}')$. By the proof of Proposition 5, it holds that $\mathbf{s}^* \neq \mathbf{s}$ with overwhelming probability. Then, from $h = G^{\mathbf{s}} = G^{\mathbf{s}^*}$, one can compute a non-trivial representation of the unit $1 \in G_q$ so that $1 = G^{\mathbf{s} - \mathbf{s}^*}$. According to Proposition 2, this is equivalent in difficulty to find a solution of the discrete logarithm problem.

If, in the witness hiding proof system, the challenge of the verifier is replaced by the value of a hash function applied to the message m, the item z and the information sent by the prover in the first move (a, b), $\mathcal{H}(m, z, a, b)$, then the interactive signature scheme can be regarded as a Fiat-Shamir type signature scheme [8]. It produces a signature on m of the form $sign(m) = (z, c, \mathbf{r} = (r_1, r_2, r_3))$. The signature is correct if $c = \mathcal{H}(m, z, G^{\mathbf{r}} h^c, m^{\hat{s}} z^c)$.

Assumption 3.1 *The hash function $\mathcal{H}(\cdot)$ is selected such that it has a pseudo-random behavior, in the sense that it is as difficult to convince a verifier who chooses $c = \mathcal{H}(m, z, a, b)$ as to convince a verifier who chooses c at random in the witness hiding proof system. Besides this requirement, the hash function is collision-resistant and it must prevent a forger from combining different signatures into a new signature.*

Proposition 7. *Under Assumption 2.1 (DLA) and Assumption 3.1, the interactive signature scheme is as secure as the witness hiding proof system and is protected against existential forgery under adaptively chosen message attacks.*

Proof. (Sketch) Firstly, we argue that it is not possible to forge signatures given only the public key of the signer. Indeed, if one assumes that $\mathcal{H}(\cdot)$ behaves as a random oracle, then the signer (prover) cannot have a better prediction of the recipient's challenge than only guessing the outcome of the hash function. Since the proof-system is sound (see Proposition 3), it is not possible to create signatures without knowing $\mathbf{s}$. If $\mathcal{H}(\cdot)$ is collision-resistant then it is also impossible to produce a valid signature for which $z \neq m^{\hat{s}}$ with a better probability than $1/q$

(according to Proposition 4). Furthermore, for a given pair (m, z), it is hopeless for the recipient to derive more information about the secret key of the signer from the execution of the proof system, considering its witness hiding property. Secondly, we analyze the possibility of forging a signature by combining various given signatures $(m_i, sign(m_i))$, where the attacker can choose m_i adaptively [9]. If one accepts $z_i = m_i^{\hat{s}}$, then there is a multiplicative relation $z_1 z_2 = (m_1 m_2)^{\hat{s}}$ that might be useful for an attacker. Therefore, the hash function must prevent the attacker from combining different signatures into a new signature. At the same time, the interactive proof system remains witness indistinguishable, even after polynomially many executions of the protocol with respect to different pairs (m_i, z_i), as a segment of the common input. Let $h = G^{\mathbf{s}}$ and $z_i = m_i^{\hat{s}}, i = 1, \ldots, l$ be the information available for an unlimitedly powerful attacker, who chooses m_i adaptively. Still, the uncertainty about the witness actually used by the signer is the same, since the rank of the matrix:

$$\begin{pmatrix} 1 & \alpha_2 & \alpha_3 \\ 1^{(1)} & 1^{(1)} & 0^{(1)} \\ \vdots & \vdots & \vdots \\ 1^{(l)} & 1^{(l)} & 0^{(l)} \end{pmatrix}$$

remains 2. Therefore, the repeated use of the interactive proof system with respect to different segments $(m_i, z_i), i = 1, \ldots, l$, of the common input $(p, q, G, h, (m, z))$ does not help the attacker to learn more information about the secret key of the signer. The above considerations prove the proposition.

4   Restrictive Blind Signature Scheme

In this section, the interactive signature scheme is transformed into a blind signature scheme by applying the technique described by Okamoto and Ohta [11]. If the recipient is allowed to determine the challenge, then the message-signature pair can be issued in a blind way.

4.1   Restrictive Blind Signing Protocol

The recipient can get a blind signature only for a restricted message set, where m has the form $m = m_0^t, m_0 \in G_q \setminus \{0, 1\}$. The blinding parameter t is a random choice of the recipient in $\mathbb{Z}_q^*$. Both the signer and the recipient know m_0 and $z_0 = m_0^{\hat{s}}$. In order to get a signature on the blind version $m = m_0^t$ of the "shape" message m_0, the recipient computes $z = m^{\hat{s}}$ from z_0 as $z = z_0^t (= (m_0^{\hat{s}})^t = (m_0^t)^{\hat{s}} = m^{\hat{s}})$. Given $(m_0, z_0 = m_0^{\hat{s}})$ the signer proves that she knows a witness of h with respect to G, the components of which also verify z_0, in such a way that the messages exchanged between the signer and the recipient are blinded. While the recipient carries out the proof system with input (p, q, G, h, m_0, z_0), he diverts the protocol with input (p, q, G, h, m, z). Instead of generating a random challenge for the signer, the recipient computes the challenge as the output of the hash function $c \leftarrow \mathcal{H}(m, z, a, b)$. Before sending it to the signer, the challenge c is additively blinded to the value c_0, through the blinding parameter $u \in \mathbf{Z}_q^*$. Finally, the recipient obtains the signature $sign(m) = (z, c, \mathbf{r})$ on m, where $\mathbf{r}$ is the response $\mathbf{r_0}$ of the signer additively blinded through the parameter $\mathbf{v}$ (which is a random choice of the recipient). The restrictive blind signing protocol is described by the following steps:

1. The signer randomly generates $\mathbf{w_0} \leftarrow (w_{01}, w_{02}, w_{03}) \in_\mathcal{R} \mathbf{Z}_q^3$ and sends to the recipient the commitment $a_0 \leftarrow G^{\mathbf{w_0}}$ and $b_0 \leftarrow m_0^{\hat{w}_0}$.
2. The recipient randomly generates $t, u \in \mathbf{Z}_q^*$, $\mathbf{v} = (v_1, v_2, v_3) \in \mathbf{Z}_q^3$, computes a new message $m = m_0^t$ and the corresponding value of a new $z = z_0^t$, both values being unknown to the signer. The recipient also computes $a \leftarrow a_0 G^{\mathbf{v}} h^u$ and $b \leftarrow (b_0 m_0^{\hat{v}} z_0^u)^t$, which are the blind versions of a_0 and b_0 respectively. He also computes $c = \mathcal{H}(m, z, a, b)$ and sends $c_0 = (c - u) \bmod q$ to the signer.
3. The signer responds to this challenge with $\mathbf{r_0} = (r_{01}, r_{02}, r_{03}) \leftarrow \mathbf{w_0} - c_0 \mathbf{s}$.
4. The recipient accepts, if and only if $a_0 = G^{\mathbf{r_0}} h^{c_0}$ and $b_0 = m_0^{\hat{r}_0} z_0^{c_0}$. The recipient also corrects the response $\mathbf{r_0}$ as $\mathbf{r} \leftarrow \mathbf{r_0} + \mathbf{v}$. Finally, the restrictive blind signature on the message $m = m_0^t$ consists of $sign(m) = (z, c, \mathbf{r})$.

4.2 Security of the Recipient

Theorem 8. *(Correctness) If the recipient follows the restrictive blind signing protocol and accepts, then $sign(m) = (z, c, \mathbf{r} = (r_1, r_2, r_3))$ is a correct signature on m.*

Proof. The signature $sign(m) = (z, c, \mathbf{r})$ is a correct signature on m if the equality $c = \mathcal{H}(m, z, G^{\mathbf{r}} h^c, m^{\hat{r}} z^c)$ is verified. If we assume that $\mathcal{H}(\cdot)$ is collision-resistant, then this is equivalent to proving that $a = G^{\mathbf{r}} h^c$ and $b = m^{\hat{r}} z^c$. The first relationship follows from:

$$G^{\mathbf{r}} h^c = G^{\mathbf{r_0} + \mathbf{v}} h^{c_0 + u} \overset{\mathbf{P_1}}{=} (G^{\mathbf{r_0}} h^{c_0}) G^{\mathbf{v}} h^u = a_0 G^{\mathbf{v}} h^u = a,$$

because $a_0 = G^{\mathbf{r_0}} h^{c_0}$ if the recipient accepts. Similarly:

$$m^{\hat{r}} z^c = m^{\hat{r}_0 + \hat{v}} z^{c_0 + u} = \left(m_0^{\hat{r}_0} z_0^{c_0} m_0^{\hat{v}} z_0^u\right)^t = \left(b_0 m_0^{\hat{v}} z_0^u\right)^t = b,$$

because we accept that $m = m_0^t$, $z = z_0^t$, and $b_0 = m_0^{\hat{r}_0} z_0^{c_0}$ if the recipient follows the protocol and accepts.

Theorem 9. *(Unconditional Unlinkability) If the recipient follows the protocol, then even a signer with unlimited computing power cannot link a message-signature pair $(m, sign(m) = (z, c, \mathbf{r}))$ to a specific execution of the restrictive blind signing protocol.*

Proof. Let m_0, z_0, a_0, b_0, c_0, and $\mathbf{r}_0$ be the view of the signer in a successful execution of the blind signing protocol, such that $a_0 = G^{\mathbf{r}_0} h^{c_0}$ and $b_0 = m_0^{\hat{r}_0} z_0^{c_0}$. It is sufficient to prove that for any valid pair $(m, sign(m) = (z, c, \mathbf{r}))$, verifying $c = \mathcal{H}(m, z, a, b)$, where $a = G^{\mathbf{r}} h^c$ and $b = m^{\hat{r}} z^c$, there exists exactly one set of values of the random variables $t, u, \mathbf{v}$ such that: $m = m_0^t$, $z = z_0^t$, $a = a_0 G^{\mathbf{v}} h^u$, $b = (b_0 m_0^{\hat{v}} z_0^u)^t$, $c = (c_0 + u) \bmod q$, and $\mathbf{r} = \mathbf{r}_0 + \mathbf{v}$.

First, given m_0 and m, the value of t can be computed as $t = \log_{m_0} m$. The values of u and $\mathbf{v}$ can be computed from c, c_0 and $\mathbf{r}, \mathbf{r}_0$ as $u = (c - c_0) \bmod q$ and $\mathbf{v} = \mathbf{r} - \mathbf{r}_0$. One still has to prove that these values of t, u and $\mathbf{v}$ verify the equalities $z = z_0^t$, $a = a_0 G^{\mathbf{v}} h^u$, and $b = (b_0 m_0^{\hat{v}} z_0^u)^t$.

According to Proposition 4, it can be assumed that $z_0 = m_0^{\hat{s}}$ and $z = m^{\hat{s}}$ (because the signer actually proves that z_0 equals $m_0^{\hat{s}}$ when making a blind signature and the recipient follows the protocol). Hence, $m = m_0^t$ implies that:

$$z = m^{\hat{s}} = \left(m_0^t\right)^{\hat{s}} = m_0^{t\hat{s}} = (m_0^{\hat{s}})^t = z_0^t.$$

Thus, the first relationship is verified. The other two relationships can be proved as follows:

$$a = G^{\mathbf{r}} h^c = G^{\mathbf{r}_0 + \mathbf{v}} h^{c_0 + u} = (G^{\mathbf{r}_0} h^{c_0}) G^{\mathbf{v}} h^u = a_0 G^{\mathbf{v}} h^u,$$
$$b = m^{\hat{r}} z^c = m^{\hat{r}_0 + \hat{v}} z^{c_0 + u} = m^{\hat{r}_0} z^{c_0} m^{\hat{v}} z^u$$
$$= m_0^{t\hat{r}_0} z_0^{tc_0} m_0^{t\hat{v}} z_0^{tu} = (m_0^{\hat{r}_0} z_0^{c_0} m_0^{\hat{v}} z_0^u)^t = (b_0 m_0^{\hat{v}} z_0^u)^t.$$

4.3 Security of the Signer

Given a signature $(z, c, \mathbf{r})$ on a message m and the view of the recipient in the protocol, $(m_0, z_0, a_0, b_0, c_0, \mathbf{r}_0)$, such that $a_0 = G^{\mathbf{r}_0} h^{c_0}$ and $b_0 = m_0^{\hat{r}_0} z_0^{c_0}$, then the pair $(u, \mathbf{v})$ can be defined by $u = (c - c_0) \bmod q$ and $\mathbf{v} = \mathbf{r} - \mathbf{r}_0$. This pair satisfies

$$a = a_0 G^{\mathbf{v}} h^u = G^{\mathbf{w}_0 + \mathbf{v} + us},$$
$$b = (b_0 m_0^{\hat{v}} z_0^u)^t = m^{(\mathbf{w}_0 + \mathbf{v} + us) \cdot (1,1,0)}.$$

The only information the recipient has about $\mathbf{w}_0$ when constructing (a, b) is that $a_0 = G^{\mathbf{w}_0}$ and $b_0 = m_0^{\hat{w}_0}$. It is conjectured that in order to construct (a, b) of this form and to derive $z = m^{\hat{s}}$, the recipient chooses $t \in \mathbb{Z}_q^*$, computes m using the formula $m = m_0^t$ and then in the blind signing protocol computes $z = z_0^t$, $a = a_0 G^{\mathbf{v}} h^u$ and $b = (b_0 m_0^{\hat{v}} z_0^u)^t$.

Furthermore, if there exists a $\mu_0 = (\mu_{01}, \mu_{02}) \in \mathbb{Z}_q^2$ that verifies $\mathcal{R}(m_0, \mu_0)$: $m_0 = g_1^{\mu_{01}} g_2^{\mu_{02}}$ and $I(\mu_0) : (\mu_{01}/\mu_{02}) \bmod q = i$, and if we accept that $m = m_0^t$, then $m = (g_1^{\mu_{01}} g_2^{\mu_{02}})^t = g_1^{t\mu_{01}} g_2^{t\mu_{02}}$. Thus, there exists a $\mu = (t\mu_{01}, t\mu_{02})$ that verifies:

$$\mathcal{R}(m, \mu) : m = g_1^{\mu_1} g_2^{\mu_2}, \text{ and}$$
$$I(\mu) : (\mu_1/\mu_2) \bmod q = (t\mu_{01}/t\mu_{02}) \bmod q = (\mu_{01}/\mu_{02}) \bmod q = i.$$

The following assumption states the restrictiveness of the blind signature scheme.

Assumption 4.1 *(Restrictiveness) The recipient obtains a signature on a message that can only be of the form* $m = m_0{}^t$, $m_0 \in G_q \setminus \{0, 1\}$, *with* $t \in \mathbb{Z}_q^*$ *randomly chosen by the recipient. In addition, if there exists a* $\mu_0 = (\mu_{01}, \mu_{02}) \in \mathbb{Z}_q^2$ *that verifies* $\mathcal{R}(m_0, \mu_0)$: $m_0 = g_1^{\mu_{01}} g_2^{\mu_{02}}$, *and the blinding-invariant function* $I(\mu_0)$: $(\mu_{01}/\mu_{02}) \bmod q = i$, *then there exists a* $\mu = (\mu_1, \mu_2) \in \mathbb{Z}_q^2$ *that verifies* $\mathcal{R}(m, \mu)$: $m = g_1^{\mu_1} g_2^{\mu_2}$ *and* $I(\mu)$: $(\mu_1/\mu_2) \bmod q = i$.

The following theorem states the resistance of the restrictive blind signature scheme against existential forgeries under adaptively chosen-message attacks. The proof of this theorem is carried out in the framework of the random oracle model, introduced by Bellare and Rogaway in [1]. In this model, both the signer and the recipient have access to a public random oracle f, which provides really random answers for each new query. The security of the restrictive blind signature scheme is analyzed by first assuming the existence of the random oracle and then replacing the oracle accesses by the computation of a collision resistant hash function $\mathcal{H}(\cdot)$. Thus, the proof of security in the random oracle model validates the design of the scheme.

Pointcheval and Stern have pointed that when this model is used to evaluate the security of Fiat-Shamir type signature schemes, a collusion between the signer, the attacker, and the random oracle allows the construction of a probabilistic polynomial-time algorithm that solves the underlying hard problem of the signature scheme [12]. Moreover, when the signature scheme is derived from an indistinguishable proof system, then the security of the scheme can be proved even against parallel attacks, when the attacker initiates new interactions with the signer before previous ones have been completed [13].

Theorem 10. *Consider the witness hiding restrictive blind signature scheme in the random oracle model. For a given pair* (m_0, z_0), *if there exists a probabilistic polynomial-time algorithm* $\mathcal{A}$ *that can get* $l + 1$ *blind signatures from* l *executions of the restrictive blind signing protocol, with non-negligible probability, even under the parallel attack, then a non-trivial representation of the unit in* G_q *can be found in polynomial time.*

Proof. (Sketch): Let $\mathcal{A}$ be a probabilistic polynomial-time algorithm with a random tape ω, which can access both the random oracle f and the signing oracle $\mathcal{S}$. The signing oracle has a random tape Ω and a knowledge tape containing the secret key $\mathbf{s} = (s_1, s_2, s_3) \in \mathbb{Z}_q^3$ corresponding to the public key h, which also verifies z_0 for a given m_0. The proof works by contradiction. Assume that after a number of l interactions $(a_{0i}, b_{0i}), c_{0i}, \mathbf{r}_{0i}$, $i = 1, \ldots, l$, based on the common input (p, q, G, h, m_0, z_0), between $\mathcal{A}$ and the signing oracle, and after a polynomial number of Q queries asked by $\mathcal{A}$ to the random oracle f, $\mathcal{Q}_1, \ldots, \mathcal{Q}_Q$, $\mathcal{A}$ returns, with non-negligible probability of success ε, $l + 1$ signatures $(m_i, z_i, (a_i, b_i), c_i, \mathbf{r}_i)$. These signatures verify that $m_i = m_0^{t_i}$, $z_i = z_0^{t_i}$ with $t_i \in_R \mathbb{Z}_q^*$, and $c_i = \mathcal{H}(m_i, z_i, a_i, b_i)$ with $a_i = G^{r_i} h^{c_i}$, $b_i = m_i^{r_i} z_i^{c_i}$, for any $i = 1, \ldots, l + 1$. Considering the randomness of the outputs produced by the random oracle, the probability that a certain "successful" query (m_i, z_i, a_i, b_i) is not asked is negligible. Hence, it can be assumed that all the tuples $(m_i, z_i, a_i, b_i), i = 1, \ldots, l + 1$, are among the queries $\mathcal{Q}_1, \ldots, \mathcal{Q}_Q$ which have been asked during the attack. Furthermore, at the cost of decreasing the success probability of $\mathcal{A}$ to the value $\rho \approx \varepsilon / Q^{l+1}$, the permutation ξ on $\{1, 2, \ldots, l + 1\}$ can be defined such that the indexes $\xi(1), \ldots, \xi(l + 1)$, corresponding to $(m_1, z_1, a_1, b_1), \ldots, (m_{l+1}, z_{l+1}, a_{l+1}, b_{l+1})$, have fixed positions in the list of queries $\mathcal{Q}_1, \ldots, \mathcal{Q}_Q$. During the attack, the signing oracle accesses the knowledge tape containing the secret key $\mathbf{s} = (s_1, s_2, s_3)$ and the random tape Ω, which determine $\mathbf{w}_{0i} = (w_{01i}, w_{02i}, w_{03i})$, such that $a_{0i} = G^{\mathbf{w}_{0i}}, b_{0i} = m_0^{w_{0i}}$ for $i = 1, \ldots, l + 1$. For any given ("shape") message m_0, the distribution of $(\mathbf{s} = (s_1, s_2, s_3), h, z_0)$, where $\mathbf{s}$ is random and $h = G^{\mathbf{s}}, z_0 = m_0^{\hat{s}}$, is the same as the distribution of $((s_1, s_2), s_3, h, z_0)$, where s_3 is random and (s_1, s_2) is the unique pair in $\mathbb{Z}_q^2$ such that $g^{-s_3} h = g_1^{s_1} g_2^{s_2}$ and $z_0 = (m_0^{\hat{s}}) = m_0^{s_1 + s_2}$. Accordingly, (s_1, s_2, s_3) can be replaced by (s_3, h, z_0), and similarly, each $\mathbf{w}_{0i} = (w_{01i}, w_{02i}, w_{03i})$ can be replaced by $(w_{03i}, a_{0i}, b_{0i})$. The variables $(\omega, (h, z_0), (a_{01}, b_{01}), \ldots, (a_{0l}, b_{0l}))$ are grouped under the variable ν and the l-tuple $(w_{031}, \ldots, w_{03l})$ is denoted by τ. The success domain of $\mathcal{A}$, denoted by $\mathcal{SD}$, is the set of quadruples (ν, s_3, τ, f) that leads $\mathcal{A}$ to success with non-negligible probability ρ. Then, $\mathrm{Pr}_{\nu, s_3, \tau, f}[(\nu, s_3, \tau, f) \in \mathcal{SD}] \geq \rho$.

The following polynomial-time reduction can be derived in order to obtain a non-trivial representation of the unit in G_q with respect to the generator-tuple G:

1. randomly choose i and compute the index $\xi(i)$;
2. randomly choose the keys and the random tapes (for the signing oracle and $\mathcal{A}$) and run the attack $\mathcal{A}$ with respect to the random oracle f, $\mathcal{A}(\nu, s_3, \tau, f)$;
3. replay the attack $\mathcal{A}$ with the same keys and random tapes, but ask queries to a different random oracle f', $\mathcal{A}(\nu, s_3, \tau, f')$, providing identical answers to the first $\xi(i) - 1$ queries;
4. with non-negligible probability, from two different outputs:

$$(m_i, z_i, (a_i, b_i), c_i, \mathbf{r}_i) \text{ in } \mathcal{A}(\nu, s_3, \tau, f), \text{ and}$$
$$(m_i, z_i, (a_i, b_i), c_i^*, \mathbf{r}_i^*) \text{ in } \mathcal{A}(\nu, s_3, \tau, f'),$$

obtained after the replay, a non-trivial representation of the unit can be derived from:

$$a_i = G^{\mathbf{r}_i} h^{c_i} = G^{\mathbf{r}_i + c_i \mathbf{s}}$$
$$= G^{\mathbf{r}_i^*} h^{c_i^*} = G^{\mathbf{r}_i^* + c_i^* \mathbf{s}},$$

with $\mathbf{r}_i + c_i \mathbf{s} \neq \mathbf{r}_i^* + c_i^* \mathbf{s}$. This inequality holds, since at least the inequality $r_{3i} + c_i s_3 \neq r_{3i}^* + c_i^* s_3$ holds with non-negligible probability. We refer to the "forking lemma" in [13], for a detailed proof of this fact.

It would be easy to get two blind signatures from one execution of the blind signing protocol, if the witness hiding proof system could be diverted to two recipients in parallel, who refer to two different pairs $(m_0^{(1)}, z_0^{(1)})$ and $(m_0^{(2)}, z_0^{(2)})$ when interacting with the signer. This kind of parallel attack should not succeed, considering the fact that a witness hiding proof of knowledge (from which the restrictive blind signature scheme is obtained) is not parallel divertible [5].

5 Concluding Remarks

We have described a restrictive blind signing protocol that is derived from a witness hiding proof system. The scheme provides enhanced security for the signer, fact which is provable to a large extent. The decrease of efficiency – the cost paid for enhanced security– is kept to a reasonable level.

References

1. M. Bellare and P. Rogaway, "Random oracles are practical: a paradigm for designing efficient protocols," *Proc. of the 1st ACM Conference on Computer and Communications Security*, 1993, pp. 62–73.
2. S. Brands, *An efficient off-line electronic cash system based on the representation problem*, Report CS-R9323, Centrum voor Wiskunde en Informatica, March 1993.
3. D. Chaum, "Blind signatures for untraceable payments," *Advances in Cryptology, Proc. Crypto'82*, D. Chaum, R.L. Rivest and A.T. Sherman, Eds., Plenum Press, New York, 1983, pp. 199–203.
4. D. Chaum and T.P. Pedersen, "Wallet databases with observers," *Advances in Cryptology, Proc. Crypto'92, LNCS 740*, E.F. Brickell, Ed., Springer-Verlag, 1993, pp. 89–105.
5. L. Chen, *Witness Hiding Proofs and Applications*, PhD thesis, Aarhus University, Computer Science Department, Aarhus (Denmark), August 1994.
6. U. Feige, A. Fiat and A. Shamir, "Zero-knowledge proofs of identity," *Journal of Cryptology*, Vol. 1, No. 2, 1988, pp. 77–94.
7. U. Feige and A. Shamir, "Witness indistinguishable and witness hiding protocols," *Proc. of the 22nd Annual ACM Symposium on Theory of Computing*, 1990, pp. 416–426.
8. A. Fiat and A. Shamir, "How to prove yourself: Practical solutions to identification and signature problems," *Advances in Cryptology, Proc. Crypto'86, LNCS 263*, A.M. Odlyzko, Ed., Springer-Verlag, 1987, pp. 186–194.
9. S. Goldwasser, S. Micali and R.L. Rivest, "A digital signature scheme secure against adaptive chosen-message attacks," *SIAM J. on Comput.*, No. 17, 1988, pp. 281–308.
10. T. Okamoto, "Provably secure and practical identification schemes and corresponding signature schemes," *Advances in Cryptology, Proc. Crypto'92, LNCS 740*, E.F. Brickell, Ed., Springer-Verlag, 1993, pp. 31–53.
11. T. Okamoto and K. Ohta, "Divertible zero knowledge interactive proofs and commutative random self-reducibility," *Advances in Cryptology, Proc. Eurocrypt'89, LNCS 434*, J.-J. Quisquater and J. Vandewalle, Eds., Springer-Verlag, 1990, pp. 134–149.
12. D. Pointcheval and J. Stern, "Security proofs for signature schemes," *Advances in Cryptology, Proc. Eurocrypt'96, LNCS 1070*, U. Maurer, Ed., Springer-Verlag, 1996, pp. 387–398.
13. D.Pointcheval and J. Stern, "Provably secure blind signature schemes," *Advances in Cryptology, Proc. Asiacrypt'96, LNCS 1163*, K. Kim and T. Matsumoto, Eds., Springer-Verlag, 1996, pp. 252–265.
14. C.P. Schnorr, "Efficient signature generation by smart cards," *Journal of Cryptology*, Vol. 4, No. 3, 1991, pp. 161–174.

A Note on the Construction and Upper Bounds of Correlation-Immune Functions

Markus Schneider

University of Hagen, Lehrgebiet Kommunikationssysteme, 58084 Hagen, Germany,
email: mark.schneider@fernuni-hagen.de

Abstract. In this paper, an algorithm for the construction of correlation-immune functions is given. It will be shown that the proposed algorithm provides a method to construct every mth order correlation-immune function. Besides correlation-immunity, also other properties of Boolean functions, like Hamming weight, can be taken into account. The complexity analysis of the proposed algorithm leads to a new upper bound for the number of specified correlation-immune functions and correlation-immune functions in general, depending on the number of input variables n and the order of correlation-immunity.

1 Introduction

Siegenthaler introduced the property of correlation-immunity for functions whose output does not give any information about their input to a certain degree (see [Sieg 84], [Sieg 85]). In this paper, we consider memoryless Boolean functions $f : GF(2)^n \rightarrow GF(2)$, $\underline{x} \rightarrow f(\underline{x})$. A memoryless function f is said to be correlation-immune of order m, with $1 \leq m \leq n$, if the output of f and any m input variables are statistically independent. To put it more formally, if the output of the binary source i, $1 \leq i \leq n$, is modelled as a random variable X_i, a function f is mth order correlation-immune if the mutual information between the function output and any subset of m elements $X_{i_1}, \ldots, X_{i_m}$ of the n input variables is zero, with $1 \leq i_1 < \ldots < i_m$ [Sieg 84]: $I\left(f(X_1, \ldots, X_n); X_{i_1}, \ldots, X_{i_m}\right) = 0$. Ghuo-Zhen and Massey pointed out that the Walsh transform of an mth order correlation-immune function has some specific properties [GuMa 88]. Proposals for the construction of correlation-immune functions were given in [Sieg 84], [CCCS 91] and [SeZh 93]. Another relevant property of Boolean functions is balance. A function f is called balanced, if its output probabilities are equal, i.e. $P\left(f(\underline{x}) = 0\right) = P\left(f(\underline{x}) = 1\right) = 0.5$.

2 Relevant properties of correlation-immune functions

The following definition and theorem are of great importance in the context of the construction method of mth order correlation-immune functions presented in the next section. Only those properties of correlation-immune functions with paper related relevance are introduced.

Definition 1. Let $n \geq 1$ be an integer. The set $\{(x_1, \ldots, x_n)\}$ with $x_i \in \{0, 1\}$ is called an $n-$cube.

Let $I = \{i_1, \ldots, i_k\}$ be a subset of the index set $\{1, \ldots, n\}$ of the $n-$cube with k elements, $1 \leq k < n$. For every $j \notin I$ let $a_j \in \{0, 1\}$ be constant. The set $\{(x_1, \ldots, x_n)\}$ with $x_i \in \{0, 1\}$ for $i \in I$ and $x_i = a_i$ for $i \notin I$ will also be called a k-cube.

Theorem 2. *The function* $f : GF(2)^n \rightarrow GF(2), \underline{x} \rightarrow f(\underline{x})$ *is mth order correlation-immune with* $1 \leq m < n$, *if and only if the binary output* a, $a \in GF(2)$, *appears in all (n-m)-dimensional subcubes of the* $n-$cube *with the same frequency* K, $0 \leq K \leq 2^{n-m}$.

Proof: $\Rightarrow$ Suppose a appears in all $(n - m)-$dimensional cubes with frequency K. Then, there is $P(f(\underline{x}) = a \mid x_{i_1} = a_1, \ldots, x_{i_m} = a_m) = K \cdot 2^{m-n}$, with $a_i \in GF(2)$. The frequency of a in the $n-$cube is obtained by summing up over all 2^m disjoint $(n - m)-$cubes. This yields $P(f(\underline{x}) = a) = 2^m \cdot K \cdot 2^{-n}$. Obviously, $P(f(\underline{x}) = a) = K \cdot 2^{m-n} = P(f(\underline{x}) = a \mid x_{i_1} = a_1, \ldots, x_{i_m} = a_m)$, which means that the value of $f(\underline{x})$ and every choice of m elements out of $\{x_1, \ldots, x_n\}$ are statistically independent. Then, f is correlation-immune of order m.
$\Leftarrow$ Let f be correlation-immune of order m. Then, one has $P(f(\underline{x}) = a \mid x_{i_1} = 0, \ldots, x_{i_m} = 0) = P(f(\underline{x}) = a \mid x_{i_1} = 0, \ldots, x_{i_m} = 1) = \cdots = P(f(\underline{x}) = a \mid x_{i_1} = 1, \ldots, x_{i_m} = 1) = P(f(\underline{x}) = a)$. By this result, we can conclude that the value a appears in all 2^m $(n - m)-$dimensional cubes with the same frequency. $\square$

In cryptographic applications, the correlation-immune functions which are also balanced are of particular interest. For those, $K = 2^{n-m-1}$. Now, we think of K to be the number of $'1's$ in an $(n - m)-$cube. The number of $'1's$ in the table of the function f is given by its Hamming weight $W(f)$. By theorem 2, it follows that a mth order correlation-immune function f with n input variables can only have certain Hamming weights $W(f)$ which will be stated in the next theorem.

Theorem 3. *Let* $0 < m < n$. *The Hamming weight* $W(f)$ *of an mth order correlation-immune function with* n *inputs and* K, *with* $K = 0, \ldots, 2^{n-m}$, *can only have the following values:*

$$W(f) = 2^m \cdot K, \text{ for } K = 0, \ldots, 2^{n-m}. \tag{1}$$

Proof: By the result of theorem 2, it is known that f is mth order correlation-immune if and only if value 1 appears exactly K times in all $(n - m)$-cubes. The union of disjoint $(n - m)$-cubes inducing the n-cube consists of 2^m such $(n - m)$-cubes. The desired result follows. $\square$

3 Construction of correlation-immune functions

In the following, an algorithm for the construction of mth order correlation-immune functions with n binary input variables is presented. The algorithm is mainly based on theorem 2. If the binary value '1' assigned to the vertices of an $(n-m)$-cube, lying in an n-cube, appears with frequency K within this $(n-m)$-cube, we will say that the K-condition is fulfilled. If K is chosen to be $K = 2^{n-m-1}$, then the constructed function is also balanced. The application of algorithm 3.1 yields always one function with the desired properties. The assignments and the selections of $(n-m)$-cubes with minimum number of free vertices are controlled by the output of a random generator. This ensures that the output is able to construct distinct functions. Within the algorithm, parameter j describes the number of taken decisions if there are different possibilities of doing the assignments, all of them fulfilling the K-condition. To apply the algorithm, it is necessary to choose appropriate n, m and K.

Algorithm 3.1

1. $j := 1$
2. *select an $(n-m)$-cube, assign binary values to its vertices in a way fulfilling the K-condition, and note the assignment in a listing*
3. *if there exists an $(n-m)$-cube not considered before with minimum number of free vertices and with no degrees of freedom to fulfill the K-condition for the assignment of the values to the vertices, then*
    (a) *assignment of the binary values to the vertices fulfilling the K-condition in the considered $(n-m)$-cube*
    (b) *if the K-condition is not fulfilled in all $(n-m)$-cubes, then*
        i. *note assignment no. j to be bad, cancel the assignment no. j, and $j := j - 1$*
    else
        i. *note assignment no. j in the listing, and go to step 3*
4. *if there still exist vertices in the n-cube without assignment, then*
    (a) *select an $(n-m)$-cube with minimum number of free vertices and $j:=j+1$*
    (b) *if there exists at least one possibility for the assignment to the vertices of the $(n-m)$-cube which is not noted in the listing to be bad, then*
        i. *assignment of the binary values to the vertices fulfilling the K-condition in the $(n-m)$-cube in this way as not noted in the listing to be bad, and note assignment no. j in the listing*
        ii. *if the K-condition is not fulfilled in the other $(n-m)$-cubes, then*
            A. *note assignment no. j to be bad, and cancel the assignment no. j, and go to step 4b*
        else
            A. *go to step 3*
    else
        i. *$j:=j-1$, note assignment no. j to be bad, and cancel the assignment no. j, delete the notes with attribute 'bad' and no. k for all $k > j$, and go to step 4b*
else end.

The result of the algorithm is an n-cube with all its vertices assigned a binary value. This can be easily translated into Boolean algebra or in the algebraic normal form defined over $GF(2)$.

Theorem 4. *Algorithm 3.1 is a method to construct all functions $f : GF(2)^n \rightarrow GF(2), \underline{x} \rightarrow f(\underline{x})$, having the property of correlation-immunity of order m with $0 < m \leq n$.*

Proof: By theorem 2, all mth order correlation-immune functions with n input variables satisfy the K-condition in all $(n - m)$−dimensional cubes with $K = 2^{n-m-1}$. Depending on the assignment in the first $(n - m)$−dimensional cube at the start of algorithm 3.1, all functions having the same vertex assignments in the first $(n - m)$−cube can be constructed obeying the K-condition in all further steps. This is valid for all combinatorial possibilities of the assignments in the first $(n - m)$−cube. Applying the method of algorithm 3.1, all these functions can be found fulfilling the K-condition in all $(n - m)$−dimensional cubes. By theorem 2, this yields all mth order correlation-immune functions. □

The condition for selecting an $(n - m)$-cube with minimum number of free vertices in step 3 of algorithm 3.1 is not necessary to guarantee the ability to construct all m-correlation-immune functions. Without this additional condition, the algorithm would work more efficiently. But this additional condition is useful for the further complexity considerations. The construction method can be looked at in a graph theoretic way. Decisions with degrees of freedom are represented as branches; states within the construction having alternative possibilities for the assignment of the binary values to the vertices of the $(n-m)$-cube are the nodes. Only those phases in the n-cube during the construction are to be understood as states (nodes), which have degrees of freedom for further assignment. Nodes with no possibility of choosing an assignment to fulfill the K-condition have no branches. The graph induced by the construction method is directed, free of cycles, and nodes can be reached by more than one path.

4 Complexity considerations

In this section, we will focus on the complexity aspects of the construction in algorithm 3.1. The parameter we are interested in is an upper bound of the frequencies to find $(n - m)$−cubes with certain numbers of free vertices in algorithm 3.1. With this parameter, the maximum number of branches at a node in the graph representation can be calculated. An exact mathematical description of algorithm 3.1 seems to be complicated. This is the reason for analyzing algorithm 3.1 in a modified manner. In the modification, the K-condition is dropped and the vertices of the n-cube are only marked instead of assigning binary values to them. The set of marked vertices will be called A.

Algorithm 4.1

1. $A = \{\}$

2. *select an $(n-m)$-cube, mark its vertices, marked vertices become elements of A*
3. *if not all vertices of the n-cube are elements of A, then*
   (a) select an $(n-m)$-cube $\not\subseteq A$ with minimum number of vertices not in A, vertices of the $(n-m)$-cube become elements of A, and go to 3
4. *end*

The gradual marking of vertices of the n-cube done in algorithm 4.1 shows all the possible paths through the n-cube obeying only the minimality condition of algorithm 3.1. Applying algorithm 3.1, there are no paths through the n-cube, which are not included in the set of all paths, applying algorithm 4.1. As already mentioned, the parameter j gives the number of decisions for the assignments applying algorithm 3.1. In algorithm 3.1, decisions are only possible if there are degrees of freedom, and these can only exist if the $(n-m)$-cube obeying the minimality condition has at least 2 vertices for assigning binary values. Applying algorithm 4.1 and counting the events when the set A is growing by at least 2 elements, one obtains an upper bound for the parameter j in algorithm 3.1.

For the sake of simplicity, we introduce an expression for the set of vertices of an n–cube, which are inducing a k–cube, with $0 < k < n$. Consider a k–cube, given by the index set $I = \{i_1 = 1, \dots, i_k = k\}$. This k–cube is identified by the set $\{(x_1, \dots, x_n) | (x_j \in GF(2)$, if $j \in I)$ and $(x_j = a_j$, with $a_j \in GF(2)$ fixed, if $j \notin I)\}$. In the further considerations, this k–cube will be described briefly by $(x_1, \dots, x_k, a_{k+1}, \dots, a_n)$.

Theorem 5. *Let $1 \le j < n$, and let the index set $I = \{i_1, \dots, i_j\} \subset \{1, \dots, n\}$ induce a j–cube $A = \{(x_1, \dots, x_n) \mid (x_i \in GF(2)$, if $i \in I)$ and $(x_i = a_i$, with $a_i \in GF(2)$ fixed, if $i \notin I)\}$. Furthermore, let $1 \le k \le j$. Given the index sets I_1, I_2 with k elements, with $I_1 \not\subseteq I$, $I_2 \not\subseteq I$, $I_1 \cap I \ne \{\}$, $I_2 \cap I \ne \{\}$ and $|I_1 \cap I| = |I_2 \cap I| + 1$, which are inducing the k–cubes A_1 and A_2. Then $|A \cup A_1| < |A \cup A_2|$.*

Proof: Let $k = k' + k''$ with $k' > 0$, $k'' \ge 0$. Consider the following cubes, without loss of generality: j-cube $A = (x_1, \dots, x_j, a_{j+1}, \dots, a_n)$, k-cube $A_1 = (a_1, \dots, a_{j-k''-1}, x_{j-k''}, \dots, x_{j+k'-1}, a_{j+k'}, \dots, a_n)$, k-cube $A_2 = (a_1, \dots, a_{j-k''}, x_{j-k''+1}, \dots, x_{j+k'}, a_{j+k'+1}, \dots, a_n)$. We have $|A| = 2^j$ and $|A_1| = |A_2| = 2^k$. A_1 and A_2 can be obtained by the union of $(k''+1)$-cubes, respectively k''-cubes:

$$A_1 = (a_1, \dots, a_{j-k''-1}, x_{j-k''}, \dots, x_{j+k'-1}, a_{j+k'}, \dots, a_n)$$
$$= (a_1, \dots, a_{j-k''-1}, x_{j-k''}, \dots, x_j, a_{j+1}, \dots, a_{j+k'-1}, a_{j+k'}, \dots, a_n)$$
$$\cup$$
$$(a_1, \dots, a_{j-k''-1}, x_{j-k''}, \dots, x_j, a_{j+1}, \dots, a_{j+k'-1}+1, a_{j+k'}, \dots, a_n)$$
$$\cup \dots \cup$$
$$(a_1, \dots, a_{j-k''-1}, x_{j-k''}, \dots, x_j, a_{j+1}+1, \dots, a_{j+k'-1}+1, a_{j+k'}, \dots, a_n)$$

The first $(k''+1)$-cube on the right side in the preceding equation is completely contained in A. All others $(k''+1)$-cubes have no elements in common with A. A $(k''+1)$-cube has $2^{k''+1}$ elements. Hence, there is $|A \cup A_1| = 2^j + 2^k - 2^{k''+1}$.

$$A_2 = (a_1, \ldots, a_{j-k''}, x_{j-k''+1}, \ldots, x_{j+k'}, a_{j+k'+1}, \ldots, a_n)$$
$$= (a_1, \ldots, a_{j-k''}, x_{j-k''+1}, \ldots, x_j, a_{j+1}, \ldots, a_{j+k'}, a_{j+k'+1}, \ldots, a_n)$$
$$\cup$$
$$(a_1, \ldots, a_{j-k''}, x_{j-k''+1}, \ldots, x_j, a_{j+1}, \ldots, a_{j+k'} + 1, a_{j+k'+1}, \ldots, a_n)$$
$$\cup \ldots \cup$$
$$(a_1, \ldots, a_{j-k''}, x_{j-k''+1}, \ldots, x_j, a_{j+1} + 1, \ldots, a_{j+k'-1} + 1, a_{j+k'}, \ldots, a_n)$$

The first k''−cube on the right side in the preceding equation is completely contained in A. All others k''−cubes have no elements in common with A. A k''−cube has $2^{k''}$ elements. Hence, there is $|A \cup A_2| = 2^j + 2^k - 2^{k''}$. Comparing the number of elements $|A \cup A_1| = 2^j + 2^k - 2^{k''+1} < |A \cup A_2| = 2^j + 2^k - 2^{k''}$, we have the desired result. $\square$

Theorem 6. *Let the vertices of an n−cube contained in the set A induce a j−cube, with $k \leq j < n$, and $k = n - m$. By the minimality condition of algorithm 4.1, a k−cube is selected, and A gets new elements. Then the smallest cube containing all elements of A has dimension $j + 1$, and the number of new elements in the set A is 2^{k-1}.*

Proof: Without loss of generality, consider the j−cube $A = (x_1, \ldots, x_j, a_{j+1}, \ldots, a_n)$. By theorem 5, it is known that the smallest cube containing the union of A and a k-cube given by the minimality condition of algorithm 4.1 is a $(j+1)$-cube, for example $(x_1, \ldots, x_{j+1}, a_{j+2}, \ldots, a_n)$. This $(j+1)$−cube can be splitted in disjoint j−cubes A and $A' = (x_1, \ldots, x_j, a_{j+1} + 1, \ldots, a_n)$. By theorem 5, the number of new elements coming to A is smallest if the k−cube is chosen in such a way that just one of the fixed $(n - j)$ components of the j−cube becomes variable. Without loss of generality, we can do the following choice for the k−cube: $(a_1, \ldots, a_{j+1-k}, x_{j+2-k}, \ldots, x_j, x_{j+1}, a_{j+2}, \ldots, a_n)$. This k−cube can be splitted in two $(k - 1)$−cubes each having 2^{k-1} vertices:

$$(a_1, \ldots, a_{j+1-k}, x_{j+2-k}, \ldots, x_j, a_{j+1}, \ldots, a_n)$$
$$\cup \tag{2}$$
$$(a_1, \ldots, a_{j+1-k}, x_{j+2-k}, \ldots, x_j, a_{j+1} + 1, \ldots, a_n).$$

The upper $(k - 1)$−cube in (2) is completely contained in A. The lower $(k - 1)$-cube in (2) and A have no elements in common. So, A gets 2^{k-1} new elements by selecting this k−cube. Such a k−cube exists for all j with $k \leq j < n$. $\square$

Theorem 7. *Consider a $(j+1)$−cube with $k \leq j < n$ in the context of algorithm 4.1, having one half completely in A, while the other half is not completely in A. Then there exists always at least one k−cube which brings not more than $2^{k-1} - 1$ new elements to A.*

Proof: Given a j-cube and a $(j + 1)$-cube: $(x_1, \ldots, x_j, a_{j+1}, \ldots, a_n) \subset A \subset$ $(x_1, \ldots, x_j, x_{j+1}, a_{j+2}, \ldots, a_n)$. Consider the vertex $(a_1, \ldots, a_j, a_{j+1}+1, a_{j+2}, \ldots,$ $a_n)$ which is supposed to be in A, but is not contained in the j−cube. There exists always a k−cube, containing this vertex and at least 2^{k-1} further elements, which are already contained in A: $(a_1, \ldots, a_{j-k+1}, x_{j-k+2}, \ldots, x_{j+1}, a_{j+2}, \ldots, a_n)$. Thus, the number of vertices becoming elements of A can not be greater than $2^{k-1} - 1$. This method can be applied to all k-cubes not completely contained in A until all elements of the $(j + 1)$−cube are contained in A. $\square$

Theorem 8. *Given an n−cube whose vertices are processed by the instructions of algorithm 4.1. Within the application of algorithm 4.1, set A runs always beside others through these states in which the number of its elements equals to a power 2^j with $k \leq j \leq n$. Furthermore, these 2^j elements of set A always induce an n−cube.*

Proof: The proof follows by induction referring to theorems 6 and 7.
$j = k$: The number of elements in A is 2^k which induce a k−cube.
$j \to j + 1$ with $j < n$: Let A contain 2^j elements inducing a j−cube. The application of algorithm 4.1 requires the selection of a k−cube described by theorem 6, where the number of new elements in A is 2^{k-1}. In the further steps, the application of algorithm 4.1 is described by theorem 7 until A induces a $(j + 1)$−cube. Then, A contains 2^{j+1} elements. $\square$

With the results of theorems 6 - 8, we have described the possible paths through the n−cube applying algorihtm 4.1. Theorem 8 says that if A contains 2^{n-1} elements, then these induce an $(n - 1)$−cube.

In the following, we will see how the consideration of the interesting complexity parameter concerning the application of algorithm 4.1 with given n and m can be decomposed into double application of algorithm 4.1 with $n' = n - 1$, $m' = m - 1$, respectively $n'' = n - 1$, $m'' = m$.

Theorem 9. *Given $1 < m < n$. Let an n−cube be processed by the instructions of algorithm 4.1, while adding vertices of k−dimensional cubes to A. Furthermore, let $k = n - m$. If the application of algorithm 4.1 is considered as long as $|A| < 2^{n-1}$, then this has the same result as application of algorithm 4.1 to an n'−cube with $n' = n - 1$ and $m' = m - 1$.*

Proof: By the result of theorem 8, it is known that the smallest cube containing A as long as $|A| < 2^{n-1}$ is an $(n - 1)$-cube. Within algorithm 4.1, with given n and m, vertices of a k−cube with $k = n - m$ become elements of A. As long as $|A| < 2^{n-1}$, all the k-cubes lie completely within the $(n - 1)$-cube. This is equivalent to applying algorithm 4.1 with n' and m'. With $k = n - m$ and $k = n' - m'$, we have $n - m = n' - m'$, and because of $n' = n - 1$, there is $m' = m - 1$. $\square$

We conclude that the application of algorithm 4.1 with n, m, as long as $|A| < 2^{n-1}$, and its application with n', m' yield the same frequencies of adding specific numbers of elements to A. In the next two theorems, it will be shown that there exist similar properties concerning the application of algorithm 4.1 with n, m when $2^{n-1} \leq |A| < 2^n$ and the application with $n'' = n - 1$, $m'' = m$.

Theorem 10. *Given $0 < m < n - 1$. Let an n–cube be processed by the instructions of algorithm 4.1 and let $2^{n-1} < |A| < 2^n$. Let A' be an $(n-1)$–cube completely contained in A. If there exists a k–cube Ψ', with $k = n - m$ and $A' \cap \Psi' = \{\ \}$, fulfilling the minimality condition while adding its elements to A, then there is always a k–cube Ψ with $|A' \cap \Psi| = 2^{k-1}$, also fulfilling the minimality condition.*

Proof: By assumption, A consists of a $(n-1)$-cube A' and further vertices, which are combined in a set Θ. Then $\Theta \cap A' = \{\ \}$, and by theorem 6, we know that $|\Theta| \geq 2^{k-1}$. Hence, the number of new elements coming to A applying algorithm 4.1 is always smaller than 2^{k-1}, by the result of theorem 7. There is $A = A' \cup \Theta \neq (x_1, \ldots, x_n)$. Suppose $A' = (x_1, \ldots, x_{n-1}, a_n)$ and $\Psi' = (a_1, \ldots, a_{n-k-1}, x_{n-k}, \ldots, x_{n-1}, a_n + 1)$, which are disjoint. Furthermore, suppose that the new elements of A resulting by adding vertices of Ψ' to A lie in a $(k-1)$-cube, given by $(a_1, \ldots, a_{n-k-1}, a_{n-k}, x_{n-k+1}, \ldots, x_{n-1}, a_n + 1)$. All other elements of Ψ' lie completely in Θ. A k–cube Ψ, with $\Psi \cap \Psi' = (a_1, \ldots, a_{n-k-1}, a_{n-k}, x_{n-k+1}, \ldots, x_{n-1}, a_n + 1)$, and half of its elements completely contained in A' can always be found with $\Psi = (a_1, \ldots, a_{n-k}, x_{n-k+1}, \ldots, x_n)$. If the growth of A by adding the elements of Ψ' is minimal, then also the growth by adding the elements of Ψ is minimal. $\square$

Theorem 11. *Given $0 < m < n - 1$. Let an n–cube be processed by the instructions of algorithm 4.1, while adding vertices of k–dimensional cubes to A. Furthermore, let $k = n - m$. If the application of algorithm 4.1 is considered from the step, when $2^{n-1} \leq |A| < 2^n$, then the result is the same as application of algorithm 4.1 to an n''–cube with $n'' = n - 1$ and $m'' = m$.*

Proof: By theorem 8, we conclude that the vertices not contained in A induce an $(n-1)$–cube, if A contains exactly 2^{n-1} elements. By theorems 6, 7, and 10, we know that there always exist such k-cubes fulfilling the minimality condition of algorithm 4.1, and having at least one half in an $(n-1)$-cube which is completely contained in A. Each half of this k-cube can be considered as a $(k-1)$–cube. With $k - 1 = n'' - m''$, $k = n - m$ and $n'' = n - 1$, we have $m'' = n'' - k + 1 = n - 1 - k + 1 = n - k = m$. $\square$

Definition 12. Let $0 < m < n$. The function $\#(c, n, m)$ describes how often c elements are added to A when applying algorithm 4.1 with n and m.

It is obvious that with given $0 < m < n$ and $k = n - m$, $\#(c, n, m) = 0$ if $c < 1$ or $c > 2^k$. Therefore, if $\#(c, n, m) \neq 0$, then $1 \leq c \leq 2^k$. We can establish a recursion for $\#(c, n, m)$.

Theorem 13. *Let $n \geq 2$ and $1 < m < n$. Then application of algorithm 4.1 to an n–cube results in:*

$$\#(c, n, m) = \#(c, n - 1, m - 1) + \#(c, n - 1, m). \tag{3}$$

Proof: The proof follows by theorems 5 - 11. □

Theorem 14. *Let $2 \leq n$ and $m = 1$. Applying algorithm 4.1 to an $n-$cube, there will be the following frequencies $\#(c, n, m = 1)$ with $1 \leq c \leq 2^{n-1}$:*

$$\#(c, n, m = 1) = \begin{cases} 1, & \text{if } c = 2^i \text{ with } i = 1 \dots n - 1, \\ 2, & \text{if } c = 1, \\ 0, & \text{otherwise.} \end{cases}$$

Proof: Without loss of generality, consider the following application of algorithm 4.1:

step	cube	growth of A
1. $(n - 1)-$ cube	$(x_1, \dots, x_{n-1}, a_n)$	2^{n-1}
2. $(n - 1)-$ cube	$(x_1, \dots, x_{n-2}, a_{n-1}, x_n)$	2^{n-2}
$\vdots$	$\vdots$	$\vdots$
(n-1). $(n - 1)-$ cube	$(x_1, a_2, x_3, \dots, x_n)$	2

Now, there are still two vertices of the n-cube not contained in A: $P_1 = (a_1, a_2 + 1, \dots, a_n + 1)$ and $P_2 = (a_1 + 1, a_2 + 1, \dots, a_n + 1)$. These vertices become elements of A in two further steps where the growth of A is 1. The desired result follows. □

Theorem 15. *Let $2 \leq n$ and $m = n - 1$. Applying algorithm 4.1 to an $n-$cube, there will be the following frequencies $\#(c, n, m = n - 1)$ with $1 \leq c \leq 2^{n-1}$:*

$$\#(c, n, m = n - 1) = \begin{cases} 1, & \text{if } c = 2, \\ 2^n - 2, & \text{if } c = 1, \\ 0, & \text{otherwise.} \end{cases}$$

Proof: Applying algorithm 4.1 with $m = n-1$, we have to consider the vertices of 1-cubes. At the start of the algorithm, there are two vertices becoming elements of A. In all further steps, the growth of A is only 1. Therefore, we have $\#(c = 2, n, m = n - 1) = 1$ and $\#(c = 1, n, m = n - 1) = 2^n - 2$. For all other values of c, there is $\#(c, n, m = n - 1) = 0$. This is the desired result. □

Theorem 16. *Let $n > 2$ and $1 \leq m < n$. Considering $c = 2^{n-m}$, applying the algorithm 4.1 yields $\#(c = 2^{n-m}, n, m) = 1$.*

Proof: At the beginning of algorithm 4.1, the growth of A is given by 2^{n-m}. In all further steps, there are only such values c for which $c = 2^{n-m-1}$ by theorem 6, or $c < 2^{n-m-1}$ by theorem 7. Therefore, $\#(c = 2^{n-m}, n, m) = 1$. □

Theorem 17. *Let $2 \leq n$ and $1 \leq m \leq n-1$. Then $\#(c, n, m) \neq 0$ only for those c, if $c = 2^j$ with $j = 0, 1, \dots, (n - m)$, otherwise $\#(c, n, m) = 0$.*

Proof: By theorems 9 and 11, applying algorithm 4.1 with given n and m can be considered in a manner of twice applying algorithm 4.1 with $n' = n - 1$, $m' = m - 1$ and $n'' = n - 1$, $m'' = m$. This can be repeated until the conditions of theorem 14 or theorem 15 are fulfilled. By these theorems, there are only such growth values c, that equal 2^j with $j = 0, 1, \ldots, (n - m)$. $\square$

With these results, we can give a general description for $\#(c = 2^j, n, m)$, which is presented in the next theorem.

Theorem 18. *Let $n \geq 2$ and $0 < m < n$. Furthermore, let $j \in \{1, 2, \ldots, (n - m)\}$. Application of algorithm 4.1 then yields the following frequencies for adding $c = 2^j$ elements to A:*

$$\#(2^j, n, m) = \binom{n - j - 1}{m - 1} \tag{4}$$

Proof: Consider the following 3 cases.

α): $n \geq 2$ and $m = 1$
In this case, $1 \leq j \leq n - m = n - 1$. For all those j, we have:

$$\#(2^j, n, m = 1) = \binom{n - j - 1}{m - 1} = \binom{n - j - 1}{0} = 1$$

This coincides with the result of theorem 14.

β): $n \geq 2$ und $m = n - 1$
Here, only $j = 1$ is relevant, because of $n - m = 1$. For $j = 1$ we have:

$$\#(2, n, m = n - 1) = \binom{n - j - 1}{m - 1} = \binom{n - 2}{n - 2} = 1$$

This coincides with the result of theorem 15.

γ): Let $m > 1$ and let $n \geq 4$ and $n > m - 1$. Therefore, $j \in \{1, 2, \ldots, (n - m)\}$. Recursion (3) can only be applied for case γ. Applying this recursion an appropriate number of times, we obtain case α and β. The proof is given by induction. The start values are given by cases α and β. It remains to prove the step from n to $n + 1$. Let $\#(c = 2^j, n, m) = \binom{n-j-1}{m-1}$ and consider $\#(c = 2^j, n + 1, m)$.

$$\#(c = 2^j, n + 1, m) = \#(c, n, m) + \#(c, n, m - 1) = \binom{n-j-1}{m-2} + \binom{n-j-1}{m-1} =$$
$$= \frac{(n-j-1)!}{(m-2)!(n-j-m+1)!} + \frac{(n-j-1)!}{(m-1)!(n-j-m)!} =$$
$$= \frac{(n-j)!}{(m-1)!(n-j-m+1)!} = \binom{n-j}{m-1}$$

$\square$

5 New upper bounds

With the result of theorem 18, we know how often $c = 2^j$ binary elements are assigned maximally to the vertices of the $n-$cube within the construction of mth order correlation-immune functions, while applying algorithm 3.1 to the $n-$cube. The number of possibilities is biggest if the $c = 2^j$ free vertices of a $k-$cube are assigned 2^{j-1} '0's and 2^{j-1} '1's, in case of $2^{j-1} \leq min(K, 2^{n-m}-K)$. Then, there are $\binom{2^j}{2^{j-1}}$ possibilities to do the assignments. In case of $2^{j-1} > min(K, 2^{n-m} - K)$, there are not more than $\binom{2^j}{min(K,2^{n-m}-K)}$ possibilities to do the assignments. This yields an upper bound for the number of mth order correlation-immune functions with Hamming weight $W(f) = K \cdot 2^m$, with $0 < K < 2^{n-m}$, which is summarized in the following theorem.

Theorem 19. *Let $1 \leq m < n$. The number of mth order correlation-immune functions with n input variables and K '1's within the $(n - m)-$cubes, with $0 < K < 2^{n-m}$, is upperbounded by:*

$$b(n,m,K) = \begin{cases} \prod_{j=1}^{n-m} \binom{2^j}{2^{j-1}}^{\binom{n-j-1}{m-1}}, & \text{if } K = 2^{n-m-1} \\ \prod_{j=1}^{1+\lfloor ld(min(K,2^{n-m}-K))\rfloor} \binom{2^j}{2^{j-1}}^{\binom{n-j-1}{m-1}} \cdot \\ \prod_{j=2+\lfloor ld(min(K,2^{n-m}-K))\rfloor}^{n-m} \binom{2^j}{min(K,2^{n-m}-K)}^{\binom{n-j-1}{m-1}}, & \text{otherwise.} \end{cases} \tag{5}$$

Theorem 19 does not include the cases $K = 0$ and $K = 2^{n-m}$. These two cases reflect the constant valued functions, which have maximum order of correlation-immunity. With theorem 19, it is also obvious that $b(n, m, K) = b(n, m, 2^{n-m} - K)$. Summing up over $b(n, m, K)$ for all K, this yields an upper bound for the number of all mth order correlation-immune functions with n inputs.

Theorem 20. *The number of mth order correlation-immune functions with n inputs, $0 < m < n$, is upperbounded by*

$$B(n,m) = \begin{cases} 2 + b(n, m, K = 1), & \text{if } n - m = 1 \\ 2\left[1 + \sum_{K=1}^{2^{n-m-1}-1} b(n, m, K)\right] + b(n, m, K = 2^{n-m-1}), & \text{otherwise.} \end{cases} \tag{6}$$

It has to be noted that $B(n, m)$ is valid for all m and n with $0 < m < n$. This result can be compared to the upper bound presented by Yang and Guo which gives the number of 1st order ($m = 1$) correlation-immune functions [YaGu 95]:

$$\sum_{k=0}^{2^{n-1}} \sum_{r=0}^{k} \binom{2^{n-2}}{r}^2 \binom{2^{n-2}}{k-r}^2. \tag{7}$$

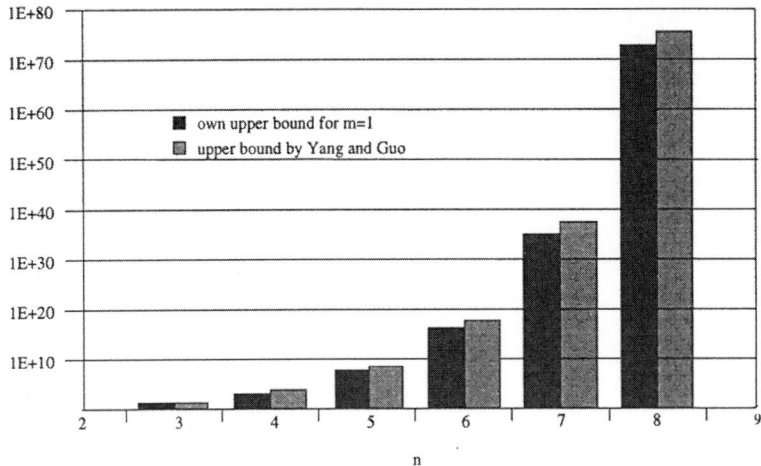

Fig. 1 : Comparison of upper bounds

6 Acknowledgements

I am grateful to Professor Firoz Kaderali and Professor Werner Poguntke for the supervision of my work. Furthermore, I would like to thank Professor Ulrich Faigle and Professor Walter Kern (both University Twente) for helpful discussions about some contents of this paper.

References

[CCCS 91] P. Camion, C. Carlet, P. Charpin, N. Sendrier: *'On Correlation-immune Functions'*, Advances in Cryptology: Crypto '91, Proceedings, Lecture Notes on Computer Science 576, 1991, p. 86-100

[GuMa 88] X. Guo-Zhen, J. Massey: *'A Spectral Characterization of Correlation-Immune Combining Functions'*, IEEE Transactions on Information Theory, Vol. 34, No.3, May 1988, p. 569-571

[SeZh 93] J. Seberry, X. Zhang, Y. Zheng: *'On Constructions and Nonlinearity of Correlation Immune Functions'*, Advances in Cryptology: Eurocrypt '93, Proceedings, Lecture Notes on Computer Science 765, 1993, p. 181-199

[Sieg 84] T. Siegenthaler: *'Correlation-Immunity of Nonlinear Combining Functions of Cryptographic Applications'*, IEEE Transactions on Information Theory, Vol. IT-30, No.5, Sep. 1984, p. 776-780

[Sieg 85] T. Siegenthaler: *'Decrypting a Class of Stream Ciphers Using Ciphertext Only'*, IEEE Transactions on Computers, Vol. C-34, No.1, Jan. 1985, p. 81-85

[YaGu 95] Y. Yang, B. Guo: *'Further Enumerating Boolean Functions of Cryptographic Significance'*, Journal of Cryptology, 1995, 8: p. 115-122

On Generalised Concatenated Codes

S. Sonander and B. Honary
Lancaster Communications Research Centre
Lancaster University
UK

Abstract

The Authors investigate a novel application of generalised concatenated coding theory to the E.S.A. Coding standard. The proposed coding and decoding system is compared to the E.S.A. coding standard, with both systems having the same coding rate. A number of RS codes are used in the system, each capable of correcting a different number of errors. The performance of the system is compared with and without feedback between the decoding modules. The system with feedback is shown to out perform the E.S.A. standard telemetry coding scheme. Different system configurations have been simulated and compared to find an optimum solution.

1. Introduction

Concatenated codes were first introduced by Forney [1]. The concatenation of convolutional and block codes has advantages of greater error correcting capability than either of the component codes, depending on the selection criteria. Concatenated coding schemes involving convolutional and RS codes [2] are extremely useful, providing good burst error correction capability [3]. The simplest concatenated block coding system comprises of two basic coding/decoding units, the inner and outer. The inner code has parameters (n_1,k_1,d_1) and the outer code has parameters (n_2,k_2,d_2), giving a combined coding system with parameters (cn_1n_2,ck_1k_2,d_3) such that $c \in I$ and d_3 is maximized for optimum code selection.

The E.S.A. telemetry coding standard consists of an outer (255,223,33) RS code and an inner convolutional code with rate 0.5 and constraint length 7, giving a total coding rate of 0.437. This paper describes the investigation into the application of Generalised Concatenated Codes (GCC) [4,5] to the ESA coding standard [6,7].

The encoding process consists of a number of different outer encoders, an interleaver and an inner encoder. The outer encoder is RS code based, such that a stream of information bits are converted into a stream of output codewords. The output RS codewords are passed to the interleaver and arranged to form the rows of an array. The RS codewords in the array are labeled as C_i, such that $(0 \le i \le h)$. Codeword C_i is a $(255,k_i,d_i)$ RS Code with $d_{i-1} > d_i > d_{i+1}$. This frame is then passed to the inner encoder, which is the same as the ESA standard convolutional encoder.

The decoding process involves an inner decoder, a de-interleaver and an outer decoder. The system relies upon a feedback mechanism in which *a posteriori* higher confidence information gained at the decoder is re-used for the decoding of the next received codeword. The higher confidence information gained at the decoder is a natural coincidence of the encoding operation. The inner decoder is a convolutional decoder and uses the Viterbi algorithm [8] on the received bit stream. The received bits are passed to the inner Viterbi decoder, with each decoded bit being stored until one complete interleaved frame of bits is available. These stored bits are then passed to the de-interleaver. The RS codewords are decoded to yield an estimate of the information. With the assumption of correct RS decoding, this information is used to modify the trellis branch metrics in the Viterbi Decoder. This process is repeated until each of the RS codewords have been decoded.

We describe a system which is based on GCC and which offers identical coding rates to the E.S.A. coding standard. However, the performance gain obtained is 1dB with a small increase in complexity. Each possible system configuration is defined by several key parameters at the encoder. The useful scope for variability in the encoder is the RS code parameters and the interleaving depth. The system encoder is equal in complexity to the ESA standard in terms of arithmetic operations, while the decoder has a slightly increased complexity.

2. Generalised Concatenated Block Codes

Fundamentally, the encoding of linear concatenated codes consists of two stages of coding. The addition of an interleaver may be as a stand alone block, or as considered here, a process of bit extraction that links the outer and inner encoders.

The performance of such concatenated block codes can be improved for carefully selected code sets. Before considering the decoding process of such codes, we first consider the outer and inner encoder.

2.1 The Encoding Process

The outer encoding process of generalised concatenated codes may be conceptualised as a series of codewords arranged into an array format. The function of the outer encoder is to generate a series of codewords, based upon a set of information vectors. These information vectors, each with k_i information symbols are denoted $\mathbf{a}_{ki}$, where $0 <= i <= n$. For each vector $\mathbf{a}_{ki}$, an associated encoding function φ_{ai} is used to generate the parity check vector $\mathbf{a}_{pi}$ with p_i parity check symbols and form the codeword vector $\mathbf{a}_i$. For the case of the information symbols, it is important that $k_{i-1} > k_i$. Therefore $\mathbf{a}_i$ is given as:

$$\mathbf{a}_i = \begin{bmatrix} \mathbf{a}_{ki} & \mathbf{a}_{pi} \end{bmatrix} = \varphi_{ai}(\mathbf{a}_{k0}) \tag{1}$$

Arranging the series of codeword vectors into array format yields the vector $\mathbf{A}$. The result of the outer encoding process is given as:

$$\mathbf{A} = \begin{bmatrix} \mathbf{a}_0 & \mathbf{a}_1 & \cdots & \mathbf{a}_i & \cdots & \mathbf{a}_{n-1} & \mathbf{a}_n \end{bmatrix}^T \tag{2}$$

and shown graphically in Fig. 1.

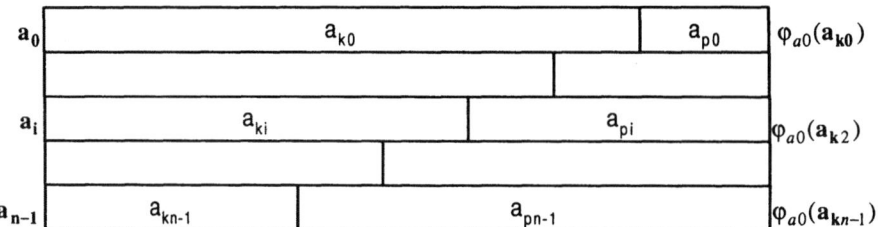

Fig. 1. Outer Encoder

The inner encoding procedure is similar to the outer encoding procedure. Again, we build up a vector **B** which is the result of the inner encoding process. As before, the information vector, parity check vector, codeword vector and associated encoding function are defined as $\mathbf{b_{ki}}$, $\mathbf{b_{pi}}$, $\mathbf{b_i}$ and φ_{bi} respectively, with $0<=i<=n$. The information vectors $\mathbf{b_{ki}}$ are derived from vector **A**, by defining $\mathbf{b_{ki}}$ to be the first k_i symbols present in the i'th column in vector **A**, again such that $k_{i-1}>k_i$. Vector $\mathbf{b_i}$ is thus defined as:

$$\mathbf{b_i} = \begin{bmatrix} \mathbf{b_{ki}} & \mathbf{b_{pi}} \end{bmatrix} = \varphi_{bi}(\mathbf{b_{k0}}) \tag{3}$$

The codeword vectors b_i are arranged into array format to form vector **A**:

$$\mathbf{B} = \begin{bmatrix} \mathbf{b_0}^T & \mathbf{b_1}^T & \cdots & \mathbf{b_i}^T & \cdots & \mathbf{b_{n-1}}^T & \mathbf{b_n}^T \end{bmatrix} \tag{4}$$

and again shown graphically in Fig. 2.

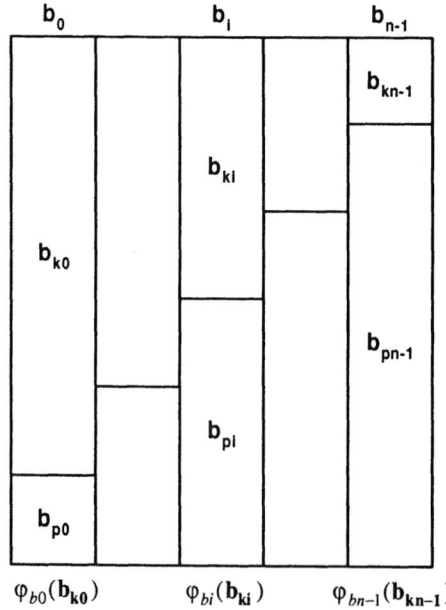

Fig. 2. Inner Encoder

Vector B is the result of both inner and outer encoding of the information vectors a_{ki}. The information rate of the code is given as eqn. 5

$$\frac{\sum_{i=0}^{n} k_i}{\sum_{i=0}^{n} p_i} = \frac{\sum_{i=0}^{n} k_i}{m(n+1)} \tag{5}$$

where $m = p_i + k_i$

2.2 The Decoding Process

In the previous section, attention was drawn to the outer and inner encoding functions φ_{ai} and φ_{bi} respectively. The introduction of the decoding functions φ^{-1}_{ai} and φ^{-1}_{bi} at this point is necessary for decoding. However, feedback between the inner and outer decoder will enhance the decoding process, due to the differing levels of error protection offered by the encoding functions. Following transmission, codeword B is distorted by noise to give a received codeword $\overline{B}$ shown as:

$$\overline{B} = \begin{bmatrix} \overline{b}_0^T & \overline{b}_1^T & \cdots & \overline{b}_i^T & \cdots & \overline{b}_{n-1}^T & \overline{b}_n^T \end{bmatrix} \tag{6}$$

From eqn. 3 and eqn. 4, an estimate of the information vectors $\overline{b}_{ki}$ input to the inner encoder is made and is expressed as eqn. 7.

$$\hat{b}_{ki} = \varphi_{bi}^{-1}(b_i) \tag{7}$$

However, the condition $k_{i-1} > k_i$ holds for b_{ki}. Therefore the vector estimate $\hat{b}_{n-1}$ is calculated most reliably, followed by $\hat{b}_i$. This information is passed to the outer decoder. With the estimate $\hat{b}_i$, the codeword vector estimates $\hat{a}_i$ are known, by extracting a_i from array $\overline{B}$. From $\overline{a}_i$, the estimate of the original information vectors $\hat{a}_{ki}$ can be calculated as:

$$\hat{A} = \begin{bmatrix} \hat{a}_0 & \hat{a}_1 & \cdots & \hat{a}_i & \cdots & \hat{a}_{n-1} & \hat{a}_n \end{bmatrix}^T \tag{8}$$

$$\hat{a}_i = \begin{bmatrix} \hat{a}_{ki} & \hat{a}_{pi} \end{bmatrix} = \varphi_{ai}^{-1}(\hat{a}_i) \tag{9}$$

The chosen coding functions φ_{ai} imply that the codewords a_i have information content $a_{ki} > a_{ki+1}$, again implying that the information vector estimate $\hat{a}_{ki}$ is known more reliably that $\hat{a}_{ki-1}$. The link between array A and b_i enables the 'fixing' of certain symbols in the information vector estimate, based upon the most reliable information vector a_i and codeword vectors b_{ki}. Further rounds of decoding and symbol replacement are performed until the least reliable of the information code vectors are found. The performance of this algorithm depends largely upon the choices made for the coding functions φ_{ai} and φ_{bi}.

3. The ESA Telemetry Channel Coding Standard

The basic components of the ESA coding standard consists of a RS outer code and a convolutional inner code. The two encoders are optionally linked with an interleaver. The convolutional code when used alone is very well suited to channels which exhibit guassian noise. Conversely, RS codes have well known burst error correction capabilities due to the inherent Galois field representation. Encoder implementation is also straightforward. The RS code used in the ESA telemetry channel coding standard has information rate 0.437 with symbols taken from **GF(256)** [6,7]. The encoding is systematic and the generator polynomial is given as:

$$g(z) = \prod_{j=112}^{143} \left(z + b^j \right) \tag{10}$$

where b is a root of the irreducible polynomial $x^8 + x^6 + x^4 + x^3 + x + 1$
The convolutional code has constraint length K=7 and connectivity polynomials **G₁** and **G₂** given by eqn. 11 and eqn. 12 respectively.

$$\mathbf{G_1} = 1 + x^2 + x^3 + x^5 + x^6 \tag{11}$$

$$\mathbf{G_2} = 1 + x + x^2 + x^3 + x^6 \tag{12}$$

The addition of an interleaver between the inner and outer codes is optional.

4. The Application of GCC to the ESA Standard

In this section, a complete coding scheme is proposed which offers improved performance over the ESA standard coding scheme. The encoder complexity is virtually unaltered. However, the decoder complexity is greater. RS codes of length 255 symbols are used, with each symbol taken from Galois field GF(2^8). It is important to note that many different variations of the proposed scheme are possible. The family of length 255 RS codes have been used here, both because of their usage in the ESA telemetry standard and also because of the consistency in Galois field arithmetic.

4.1 The Encoding Process

The encoding process consists of a number of outer encoders, an interleaver and an inner encoder. The outer encoder is RS code based, such that a stream of information symbols are converted into a stream of output code words. The output RS codewords are passed to the interleaver and arranged to form the rows of an array. The RS codewords in the array are labeled C_i, where ($0 \le i \le h$). Code C_i is a (255,k_i,p_i) RS Code. The interleaving depth of the system is also equal to the number of RS codewords, or $h+1$. At this point, we introduce a new parameter, λ , which gives the difference in number of information symbols per adjacent code words. Thus the following holds:

$$k_0 = k_0 \tag{13}$$

$$k_1 = k_0 + \lambda \tag{14}$$

$$k_i = k_0 + i\lambda \tag{15}$$

$$k_h = k_0 + h\lambda \tag{16}$$

where k_i gives the number of information symbols in C_i. Hence the number of information symbols in each RS code C_i increases, hence the error correcting capability of each code decreases. The coding parameters governing the array formatted codes are as follows:

Number of codes in array	$h+1$
Information Symbols in Row 'i'	$k_i=k_0+i\lambda$
Code Symbols in Row 'i'	$n=2^8-1$
Total No. of Information Symbols in array	$k(h+1)$

Thus to satisfy the above parameter specification and system constraints $k_i=k+\lambda(i-h/2)$ yielding eqn. 17, proving that the information rate is as stated.

$$\frac{\sum_{i=0}^{h} k+\lambda\left(i-\frac{h}{2}\right)}{kn} = \frac{k}{n} \tag{17}$$

The process of bit extraction is performed by the interleaver. The bits are extracted on a column by column basis and are passed to the convolutional encoder. Fig. 3 shows the interleaving operation.

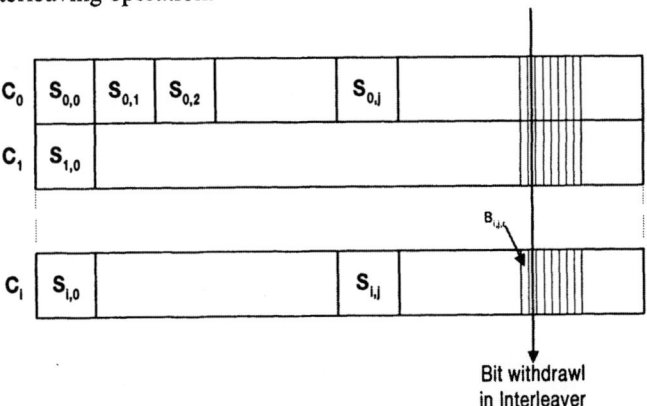

Bit withdrawl
in Interleaver

Fig. 3   Interleaver Operation

Considering Fig. 3, the codeword C_i is a vector consisting of n symbols vectors $S_{i,j}$, such that

$$C_i = \begin{bmatrix} S_{i,0} & S_{i,1} & \cdots & S_{i,j} & \cdots & S_{i,n-2} & S_{i,n-1} \end{bmatrix} \qquad (18)$$

Each symbol vector $S_{i,j}$ consists of m bits, $B_{i,j,r}$ such that

$$S_{i,j} = \begin{bmatrix} B_{i,j,0} & B_{i,j,1} & \cdots & B_{i,j,r} & \cdots & B_{i,j,m-2} & B_{i,j,m-1} \end{bmatrix} \qquad (19)$$

now the vertical bit extraction is given as
$$j = 0,...,n-1$$
$$r = 0,...,m-1$$
$$i = 0,...,h$$
$$\text{Output bit } B_{i,j,r}$$

Such that the number of bits extracted from the array is $nm(h+1)$. The interleaved $h+1$ RS codewords or $nm(h+1)$ bits form one frame. This frame is passed to the inner encoder.

4.2 The Decoding Process

The technique used for the decoding of the concatenated codes is shown in Fig. 4.

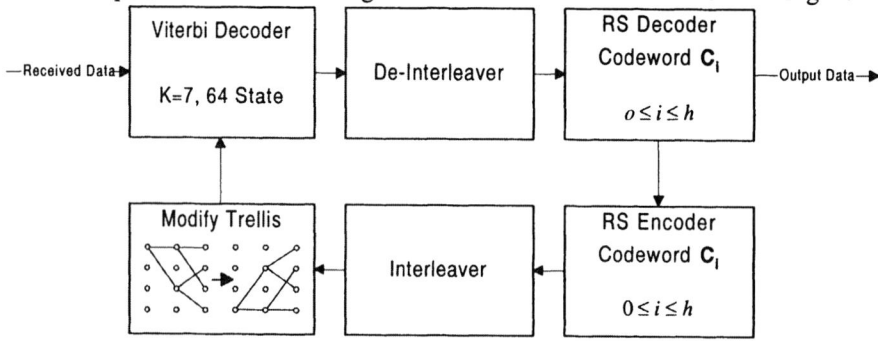

Fig. 4   The Decoding Algorithm

The decoding algorithm has similarities with Turbo Decoding [9]. However, the omission of log likelihood Algebra in the decoding process distinguishes it from Turbo Decoding. The inner decoder uses the Viterbi Algorithm on the received bit stream. Each decoded bit is stored until a total of $nm(h+1)$ bits are available. These stored bits are then passed to the de-interleaver and $\hat{C}_0$ is extracted. $\hat{C}_0$ is decoded to form an estimate of the information $\hat{k}_0$. The encoding process ensures that $\hat{k}_0$ is known to a high degree of certainty and is therefore re-encoded and passed to the interleaver. The inner decoder is now able to modify the trellis structure using this information. Each branch metric on the trellis corresponds directly to one of the $h+1$

RS codewords, and these trellis branch metrics are modified. This process is repeated h times until each $\hat{C}_0$ has been decoded.

The Decoding Algorithm:

Step 1)   The received frame is decoded and de-interleaved, thus extracting all the RS codewords $\hat{C}_0$, $\hat{C}_1$, ..., $\hat{C}_h$

Step 2) $\hat{C}_0$ is decoded using the RS decoder and the information estimate $\hat{k}_0$ is obtained.

Step 3)   Assuming that the decoding operation resulted in a correct calculation, we set $\hat{k}_0 = k_0$.

Step 4)   The information vector k_0 is RS encoded, giving C_0

Step 5)   Codeword C_0 is convolutional encoded and is used to modify the received frame trellis information.

Step 6)   The received modified coded signal is Viterbi decoded.

Step 7)   Vector $\hat{C}_1$ is decoded, again using the RS decoder. The process then repeats from step 2 until all of the information is extracted from the received signal.

The *h+1* RS decoders can be seen in Fig. 4, together with the feedback path and feedback decoding units.

5. Simulation Results

The results shown in this section were derived by simulation under gaussian noise. A different number of system configurations have been considered for simulation, with only those deemed suitable for comparison with the E.S.A. standard shown.

Each different system configuration is defined by several key parameters at the encoder. The decoding parameters are a natural consequence of the encoding algorithm and have been described above. The useful scope for variability in the encoder system is the number of RS encoders, the RS code parameters and the interleaving depth.

For the decoder the performance with and without feedback is also considered. The complete simulation results obtained under additive white Guassian noise conditions are shown as Fig. 5. The system parameters governing the simulation results are shown in Table 1.

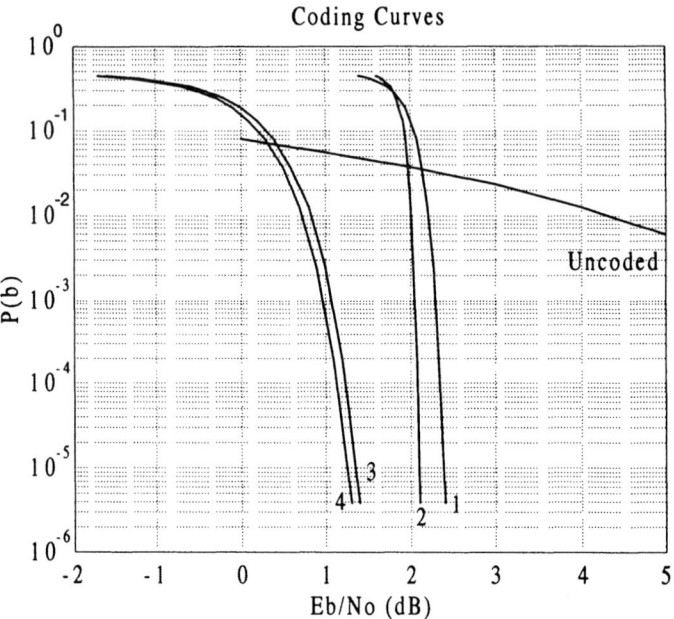

Fig. 5   Simulation Results

The E.S.A. standard coding performance without feedback is depicted as 'curve 2.' The inclusion of results without feedback have been included for a direct comparison to those with feedback.

Curve	RS Codes Used	Interleaving Depth	Decoder Feedback	Code Rate
1	C_0=RS(255,207,49) C_1=RS(255,223,33) C_2=RS(255,239,17)	3	No	0.437
2	C_0=RS(255,207,49) C_1=RS(255,223,33) C_2=RS(255,239,17)	3	Yes	0.437
3	C_0=RS(255,207,49) C_1=RS(255,215,41) C_2=RS(255,223,33) C_3=RS(255,231,25) C_4=RS(255,239,17)	5	Yes	0.437
4	C_0=RS(255,207,49) C_1=RS(255,215,41) C_2=RS(255,223,33) C_3=RS(255,231,25)	4	Yes	0.437

Table 1 System Simulations

6. Conclusion

We have described a system which offers a coding system with identical key coding parameters to the E.S.A. coding standard. These key parameters are coding rate, RS code block length, convolutional constraint length, convolutional connectivity vectors and variable interleaving depth. The system encoder is equal in complexity to the ESA standard in terms of arithmetic operations. However, the number of different RS encoders may require additional storage requirements. The system decoder complexity in increased by a greater degree. The introduction of feedback into the decoder requires an extra round of Viterbi decoding per interleaved code block to be completed, per code word.

7. References

[1]   G. D. Forney, Jr., 'Concatenated Codes', MIT Press, Cambridge, Mass., USA, 1966

[2]   P. Elias, 'Coding for Noisy Channels - IRE Convention Record, Part 4, pp. 37-46, March 1955.

[3]   D. W. Hagelbarger, 'Recurrent Codes: Easily Mechanized, Burst - Correcting Binary Codes', BSTJ, Vol. 38, No. 4, pp. 969-984, July 1959

[4]   B. B. Zayblov and E.L. Blokh: 'Linear Concatenated Codes', Moscow, 1982

[5]   Martin Bossert, 'Concatenation of Block Codes', DFG Report, 1988

[6]   ESTEC: 'Telemetry Channel Coding Standard', European Space Agency, ESA PSS-04-103 Issue 1, September 1989

[7]   ESTEC, 'Telemetry Channel Coding', European Space Agency, ESA CCSDS 101.0-B-2, January 1987

[8]   Viterbi, A. J., 'Error Bounds for Convolutional Codes and an Asymptotically Optimum Decoding Algorithm", IEEE Trans. Inf. Theory, vol. IT13, April 1967, pp.260-269

[9]   Joachim Hagenauer, 'Iterative Decoding of Binary Block and Convolutional

8. Acknowledgments

The authors with to thank the EPSRC for their help in funding this Research Project.

Error Performance Analysis of Different Interleaving Strategies Applied to Eight Track Digital Tape Systems

A. Tandon and P. G. Farrell

Communications Research Group
School of Engineering
University of Manchester
Manchester, UK

1 Introduction

In the design of powerful error detection and correction (EDC) codes, for multi-track recording channels, Reed Solomon block symbol codes are used [1]. It is particularly effective with Reed Solomon codes to use a powerful technique known as interleaving [2] which alters the number of error symbols within a block and can therefore improve error performance. Important parameters that influence the efficiency and cost effectiveness of such codes include block size, interleave depth and the power of the symbol correcting code. Using error performance estimation techniques allows a designer to experiment with these parameters to produce efficient EDC codes. For the analysis of performance estimation schemes, a simulation [3] has been developed that incorporates the effects of multi-tracks and drop-outs.

2 Reed Solomon Codes

Reed Solomon (RS) codes [1] have qualities, applicable to multi-track tape systems, that make use of these codes more attractive than other coding schemes. In particular, RS codes are symbol oriented rather than bit oriented in nature. This is ideal for a multi-track tape system which is organised in terms of groups of bit streams rather than a single stream of bits. Another attractive feature of RS codes is the ability to efficiently handle burst errors that span a number of adjacent symbols, as often occurs in multi-track tape systems [4]. The parameters of RS codes are as follows:

$$n = 2^m - 1 \text{ or } 2^m \text{ or } 2^m + 1 \tag{1}$$

$$n - k = 2t \tag{2}$$

$$d = 2t + 1(= (n - k) + 1) \qquad (3)$$

$$R = \frac{k}{n} = \frac{(n - 2t)}{n} \qquad (4)$$

Equations (1) - (4) are the parameters of a t - error correcting RS code with symbols drawn from GF (2m)[1]. The equations (1) - (4) use m bit words, n symbol codewords (i.e. a block size of n), k message symbols and a minimum distance between codewords of d. RS codes have good performance as they are maximum distance separable, since $d = n - k + 1$ and this is the highest possible value that d can have, for a given n and k. The values of n in equation (1) of $2m$ and $2m + 1$ refer to extended and doubly extended RS codes respectively.

The codes used below all have 8-bit symbols (i.e. $m = 8$). The maximum block length used is 256 symbols, corresponding to the extended case. The other block lengths used correspond to shortened RS codes, where the appropriate number of information symbols are set to zero to reduce the block length desired. A wide range of values of t are used, with corresponding effect on the rate, R, of the code. Increasing the value of t reduces the rate R, for a constant value of n, as shown in equation (4). A suitable generator polynomial used in a t error correcting RS code is:

$$g(x) = (x + \alpha^1)(x + \alpha^2)(x + \alpha^3)(x + \alpha^4) \ldots (x + \alpha^{2t}) \qquad (5)$$

3 Interleaving Techniques

3.1 Horizontal Symbol Stack Arrangement

Figure 1 shows the arrangement of symbols required to produce a horizontal symbol stack on a multi-track tape system. It can be seen that the symbols are positioned vertically across the tape tracks but the stack is horizontal. If a burst of errors occurs on the tape medium due to for example, a drop-out, the errors are contained within the symbols. With a powerful enough EDC code, these symbols can be corrected and therefore the effects of the burst of errors can be removed or at least diminished.

An alternative stacking arrangement, shown in figure 2, is vertical stacking where symbols are positioned horizontally along the tape track but the stack is vertical. This stacking arrangement is able to reduce the possibility of long burst errors from occurring. Figure 2 also illustrates an alternative interleaving strategy where symbol and block interleaving are combined. This interleaving strategy produces the mixing of symbols that can be seen in figure 2.

	Block		Symbol		Bit	
Symbols corrected (t)	number of errors	error rate	number of error	error rate	number of errors	errorrate
0	676	1.3798e-1	1560	2.4384e-3	2013	4.0126e-4
2	218	4.4498e-2	1145	1.7897e-3	1367	2.7249e-4
3	141	2.8781e-2	929	1.4521e-3	1071	2.1349e-4
4	79	1.6125e-2	398	6.2212e-4	710	1.4153e-4
5	45	9.1855e-3	230	3.5951e-4	483	9.6280e-5
6	27	5.5113e-3	132	2.0633e-4	304	6.0599e-5
8	9	1.8371e-3	84	1.3130e-4	116	2.3123e-5
10	2	4.0824e-4	21	3.2825e-5	31	6.1795e-6
11	1	2.0412e-4	11	1.7194e-5	17	3.3887e-6
12	0	0	0	0	0	0

Table 1. Analysis of block, symbol and bit errors and error rates at interleaving depth 100 and block size 128 using horizontal stack symbol interleaving

4 Error Performance Estimation Results

4.1 Block, Symbol and Bit Errors and Error Rate Analysis

Table 1 shows the analysis of bit, block symbol errors and error rates at an interleaving depth of 100 and a block size of 128 symbols. The stacking arrangement used in table 1 was horizontal and only symbol interleaving was performed. The number of errors are those occurring in a sample of $10,000,384$ bits ($= 1,250,048$ symbols). The table shows how many symbols the code must be capable of correcting in order to remove all the errors in the sample.

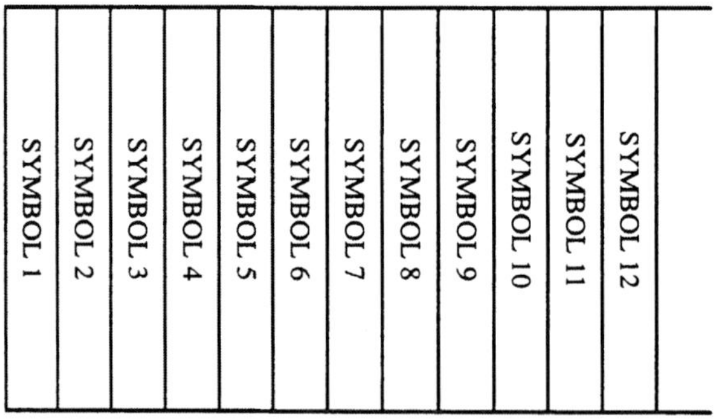

Fig. 1. The horizontal stacking arrangement of symbols

Interleaving					
Symbols corrected (t)	horiz'al stack-symbol	vert'al stack-symbol	horiz'al stack-block	vert'al stack-block	horiz'al stack-both	vert'al stack-both
0	1560	1015	1560	1015	1560	1015
3	745	548	1285	709	779	522
5	327	265	1040	627	259	172
6	219	151	986	573	91	112
7	128	95	944	566	56	42
8	88	47	928	558	0	18
9	43	20	901	558	0	0
10	23	0	841	548	0	0
11	12	0	841	526	0	0
12	0	0	817	514	0	0
18	0	0	676	257	0	0
25	0	0	343	26	0	0
26	0	0	291	0	0	0
34	0	0	35	0	0	0

Table 2. Analysis of number of symbols to be corrected at an interleaving depth 100 and a block size of 128 for different stack and interleaving scenarios

4.2 Analysis of the Number of Symbols to be Corrected

Table 2 shows the analysis of the number of symbols to be corrected at an interleaving depth of 100 and a block size of 128. The analysis was performed with both the horizontal and vertical stacking arrangements and with the three interleaving techniques. It is observed that without any symbols being corrected there is a difference in the number of symbols to be corrected in the vertical stack compared with the horizontal stack. This difference is due to the nature of bursts

SYMBOL 1	SYMBOL 5	
SYMBOL 9	SYMBOL 13	
SYMBOL 2	SYMBOL 6	
SYMBOL 10	SYMBOL 14	
SYMBOL 3	SYMBOL 7	
SYMBOL 11	SYMBOL 15	
SYMBOL 4	SYMBOL 8	
SYMBOL 12	SYMBOL 16	

Fig. 2. Combined interleaving for the vertical stack arrangement

in tape systems. A burst of 6 bit errors along a single track, if handled using horizontal stacking, produces six error symbols with each symbol containing one error. If, however, vertical stacking is used then only one or two error symbols are produced. Therefore it is not surprising that the vertical stack has fewer error symbols.

In table 2, it is clearly demonstrated that doing the block interleaving process on its own produces the worst results, and this is true in all cases. Thus block interleaving only analysis is no longer considered.

4.3   Trends that occur using both block and symbol interleaving

Figures 3 and 4 show the trends produced using both block and symbol interleaving with the horizontal and vertical stacking arrangement respectively, for a block size of 128. High interleaving depths produce the lowest t EDC symbol code required to remove all the symbol errors from the system. The best trend occurs at a depth of 5000 where an EDC symbol code with $t = 3$ symbols (i.e. a RS code that can correct 3 symbols) can remove all the symbol errors from the system. This is significantly better performance than was achievable using symbol interleaving only (which required $t = 7$). Thus it can be seen that using high interleaving depths is a significant technique for producing the low EDC symbol codes required to remove all the symbol errors out of the simulation. It is also found that horizontal stacking produces similar results to vertical stacking at a depth of 5000 where, for both stacking arrangements, a $t = 3$ EDC symbol code removes all the symbol errors from the sample stream.

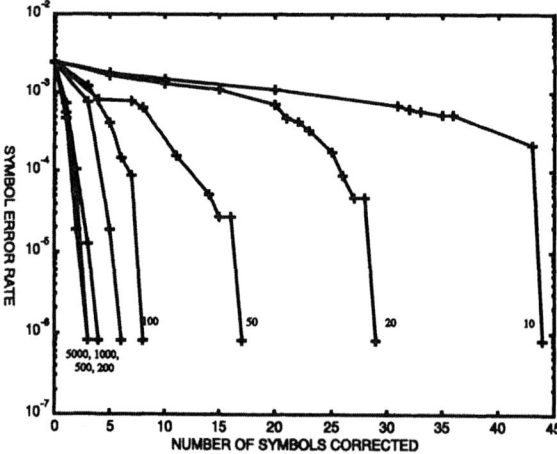

Fig. 3. Interleaving depth trends using both block and symbol interleaving with horizontal stacking arrangement for block size 128

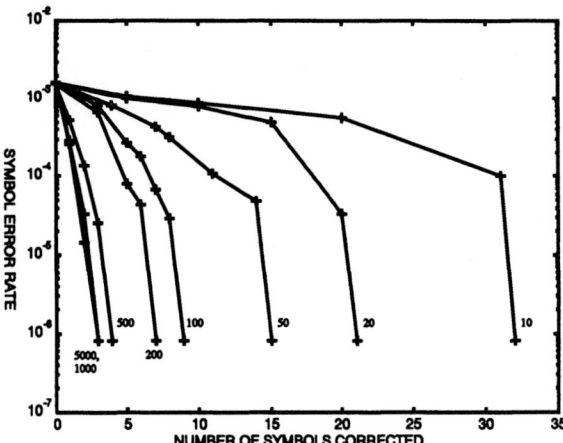

Fig. 4. Interleaving depth trends using both block and symbol interleaving with vertical stacking arrangement for block size 128

5 Concluding Remarks

It was found that, for the best possible results, a designer should use large interleaving depths, large block sizes and a strategy of using both block and symbol interleaving. The use of large interleaving depths is the significant factor in reducing the power of the EDC symbol code required to remove all the error symbols in the system. If the designer has a restriction on using large interleaving depths (for example, on the amount of memory available), then the following rules should be used:

- If the designer is able to use an interleave depth above 500 then the strategy to use is large block sizes and both block and symbol interleaving.
- If the designer is required to use an interleave depth of between 50 and 500 then the designer must manually work out which strategy to use.
- If the designer is required to use an interleave depth below 50 then the strategy to use is small block sizes and symbol interleaving only.

References

1. SWEENEY, P. *Error Control Coding - An Introduction*, Prentice Hall Ltd, 1991.
2. WADE G. *Signal Coding and Processing*, Cambridge University Press, 1994.
3. TANDON, A., MIDDLETON, B.K., FARRELL, P.G., AND MILES, J.J., A Simulation of a Noisy and Drop-out Infected Multi-track Digital Magnetic Recording Channel for 10 Million bit Data Trains, J. Inf. Recording, Vol. 23, pp469-487, 1997.
4. ABDEL-GHAFFER, K. A. S., AND HASSNER, M., Multilevel Error-Control Codes for Data Storage Channels, IEEE Trans. Magn., MAG - 37, p735 - 741, May 1991.

Efficient Error-Propagating Block Chaining

André Zúquete and Paulo Guedes

IST / INESC
R. Alves Redol 9, 1000 Lisboa, PORTUGAL
(Andre.Zuquete, Paulo.Guedes)@inesc.pt

Abstract. This document presents EPBC, Efficient Error-Propagating Block Chaining, a new and efficient block encryption mode using both plaintext and ciphertext feedback. This encryption mode is similar to another one, IOBC, and was likewise designed to propagate erroneous decryptions of tampered blocks of ciphered data to all following blocks, hence allowing to validate the integrity of that data using a predefined trailing value. However, EPBC is more secure than IOBC, as it is not vulnerable to any known-plaintext attacks, and is more efficient than IOBC. Performance tests ran on a SPARCstation 10/40 show that EPBC is in average 1.2 times faster than IOBC, and 6.3 to 10.9 times faster than a common combination of an encryption mode and a one-way hash function (CBC and MD5).

1 Introduction

The integrity control of encrypted data requires data to carry an extra integrity control value. This integrity control value allows legitimate principals to detect, after decryption, modifications on the original data contents. There are two basic ways to handle integrity control values:

1. They are generated using a one-way hash function (e.g. MD5 [9] or SHA [11]) from all the bits of the original or recovered plaintext data independently of the encryption/decryption algorithm [13, 3, 2, 1].
2. They are predefined values, which may be set up in many different ways: agreed between interacting peers, derived from a secret value, like the encryption key, or efficiently computed from some bits, but not necessarily all, of the plaintext data. In this case the encryption/decryption algorithm must guarantee that any modifications of the ciphertext will propagate erroneous decryptions until the end of the ciphertext, thus affecting the resulting integrity control value [12, 14].

The second way to handle the integrity control of encrypted data is attractive because one may save the time expended in the generation of data's hash values by slightly increasing the complexity of the encryption mode. However, most commonly used block encryption modes, like Electronic Code Book (ECB) or Cipher Block Chaining (CBC) [4], do not propagate erroneous decryptions of a modified ciphertext block to all following blocks. There are several examples of encryption modes providing error propagation, like the Kerberos' Propagating CBC [12]. Unfortunately, they have weaknesses, such as allowing the addition of arbitrary values to ciphertext blocks, swapping of ciphertext blocks, or the replacement of ciphertext blocks by new ones using known-plaintext attacks.

This document presents the Efficient Error-Propagating Block Chaining (EPBC), a new encryption mode providing error propagation without suffering from the weaknesses of other encryption modes with a similar functionality. In particular, it resists to all the attacks previously referred. Note that EPBC is not intended to be used as a keyed hash function. This because it was designed to propagate ciphertext modifications to a trailing, predefined integrity control value, and not to produce good hash values from plaintext data.

EPBC conceals plaintext patterns by randomising the input of the block cipher with previous outputs of it. Similarly, ciphertext blocks result from the randomisation of the output of the block cipher with previous inputs of if. This double randomisation prevents attackers from gathering pairs of input and output blocks of the cipher in order to guess the cipher key. EPBC is similar to another error-propagating cipher mode, IOBC [8], but uses a different function in the plaintext feedback path. Such difference makes EPBC completely immune against known-plaintext attacks, thus more secure than IOBC, and makes it also faster than IOBC.

To assess the security qualities of EPBC, we show that EPBC guarantees confidentiality and integrity control of encoded data. Concerning confidentiality, we show that attackers cannot compute particular plaintext blocks even knowing all other plaintext and ciphertext blocks. Concerning integrity control, we show that attackers are unable to derive the correct tampering of the ciphertext produced by EPBC in order to perform a limited modification of the resulting plaintext. As a consequence, attackers can only try to tamper ciphertext blocks without any guaranties of success, and the probability of success is only given by the number of bits of the trailing integrity values used with EPBC.

Performance tests run on a SPARCstation 10/40 showed that, without considering the time expended by the block cipher, EPBC is 1.2 times faster than IOBC, in average, and 6.3 to 10.9 times faster than a combination of CBC and MD5.

The rest of the paper is structured as follows. The next section presents related work. Section 3 describes the algorithm of EPBC and how it guarantees confidentiality and integrity control of encoded data. In section 4 we evaluate the performance of EPBC. Finally, in section 5 we draw some conclusions.

2 Related Work

This section overviews some encryption modes providing error propagation and their weaknesses. These algorithms are the Block Chaining (BC [10]), the Cipher Block Chaining with Checksum (CBCC [7,10]), the Propagating CBC (PCBC [6]), the PES_PCBC [14], and the Input and Output Block Chaining (IOBC [8]).

Hereafter C_i and P_i represent genuine ciphertext and plaintext blocks on the i-th iteration, c_i represents a tampered ciphertext block, p_i represents a plaintext block resulting from the decryption of a block from a tampered ciphertext, and $\mathbf{E}_K()$ and $\mathbf{D}_K()$ represent block encryption and decryption functions using key K.

Block Chaining (BC): The BC encryption mode uses all previous ciphertext blocks as feedback prior to encrypt a plaintext block [10].

$$\text{Encryption:} \quad C_i = \mathbf{E}_K(P_i \oplus F_{i-1})$$
$$\text{Decryption:} \quad P_i = \mathbf{D}_K(C_i) \oplus F_{i-1}$$
$$\text{Encryption \& Decryption:} \quad F_i = \bigoplus_{k=1}^{i} C_k$$

The initial value of F_{i-1} is a secret initialisation vector. This encryption mode is very weak in detecting ciphertext tampering: all ciphertext blocks before the ones containing integrity control values can be shuffled, or pairs of ciphertext blocks can be XORed with an arbitrary value, without propagating erroneous decryptions to the remaining blocks.

Cipher Block Chaining with Checksum (CBCC): The CBCC encryption mode is a variant of CBC that keeps a XOR of all plaintext blocks, and XOR that with the last plaintext block before encryption; the last plaintext block is intended for integrity control using a constant value [7, 10]. This encryption mode is stronger than BC, but shuffling all ciphertext blocks, except the last one, does not affect the plaintext containing the integrity control value. The following example shows that we can recover the original value of the last plaintext block P_n by swapping C_1 with C_{n-1}:

$$\begin{cases} c_1 = C_{n-1} \\ c_{n-1} = C_1 \end{cases} \implies \begin{cases} p_1 = P_{n-1} \oplus C_{n-2} \oplus IV \\ p_2 = P_2 \oplus C_1 \oplus C_{n-1} \\ p_{n-1} = P_1 \oplus C_{n-2} \oplus IV \\ p_n = P_n \end{cases}$$

Propagating CBC (PCBC): The PCBC encryption mode is similar to CBC and was used in the Kerberos Version 4 in order to simultaneously provide encryption and integrity control of data exchanged between Kerberos' principals and services [6, 12]. PCBC uses both plaintext and ciphertext feedback in order to propagate erroneous decryptions of a tampered encrypted message until the end of the message, rendering the entire message useless.

$$\text{Encryption:} \quad C_i = \mathbf{E}_K(P_i \oplus P_{i-1} \oplus C_{i-1})$$
$$\text{Decryption:} \quad P_i = \mathbf{D}_K(C_i) \oplus P_{i-1} \oplus C_{i-1}$$

The initial value of $P_{i-1} \oplus C_{i-1}$ is a secret initialisation vector. Like for CBCC, it is possible to shuffle encrypted blocks without propagating the erroneous decryption of those blocks until the last encrypted block; it only affects the corresponding plaintext blocks recovered after decryption and, possibly, the immediately following blocks [5]:

$$\begin{cases} c_i = C_{i+1} \\ c_{i+1} = C_{i+2} \\ c_{i+2} = C_i \end{cases} \implies \begin{cases} p_i = P_{i-1} \oplus P_i \oplus P_{i+1} \oplus C_{i-1} \oplus C_i \\ p_{i+1} = P_{i-1} \oplus P_i \oplus P_{i+2} \oplus C_{i-1} \oplus C_i \\ p_{i+2} = P_{i+2} \oplus C_i \oplus C_{i+2} \\ p_{i+3} = P_{i+3} \end{cases}$$

PES_PCBC: The PES_PCBC encryption mode was introduced in the Privacy Enhanced Sockets (PES) subsystem to simultaneously provide encryption and integrity control of data exchanged between client-server applications [14]. Like PCBC, PES_PCBC uses both plaintext and ciphertext feedback to achieve the error propagation effect.

Encryption: $\begin{cases} C_i = F_i \oplus G_{i-1} \\ F_i = \mathbf{E}_K(G_i) \\ G_i = P_i \oplus F_{i-1} \end{cases}$ Decryption: $\begin{cases} P_i = F_{i-1} \oplus G_i \\ G_i = \mathbf{D}_K(F_i) \\ F_i = C_i \oplus G_{i-1} \end{cases}$

The initial values of F_{i-1} and G_{i-1} are distinct, secret initialisation vectors.

The PES_PCBC encryption mode resists to attacks changing the order of ciphertext blocks but is weak against known-plaintext attacks. It is possible to compute tampered ciphertext blocks, resulting from the combination of plaintext and genuine ciphertext blocks, that defeat the desired error propagation effect. For example:

$$\begin{cases} c_i = P_{i-1} \\ c_{i+1} = P_i \oplus C_{i-1} \oplus C_{i+1} \\ c_{i+2} = C_{i+2} \end{cases} \implies \begin{cases} p_i = C_{i-1} \\ p_{i+1} = P_{i-1} \oplus P_{i+1} \oplus C_i \\ p_{i+2} = P_{i+2} \end{cases}$$

Input and Output Block Chaining (IOBC): The IOBC encryption mode is similar to PES_PCBC but stronger concerning known-plaintext attacks [8]. Comparing with PES_PCBC, IOBC has an extra function in the plaintext feedback path that rotates feedback values before using them.

Encryption: $\begin{cases} G_i = P_i \oplus F_{i-1} \\ F_i = \mathbf{E}_K(G_i) \\ C_i = F_i \oplus \mathbf{f}(G_{i-1}) \end{cases}$ Decryption: $\begin{cases} F_i = C_i \oplus \mathbf{f}(G_{i-1}) \\ G_i = \mathbf{D}_K(F_i) \\ P_i = G_i \oplus F_{i-1} \end{cases}$

The initial values of F_{i-1} and G_{i-1} are distinct, secret initialisation vectors. The function $\mathbf{f}()$ makes two different rotations on the bits of G: one is applied to the most significative $b/2 - 1$ bits, and the other one is applied to the less significative $b/2 + 1$ bits; b is the number of bits of G and is assumed to be even.

The security of IOBC depends on the length of encrypted data; if it is longer than a given threshold length then it is subject to known-plaintext attacks. However, these attacks are much difficult to achieve than with PES_PCBC because the attacker must know many plaintext blocks to perform it. For example, a tampered ciphertext block to start a known-plaintext attack is computed as follows:

$$\begin{cases} n = m \times (\frac{b}{2} + 1) \times (\frac{b}{2} - 1) = m \times (\frac{b^2}{4} - 1) \quad \forall m \in \mathcal{N} \\ c_i = \mathbf{f}(P_{i-1}) \oplus \bigoplus_{k=1}^{n-1} \mathbf{f}^k (C_{i-2k}) \oplus \mathbf{f}^{k+1} (P_{i-2k-1}) \quad \text{for} \quad i, n \in \mathcal{N}, \ i \geq 2n \end{cases}$$

For $m = 1$ and a typical value of $b = 64$, we get

$$n = 1023 \quad \Rightarrow \quad i \geq 2046$$

which means that attackers need to know $n = 1023$ specific plaintext blocks in order to start a known-plaintext attack. Therefore, for encrypted data shorter than 2046 64-bit blocks ($\approx$ 16 Kbytes) it is impossible to start such an attack. For further details regarding the strength of IOBC against known-plaintext attacks see [8].

3 Efficient Error-Propagating Block Chaining (EPBC)

EPBC is a new block encryption mode resulting from an improvement of IOBC: it has a different function, $\mathbf{g}\,()$, in the plaintext feedback path (see Figure 1).

$$\text{Encryption:} \quad \begin{cases} G_i = P_i \oplus F_{i-1} & (a) \\ F_i = \mathbf{E}_K(G_i) \\ C_i = F_i \oplus \mathbf{g}\,(G_{i-1}) & (b) \end{cases} \tag{1}$$

$$\text{Decryption:} \quad \begin{cases} F_i = C_i \oplus \mathbf{g}\,(G_{i-1}) & (a) \\ G_i = \mathbf{D}_K(F_i) \\ P_i = G_i \oplus F_{i-1} & (b) \end{cases} \tag{2}$$

The initial values of F_{i-1} and G_{i-1} are distinct, secret initialisation vectors. Like PES_PCBC and IOBC, EPBC conceals plaintext patterns by randomising the input of the block cipher G_i with previous outputs of it (F_{i-1}). Similarly, ciphertext blocks C_i result from the randomisation of the output of the block cipher F_i with values derived from previous inputs of it ($\mathbf{g}\,(G_{i-1})$). This double

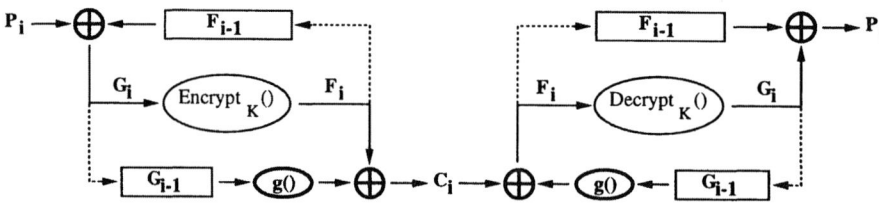

Fig. 1. Encryption/decryption of data blocks using the EPBC cipher mode and a block cipher. Dashed arrows represent value transfers at the end of each iteration, while solid arrows represent value transfers during each iteration.

randomisation prevents attackers from gathering (G_i, F_i) pairs in order to guess the encryption key K.

The function $\mathbf{g}()$ operates as follows, assuming that G values have an even number of bits, and that $G \equiv \langle G_H, G_L \rangle$, where G_H and G_L are the high and low order halves of G, respectively:

$$g(G) = \langle G_H + \overline{G_L}, G_H \cdot \overline{G_L} \rangle \tag{3}$$

where the operators "$+$" and "$\cdot$" represent the bitwise OR and AND operations, respectively, and $\overline{G_L}$ is the bitwise inverse of G_L. It is easy to show that the function $\mathbf{g}()$ is not injective, i.e. there are different arguments of $\mathbf{g}()$ that produce the same outcome (see Appendix A). As we will see below, that is an advantage because $\mathbf{g}()$ has no inverse.

Empirically, we found that for a domain with 2^b elements, each with b bits, the image of $\mathbf{g}()$ contains $3^{b/2}$ elements. For a typical value of $b = 64$, and a domain with 2^{64} elements, the image of $\mathbf{g}()$ includes $3^{32} \approx 2^{51}$ values. The fact that $\mathbf{g}()$ has a range smaller than its domain is not a security problem since such range is large enough for providing a good randomisation of F values.

The function $\mathbf{g}()$ is more efficient than the function $\mathbf{f}()$ of IOBC (see section 4), and has several properties that make it suitable for preventing known-plaintext attacks on EPBC. These properties, which are fully demonstrated in Appendix A, are the following:

$$\begin{cases} \forall x \in D_g & \mathbf{g}(x) \neq x & (a) \\ \nexists x \in D_g, \mathbf{h}() \quad \forall y \in D_g \; \mathbf{g}(x \oplus y) = \mathbf{h}(x) \oplus y & (b) \\ \nexists \mathbf{h}() \quad \forall x, y \in D_g & \mathbf{g}(x \oplus y) = \mathbf{h}(x) \oplus \mathbf{h}(y) & (c) \end{cases} \tag{4}$$

where D_g is the domain of $\mathbf{g}()$. Expression (4a) is straightforward and needs no further explanation. Expression (4b) says that there is no x value in D_g and another function $\mathbf{h}()$ so that, for all y values in D_g, $\mathbf{g}(x \oplus y)$ is equal to $\mathbf{h}(x) \oplus y$. In other words, it means that even knowing the value of x, one cannot expand $\mathbf{g}(x \oplus y)$ in terms of another known value $\mathbf{h}(x)$ XORed with y. Expression (4c) says that there is no function $\mathbf{h}()$ so that, for all values of x and y in D_g, $\mathbf{g}(x \oplus y)$ is equal to $\mathbf{h}(x) \oplus \mathbf{h}(y)$.

Other similar functions, with an equal number of elements in the image and similar properties, could be used as $\mathbf{g}()$. These functions are:

$$\mathbf{g'}(G) = \langle \overline{G_H} + G_L, \overline{G_H} \cdot G_L \rangle$$
$$\mathbf{g''}(G) = \langle \overline{G_H} + \overline{G_L}, \overline{G_H} \cdot G_L \rangle$$

3.1 Confidentiality

Concerning confidentiality, it is necessary to prove that EPBC does not enable attackers to compute a specific plaintext block (P_i) even knowing all other plaintext and ciphertext blocks. Resolving equation (2b) to compute P_i we obtain:

$$P_i = G_i \oplus F_{i-1} \tag{5}$$
$$= G_i \oplus C_{i-1} \oplus \mathbf{g}\,(G_{i-2}) \tag{6}$$
$$= G_i \oplus C_{i-1} \oplus \mathbf{g}\,(P_{i-2} \oplus F_{i-3})$$
$$= G_i \oplus C_{i-1} \oplus \mathbf{g}\,(P_{i-2} \oplus C_{i-3} \oplus \mathbf{g}\,(G_{i-4})) \tag{7}$$
$$= \cdots$$

As the function $\mathbf{g}\,()$ is not injective, it has no inverse and the term G_i of equation (5) cannot be further expanded in terms of C_{i+1} and F_{i+1}. Since attackers ignore G values, and due to the characteristics of function $\mathbf{g}\,()$ given by expressions (4), they cannot compute or simplify, in order to isolate G values, expressions $\mathbf{g}\,(G_{i-2})$ in equation (6), and $\mathbf{g}\,(P_{i-2} \oplus C_{i-3} \oplus \mathbf{g}\,(G_{i-4}))$ in equation (7). Consequently, they cannot remove G terms from equations (6) and (7) and, thus, they cannot compute the value of the plaintext block P_i.

3.2 Integrity Control

Concerning integrity control, to tamper the ciphertext without affecting the decryption of trailing integrity control values attackers must perform a controlled modification of the ciphertext. This means that attackers must provide a correct sequence of false ciphertext blocks (c values) in order to modify and latter recover the correct internal state (G and F values) of the decryption engine before the actual decryption of the integrity control values. This implies that all f values resulting from c values during decryption must be correct F values. Otherwise, f values would decrypt to something unpredictable, instead of correct G values, and attackers would loose control of the tampering. Therefore, the possible values of a c_i block are:

$$c_i = F_j \oplus \mathbf{g}\,(G_{i-1}) \quad \wedge \quad i \neq j$$
$$= F_j \oplus \mathbf{g}\,(P_{i-1} \oplus F_{i-2}) \tag{8}$$
$$= C_j \oplus \mathbf{g}\,(G_{j-1}) \oplus \mathbf{g}\,(P_{i-1} \oplus F_{i-2})$$
$$= C_j \oplus \mathbf{g}\,(P_{j-1} \oplus F_{j-2}) \oplus \mathbf{g}\,(P_{i-1} \oplus F_{i-2}) \tag{9}$$
$$= \cdots$$

We could further expand equation (9) but it would not simplify the task of attackers. Since attackers ignore F values, and due to the characteristics of function $\mathbf{g}\,()$ given by expressions (4), they cannot compute or simplify, in order to isolate F values, expressions $\mathbf{g}\,(P_{i-1} \oplus F_{i-2})$ in equations (8) and (9), and $\mathbf{g}\,(P_{j-1} \oplus F_{j-2})$ in equation (9). Consequently, they cannot remove F terms from equations (8) and (9) and, thus, they cannot compute a false ciphertext block c_i in order to start a controlled modification of the encrypted data.

In conclusion, the EPBC encryption mode does not suffer from the weaknesses of all other modes presented in section 2. Attackers are unable to take control of the EPBC decryption engine by providing tampered ciphertext blocks. Therefore, they can only try to tamper ciphertext blocks without any guaranties of success, and the probability of success is only given by the number of bits of the trailing integrity values used with EPBC.

4 Performance Evaluation

In this section we evaluate the performance of EPBC, and we compare it with other techniques used to achieve confidentiality and integrity control: the IOBC error-propagating encryption mode and a common combination of an encryption mode (CBC) and a one-way hash function (MD5). The performance tests were executed on a Sun SPARCstation 10/40.

In order to do a fair comparison between different encryption modes, we used our own optimised implementations. Our implementations do not modify source bytes, either plaintext or ciphertext, and were optimised in order to reduce the number of function calls and maximise the use of CPU registers instead of memory accesses. Encryption/decryption cycles were unrolled 4 times. In Appendix B we present our implementation of CBC, IOBC, and EPBC.

Table 1 presents the average elapsed time per block expended by CBC, IOBC and EPBC when encoding and decoding arrays of 64-bit blocks with two different lengths: 128 blocks and 1 M blocks. These two different lengths allow us to assess the impact of the memory cache status, either warm or cold, in the performance of the algorithms. The time measurements do not include the time expended by the block cipher function and were obtained by dividing the total elapsed time expended in encoding and decoding a total of 160 M blocks (1280 Mbytes). These values show that, on average, the EPBC encryption mode is 1.2 times faster than IOBC and 1.5 times slower than CBC.

Cache status	Algorithm	Encryption (ns)	Decryption (ns)
Warm	CBC	166	177
	IOBC	344	369
	EPBC	268	294
	CBC & MD5	2 910	2 918
Cold	CBC	334	337
	IOBC	519	529
	EPBC	455	458
	CBC & MD5	2 894	2 879

Table 1. Average user time per block expended on a Sun SPARCstation 10/40 by three encryption modes (CBC, IOBC and EPBC) and a combination of an encryption mode and a one-way hash function (CBC & MD5), when encoding and decoding arrays of 64-bit blocks (not including the time expended by the block cipher function).

We did the same performance evaluations for a combination of the CBC encryption mode and the MD5 one-way hash function, using our implementation of CBC and the MD5 implementation presented in the document describing it [9]. As input for MD5 we used all the blocks of the arrays except the last two; these were used to store the resulting hash value. The average user time expended in processing each block with CBC and MD5 is also presented in Table 1.

The values in Table 1 show that the combination of CBC and MD5 is almost insensible to the cache status. The small speedup that is observed in the measurements obtained in the simulations with the cold cache is due to the fact that we execute less function calls to encode/decode all the 160 M blocks. Comparing with EPBC, we can see that this is 6.3 to 10.9 times faster than the combination of CBC and MD5. If we had used SHA [11] instead of MD5, the performance gain would probably be greater. According to a table presented by Schneier [10] for a 486 SX processor, the SHA function is about two times slower than MD5.

5 Conclusions

In this document we presented a new encryption mode, Efficient Error-Propagating Block Chaining, that propagates erroneous decryptions of tampered ciphertext blocks to all following blocks. This encryption mode allows integrity validation of the recovered plaintext by checking a predefined value at the end of the plaintext. This value may be set up in many different ways: agreed between interacting peers, derived from a secret value, like the encryption key, or efficiently computed from some bits, but not necessarily all, of the plaintext data.

Unlike other encryption modes also providing error propagation, such as BC, CBCC, PCBC, PES_PCBC, or IOBC, the EPBC encryption mode is not vulnerable to attacks changing the order of ciphertext blocks or known-plaintext attacks. It also conceals plaintext patterns by randomising the input of the block cipher with previous outputs of it, and prevents attackers from gathering pairs of cipher input and output blocks in order to guess the encryption key.

Performance tests run on a SPARCstation 10/40 showed that, without considering the time expended by the block cipher, EPBC is in average 1.2 times faster than IOBC, and 6.3 to 10.9 times faster than a combination of CBC and MD5. The performance gain of EPBC when compared with the combination of CBC and MD5 can be very important if we are concerned with providing together data confidentiality and integrity control. This is the case of most secure communication protocols, such as SSL [3], PES [14], or SKIP [1], where interacting applications or operating systems can use constant or easy-to-compute, secret integrity control values for one or more secure communication channels. This way they could save a significative amount of processing time when exchanging large quantities of confidential information through secure communication channels.

References

1. Ashar Aziz, Tom Markson, and Hemma Prafullchandra. Simple Key-Management For Internet Protocols (SKIP). Internet Draft, Sun Microsystems, Inc., December 1995.

2. D. Balenson. Privacy Enhancement for Internet Electronic Mail (Part III): Algorithms, Modes, and Identifiers. RFC 1423, IAB IRTF PSRG, IETF PEM WG, February 1993.

3. Alan O. Freier, Philip Karlton, and Paul C. Kocher. SSL Protocol Version 3.0. Internet Draft, Netscape Communications Corp., March 1996.

4. Information Processing - Modes of Operation for an n-bit Block Cipher Algorithm. ISO IEC/DIS 10116, 1989.

5. J. T. Kohl. The Use of Encryption in Kerberos for Network Authentication. In *Advances in Cryptology – CRYPTO '89 Proceedings*, pages 35–43. Springer-Verlag, 1990.

6. C. H. Meyer and S. M. Matyas. *Cryptography: A New Dimension in Computer Data Security*. John Wiley & Sons, Inc., New York, 1982.

7. Xerox Network System (XNS) Authentication Protocol. XSIS 098404, Xerox Corporation, April 1984.

8. Francisco Recacha. IOBC: Un nuevo modo de encadenamiento para cifrado en bloque. In *Proc. of the IV Reunion Espanyola sobre Criptologia*, Valladolid, September 1996.

9. R. Rivest. The MD5 Message-Digest Algorithm. RFC 1321, MIT Laboratory for Computer Science and RSA Data Security, Inc., April 1992.

10. Bruce Schneier. *Applied Cryptography: Protocols, Algorithms and Source Code in C*. John Wiley & Sons, Inc., second edition, 1996.

11. Secure Hash Standard. NIST FIPS PUB 180, April 1993.

12. Jennifer G. Steiner, Clifford Neuman, and Jeffrey I. Schiller. Kerberos: An Authentication Service for Open Network Systems. In *Proc. of the USENIX Winter Conf.*, pages 191–202, Dallas, Texas, USA, February 1988.

13. Philip Zimmermann. *The Official PGP User's Guide*. MIT Press, 1995.

14. André Zúquete and Paulo Guedes. Transparent Authentication and Confidentiality for Stream Sockets. *IEEE Micro*, 16(3):34–41, June 1996.

A   Demonstrations

In this Appendix we will demonstrate some of the properties of the function $\mathbf{g}\,()$ previously introduced in section 3. As a reminder, the function $\mathbf{g}\,()$ operates as follows:

$$\mathbf{g}(x) = \langle x_H + \overline{x_L}, x_H \cdot \overline{x_L} \rangle \tag{10}$$

where x is a value with an even number of bits (b), and $x \equiv \langle x_H, x_L \rangle$, where x_H and x_L are the high and low order halves of x, respectively. The operators "+" and "·" represent the bitwise OR and AND operations, respectively, and $\overline{x_L}$ is the bitwise inverse of x_L.

1st property: $\mathbf{g}\left(\right)$ is not injective

Demonstration: The function $\mathbf{g}\left(\right)$ is not injective if:

$$\exists x, y \in D_g, x \neq y \quad \mathbf{g}\left(x\right) = \mathbf{g}\left(y\right)$$

Expanding $\mathbf{g}\left(x\right) = \mathbf{g}\left(y\right)$ using equation (10), we get:

$$\mathbf{g}\left(x\right) = \mathbf{g}\left(y\right) \Longleftrightarrow \begin{cases} x_H + \overline{x_L} = y_H + \overline{y_L} \\ x_H \cdot \overline{x_L} = y_H \cdot \overline{y_L} \end{cases}$$

This two equations can be resolved in several ways. One solution, for example, is $y_H = \overline{x_L}$ and $y_L = \overline{x_H}$. Therefore, $\mathbf{g}\left(\right)$ is not injective.

2nd property: $\forall x \in D_g \quad \mathbf{g}\left(x\right) \neq x$

Demonstration: If the opposite is true, then:

$$\exists x \in D_g \quad \mathbf{g}\left(x\right) = x \Longleftrightarrow \begin{cases} x_H + \overline{x_L} = x_H \\ x_H \cdot \overline{x_L} = x_L \end{cases}$$

Expanding these two equations in terms of the individual bits of each of the halves of x, we have:

$$\forall i \in \{1, 2, \cdots, b/2\} \begin{cases} x_{Hi} + \overline{x_{Li}} = x_{Hi} \Leftrightarrow x_{Li} = 1 \vee x_{Hi} = 1 \\ x_{Hi} \cdot \overline{x_{Li}} = x_{Li} \Leftrightarrow x_{Li} = 0 \wedge x_{Hi} = 0 \end{cases}$$

As there are no suitable values for the pair of bits (x_{Li}, x_{Hi}), the opposite is false and, thus, the property holds.

3th property: $\nexists x \in D_g, \mathbf{h}\left(\right) \quad \forall y \in D_g \quad \mathbf{g}\left(x \oplus y\right) = \mathbf{h}\left(x\right) \oplus y$

Demonstration: If the opposite is true, then:

$$\exists x \in D_g, \mathbf{h}\left(\right)$$

$$\mathbf{g}\left(x \oplus y\right) = \mathbf{h}\left(x\right) \oplus y \Longleftrightarrow \begin{cases} (x_H \oplus y_H) + \overline{(x_L \oplus y_L)} = \mathbf{h}\left(x\right)_H \oplus y_H \\ (x_H \oplus y_H) \cdot \overline{(x_L \oplus y_L)} = \mathbf{h}\left(x\right)_L \oplus y_L \end{cases}$$

As the first equation depends on y_L and the second equation depends on y_H, that implies that $\mathbf{h}\left(x\right)$ depends on y, which cannot not happen. Therefore, the property holds.

4th property: $\nexists \mathbf{h}\left(\right) \quad \forall x, y \in D_g \quad \mathbf{g}\left(x \oplus y\right) = \mathbf{h}\left(x\right) \oplus \mathbf{h}\left(y\right)$

Demonstration: If we have $x = y$, then:

$$\begin{cases} \mathbf{g}\left(x \oplus y\right) = \mathbf{g}\left(0\right) \neq 0 \\ \mathbf{h}\left(x\right) \oplus \mathbf{h}\left(y\right) = 0 \end{cases} \Longrightarrow$$

$$\Longrightarrow \forall x, y \in D_g, x = y \quad \forall \mathbf{h}\left(\right) \quad \mathbf{g}\left(x \oplus y\right) \neq \mathbf{h}\left(x\right) \oplus \mathbf{h}\left(y\right)$$

Consequently, the property holds.

B   Source Code of CBC, IOBC and EPBC

```
/*  Global macros  */
#define w32              unsigned int    /* 32 bit word */
#define CIPHER(to,from)                  /* nothing */
#define DECIPHER(to,from)                /* nothing */
#define DO_4_TIMES(op)  {op op op op}
```

```
#define CBC_ENC_BLOCK \
  I[0] = *P++ ^ c0; I[1] = *P++ ^ c1;  \
  CIPHER(C,I);                          \
  c0 = *C++; c1 = *C++;

cbcEnc ( int n, w32 *C, w32 *P, w32 *C_1 )
{
  register w32 c0 = C_1[0], c1 = C_1[1];
  w32 I[2];

  for (; n >= 4; n -= 4)
    DO_4_TIMES( CBC_ENC_BLOCK )
  while (n--) { CBC_ENC_BLOCK }
}
```

```
#define CBC_DEC_BLOCK \
  DECIPHER(P,C);                        \
  *P++ ^= c0; *P++ ^= c1;               \
  c0 = *C++; c1 = *C++;

cbcDec ( int n, w32 *C, w32 *P, w32 *C_1 )
{
  register w32 c0 = C_1[0], c1 = C_1[1];

  for (; n >= 4; n -= 4)
    DO_4_TIMES( CBC_DEC_BLOCK )
  while (n--) { CBC_DEC_BLOCK }
}
```

```
#define ROT_L(h,l) (((l & 0xFFFFFFFC) >> 1) | ((l & 2) << 30) | (h & 1))
#define ROT_H(h,l) ((h >> 1) | (l << 31))
```

```
#define IOBC_ENC_BLOCK \
  c0 = ROT_L(g1,g0); c1 = ROT_H(g1,g0);  \
  G[0] = g0 = *P++ ^ f0;                  \
  G[1] = g1 = *P++ ^ f1;                  \
  CIPHER(F,G);                            \
  f0 = F[0]; f1 = F[1];                   \
  *C++ = c0 ^ f0; *C++ = c1 ^ f1;

iobcEnc ( int n, w32 *C, w32 *P,
                 w32 *F, w32 *G )
{
  register w32 f0 = F[0], f1 = F[1];
  register w32 g0 = G[0], g1 = G[1];
  register w32 c0, c1;

  for (; n >= 4; n -= 4)
    DO_4_TIMES( IOBC_ENC_BLOCK )
  while (n--) { IOBC_ENC_BLOCK }
}
```

```
#define IOBC_DEC_BLOCK \
  p0 = f0; p1 = f1;                       \
  F[0] = f0 = *C++ ^ ROT_L(g1,g0);        \
  F[1] = f1 = *C++ ^ ROT_H(g1,g0);        \
  DECIPHER(G,F);                          \
  g0 = G[0]; g1 = G[1];                   \
  *P++ = p0 ^ g0; *P++ = p1 ^ g1;

iobcDec ( int n, w32 *C, w32 *P,
                 w32 *F, w32 *G )
{
  register w32 f0 = F[0], f1 = F[1];
  register w32 g0 = G[0], g1 = G[1];
  register w32 p0, p1;

  for (; n >= 4; n -= 4)
    DO_4_TIMES( IOBC_DEC_BLOCK )
  while (n--) { IOBC_DEC_BLOCK }
}
```

```
#define EPBC_ENC_BLOCK \
  c0 = ~g0 & g1; c1 = ~g0 | g1;           \
  G[0] = g0 = *P++ ^ f0;                  \
  G[1] = g1 = *P++ ^ f1;                  \
  CIPHER(F,G);                            \
  f0 = F[0]; f1 = F[1];                   \
  *C++ = c0 ^ f0; *C++ = c1 ^ f1;

epbcEnc ( int n, w32 *C, w32 *P,
                 w32 *F, w32 *G )
{
  register w32 f0 = F[0], f1 = F[1];
  register w32 g0 = G[0], g1 = G[1];
  register w32 c0, c1;

  for (; n >= 4; n -= 4)
    DO_4_TIMES( EPBC_ENC_BLOCK )
  while (n--) { EPBC_ENC_BLOCK }
}
```

```
#define EPBC_DEC_BLOCK \
  p0 = f0; p1 = f1;                       \
  F[0] = f0 = *C++ ^ (~g0 & g1);          \
  F[1] = f1 = *C++ ^ (~g0 | g1);          \
  DECIPHER(G,F);                          \
  g0 = G[0]; g1 = G[1];                   \
  *P++ = p0 ^ g0; *P++ = p1 ^ g1;

epbcDec ( int n, w32 *C, w32 *P,
                 w32 *F, w32 *G )
{
  register w32 f0 = F[0], f1 = F[1];
  register w32 g0 = G[0], g1 = G[1];
  register w32 p0, p1;

  for (; n >= 4; n -= 4)
    DO_4_TIMES( EPBC_DEC_BLOCK )
  while (n--) { EPBC_DEC_BLOCK }
}
```

Author Index

Springer
and the
environment

At Springer we firmly believe that an international science publisher has a special obligation to the environment, and our corporate policies consistently reflect this conviction.

We also expect our business partners – paper mills, printers, packaging manufacturers, etc. – to commit themselves to using materials and production processes that do not harm the environment. The paper in this book is made from low- or no-chlorine pulp and is acid free, in conformance with international standards for paper permanency.

 Springer

Lecture Notes in Computer Science

For information about Vols. 1–1275

please contact your bookseller or Springer-Verlag

Vol. 1313: J. Fitzgerald, C.B. Jones, P. Lucas (Eds.), FME '97: Industrial Applications and Strengthened Foundations of Formal Methods. Proceedings, 1997. XIII, 685 pages. 1997.

Vol. 1314: S. Muggleton (Ed.), Inductive Logic Programming. Proceedings, 1996. VIII, 397 pages. 1997. (Subseries LNAI).

Vol. 1315: G. Sommer, J.J. Koenderink (Eds.), Algebraic Frames for the Perception-Action Cycle. Proceedings, 1997. VIII, 395 pages. 1997.

Vol. 1316: M. Li, A. Maruoka (Eds.), Algorithmic Learning Theory. Proceedings, 1997. XI, 461 pages. 1997. (Subseries LNAI).

Vol. 1317: M. Leman (Ed.), Music, Gestalt, and Computing. IX, 524 pages. 1997. (Subseries LNAI).

Vol. 1318: R. Hirschfeld (Ed.), Financial Cryptography. Proceedings, 1997. XI, 409 pages. 1997.

Vol. 1319: E. Plaza, R. Benjamins (Eds.), Knowledge Acquisition, Modeling and Management. Proceedings, 1997. XI, 389 pages. 1997. (Subseries LNAI).

Vol. 1320: M. Mavronicolas, P. Tsigas (Eds.), Distributed Algorithms. Proceedings, 1997. X, 333 pages. 1997.

Vol. 1321: M. Lenzerini (Ed.), AI*IA 97: Advances in Artificial Intelligence. Proceedings, 1997. XII, 459 pages. 1997. (Subseries LNAI).

Vol. 1322: H. Hußmann, Formal Foundations for Software Engineering Methods. X, 286 pages. 1997.

Vol. 1323: E. Costa, A. Cardoso (Eds.), Progress in Artificial Intelligence. Proceedings, 1997. XIV, 393 pages. 1997. (Subseries LNAI).

Vol. 1324: C. Peters, C. Thanos (Eds.), Research and Advanced Technology for Digital Libraries. Proceedings, 1997. X, 423 pages. 1997.

Vol. 1325: Z.W. Raś, A. Skowron (Eds.), Foundations of Intelligent Systems. Proceedings, 1997. XI, 630 pages. 1997. (Subseries LNAI).

Vol. 1326: C. Nicholas, J. Mayfield (Eds.), Intelligent Hypertext. XIV, 182 pages. 1997.

Vol. 1327: W. Gerstner, A. Germond, M. Hasler, J.-D. Nicoud (Eds.), Artificial Neural Networks – ICANN '97. Proceedings, 1997. XIX, 1274 pages. 1997.

Vol. 1328: C. Retoré (Ed.), Logical Aspects of Computational Linguistics. Proceedings, 1996. VIII, 435 pages. 1997. (Subseries LNAI).

Vol. 1329: S.C. Hirtle, A.U. Frank (Eds.), Spatial Information Theory. Proceedings, 1997. XIV, 511 pages. 1997.

Vol. 1330: G. Smolka (Ed.), Principles and Practice of Constraint Programming – CP 97. Proceedings, 1997. XII, 563 pages. 1997.

Vol. 1331: D. W. Embley, R. C. Goldstein (Eds.), Conceptual Modeling – ER '97. Proceedings, 1997. XV, 479 pages. 1997.

Vol. 1332: M. Bubak, J. Dongarra, J. Waśniewski (Eds.), Recent Advances in Parallel Virtual Machine and Message Passing Interface. Proceedings, 1997. XV, 518 pages. 1997.

Vol. 1333: F. Pichler. R.Moreno-Díaz (Eds.), Computer Aided Systems Theory – EUROCAST'97. Proceedings, 1997. XII, 626 pages. 1997.

Vol. 1334: Y. Han, T. Okamoto, S. Qing (Eds.), Information and Communications Security. Proceedings, 1997. X, 484 pages. 1997.

Vol. 1335: R.H. Möhring (Ed.), Graph-Theoretic Concepts in Computer Science. Proceedings, 1997. X, 376 pages. 1997.

Vol. 1336: C. Polychronopoulos, K. Joe, K. Araki, M. Amamiya (Eds.), High Performance Computing. Proceedings, 1997. XII, 416 pages. 1997.

Vol. 1337: C. Freksa, M. Jantzen, R. Valk (Eds.), Foundations of Computer Science. XII, 515 pages. 1997.

Vol. 1338: F. Plášil, K.G. Jeffery (Eds.), SOFSEM'97: Theory and Practice of Informatics. Proceedings, 1997. XIV, 571 pages. 1997.

Vol. 1339: N.A. Murshed, F. Bortolozzi (Eds.), Advances in Document Image Analysis. Proceedings, 1997. IX, 345 pages. 1997.

Vol. 1340: M. van Kreveld, J. Nievergelt, T. Roos, P. Widmayer (Eds.), Algorithmic Foundations of Geographic Information Systems. XIV, 287 pages. 1997.

Vol. 1341: F. Bry, R. Ramakrishnan, K. Ramamohanarao (Eds.), Deductive and Object-Oriented Databases. Proceedings, 1997. XIV, 430 pages. 1997.

Vol. 1342: A. Sattar (Ed.), Advanced Topics in Artificial Intelligence. Proceedings, 1997. XVII, 516 pages. 1997. (Subseries LNAI).

Vol. 1343: Y. Ishikawa, R.R. Oldehoeft, J.V.W. Reynders, M. Tholburn (Eds.), Scientific Computing in Object-Oriented Parallel Environments. Proceedings, 1997. XI, 295 pages. 1997.

Vol. 1344: C. Ausnit-Hood, K.A. Johnson, R.G. Pettit, IV, S.B. Opdahl (Eds.), Ada 95 – Quality and Style. XV, 292 pages. 1997.

Vol. 1345: R.K. Shyamasundar, K. Ueda (Eds.), Advances in Computing Science - ASIAN'97. Proceedings, 1997. XIII, 387 pages. 1997.

Vol. 1346: S. Ramesh, G. Sivakumar (Eds.), Foundations of Software Technology and Theoretical Computer Science. Proceedings, 1997. XI, 343 pages. 1997.

Vol. 1347: E. Ahronovitz, C. Fiorio (Eds.), Discrete Geometry for Computer Imagery. Proceedings, 1997. X, 255 pages. 1997.

Vol. 1348: S. Steel, R. Alami (Eds.), Recent Advances in AI Planning. Proceedings, 1997. IX, 454 pages. 1997. (Subseries LNAI).

Vol. 1349: M. Johnson (Ed.), Algebraic Methodology and Software Technology. Proceedings, 1997. X, 594 pages. 1997.

Vol. 1350: H.W. Leong, H. Imai, S. Jain (Eds.), Algorithms and Computation. Proceedings, 1997. XV, 426 pages. 1997.

Vol. 1351: R.T. Chin, T.-C. Pong (Eds.), Computer Vision – ACCV'98. Proceedings Vol. I, 1998. XXIV, 761 pages. 1997.

Vol. 1352: R.T. Chin, T.-C. Pong (Eds.), Computer Vision – ACCV'98. Proceedings Vol II, 1998. XXIV, 757 pages. 1997.

Vol. 1355: M. Darnell (Ed.), Cryptography and Coding. Proceedings, 1997. IX, 335 pages. 1997.